Diagnosis of Pathogenic Microorganisms Causing Infectious Diseases

Editor

Irshad M. Sulaiman

U.S. Food and Drug Administration
Atlanta, GA
USA

CRC Press
Taylor & Francis Group
Boca Raton London New York

CRC Press is an imprint of the
Taylor & Francis Group, an **informa** business

A SCIENCE PUBLISHERS BOOK

Cover illustrations reproduced by kind courtesy of Dr. Ynes Ortega, Associate Professor, Center for Food Safety, University of Georgia, USA.

First edition published 2024
by CRC Press
2385 NW Executive Center Drive, Suite 320, Boca Raton FL 33431

and by CRC Press
4 Park Square, Milton Park, Abingdon, Oxon, OX14 4RN

CRC Press is an imprint of Taylor & Francis Group, LLC

Library of Congress Cataloging-in-Publication Data (applied for)

ISBN: 978-0-367-02926-5 (hbk)
ISBN: 978-1-032-70850-8 (pbk)
ISBN: 978-0-429-00119-2 (ebk)

DOI: 10.1201/9780429001192

Typeset in Times New Roman
by Radiant Productions

Preface

Infectious diseases are the sicknesses caused by the pathogenic microbial organisms that include bacteria, viruses, fungi, or parasites. These infectious microorganisms can multiply and cause an infection once they enter the body. Several infectious diseases are known to be transmissible, that spread from person-to-person (human-to-human transmission). Infectious diseases can also be transmitted by the vectors or from animals to humans (zoonotic transmission). In addition, the pathogens causing mild to life-threatening infectious diseases can be airborne, waterborne, foodborne, and soilborne. Several new microbial pathogens have emerged to cause several infectious diseases, sporadic cases, and deadly outbreaks worldwide.

The biology of microorganisms is complex as every single microbial organism (species/genotype/sub genotype) occupies a unique ecosystem and a distinctive niche. Furthermore, it has been recognized that the strains and species of microbial organisms are not only typically restricted together but also compete for limited nutrients and space. These severe challenging environments frequently result in the evolution of unique phenotypes of microorganisms that can outcompete and displace the co-existing genotype/strain/species. Therefore, to control and prevent any infectious disease, it is warranted to characterize more known, as well as, unknown/little less known/less prevalent microorganisms, recovered from clinical, food, environmental, and other microbiological settings to understand the apparent biology of microbial organisms and to better predict the performance of microbes causing various infectious disease. It will also help in the development of more robust high-throughput diagnostic techniques for the rapid diagnosis of infectious microbial pathogens of public health importance.

Thus, considering the above facts, this book has been compiled on the advancement that has been made in the recent past for the diagnosis of various human-pathogenic microorganisms of public health significance. This book includes nine chapters from experts in the field on the diagnosis of the following pathogenic microorganisms known to cause infectious diseases worldwide: (i) Diagnosis of Human-pathogenic Gram-Negative Bacteria (*Cronobacter* species, *Campylobacter*), (ii) Diagnosis of Human-pathogenic Gram-Positive Bacteria (*Listeria monocytogenes, Staphylococcus*), (iii) Diagnosis of Human-pathogenic Virus (Hepatitis B Virus and Hepatitis C Virus), and (iv) Diagnosis of Human-pathogenic Gastrointestinal Parasites (Intestinal Amebiasis, *Toxoplasma gondii, Cyclospora cayetanensis* Coccidian parasites). This book will help the scientific community to better understand the diagnosis of several disease-causing microbial pathogens of public health significance.

Contents

Part IV: Diagnosis of Human-Pathogenic Gastrointestinal Parasites

Part I

Diagnosis of Human-Pathogenic Gram-Negative Bacteria

Chapter 1

Recent Advancement in the Diagnosis of *Cronobacter* Species and Related Species Causing Foodborne Disease and Outbreaks#

Hyein Jang*, Jayanthi Gangiredla, Isha R. Patel, Flavia Negrete,
Leah M. Weinstein, Katie Ko, Ben D. Tall and *Gopal R. Gopinath*

Taxonomy, Phenotypic Characteristics, Natural History of *Cronobacter* and Related Public Health Trends

Cronobacter are Gram-negative, non-sporulating, mesophilic, facultatively anaerobic bacteria that belong to the phylum *Proteobacteria*, class *Gammaproteobacteria*, order *Enterobacterales*, and family *Enterobacteriaceae*. Bacterial cells are rod-shaped measuring approximately 3 by 1 μm when growing exponentially. Cells are motile by peritrichously-expressed flagella (Iversen et al. 2008a). Figure 1 shows the morphological features of typical *Cronobacter* cells. The genus *Cronobacter* was first proposed in 2007 by Iversen et al. (2007a) with further clarification of the taxonomic relationships of the biogroups of *Enterobacter sakazakii* as described by Farmer et al. (1980) and Iversen et al. (2006) in a second report published in 2008 (Iversen et al. 2008a). The species described by Iversen et al. (2008a)

Center for Food Safety and Applied Nutrition, U.S. Food and Drug Administration, Laurel, MD 20708, USA.
* Corresponding author: hyein.jang@fda.hhs.gov
# Disclaimer: This book chapter reflects the views of the authors and should not be construed to represent FDA's views or policies.

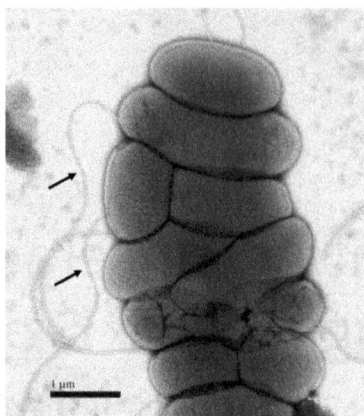

Figure 1. Transmission electron photomicrograph of a typical ST 4 *Cronobacter sakazakii* strain (LR632, synonym ES632) grown on TSA supplemented with 1% sodium chloride and incubated at 37°C for 22 h. *C. sakazakii* LR632 was isolated from USA dairy plant environment in 2005. The cells were negatively stained with 0.5% sodium phosphotungstate (pH 6.8). Note the typical undulating nature of the cell surface (outer membrane) and the many peritrichously-expressed flagella (arrows). The cells shown in the electron photomicrograph are in a clump which typically is found in cultures that are undergoing auto aggregation. Bar represents 1 µm. The image was adapted from Jang et al. (2018a).

included *Cronobacter sakazakii, Cronobacter malonaticus, Cronobacter muytjensii, Cronobacter turicensis, Cronobacter dublinensis* (three subspecies; *lactaridi, dublinensis,* and *lausannensis*) and Genomospecies 1. In 2012, Joseph et al. (2012a) renamed Genomospecies 1 as *Cronobacter universalis* and described *Cronobacter condimenti.* Of the seven species, the primary pathogens are *C. sakazakii, C. malonaticus,* and *C. turicensis* (Stephan et al. 2011; Joseph et al. 2012; 2012b; Holý et al. 2014; Alsonosi et al. 2015).

However, the natural history of *Cronobacter* has had an arduous journey primarily because phenotypically this group of organisms was difficult to taxonomically place within the family *Enterobacteriaceae,* and there was an enormous lack of knowledge about their microbiological characteristics, clinical prevalence, and ecological distribution. Table 1 shows the key phenotypic traits possessed by *Cronobacter* species and similar phylogenetically related species.

The first noted isolate dates back only to 1950 with NCTC 8155, a *C. sakazakii* strain that was obtained from a can of dried milk, and the first strain isolated from a human clinical specimen also dated in 1950 when *C. sakazakii* NCTC 9238 was isolated from an abdominal abscess (Farmer et al. 2015). The first clinical neonatal meningitis isolates were those that Urmenyi and White-Franklin described in their 1958 case report published in 1961 (Urmenyi and White-Franklin 1961). For more information about the history of outbreaks and unique cases, refer to the review by Henry and Fouladkhah (2019).

The isolates in these early cases were presumptively identified as either pigmented coliforms or yellow-pigmented *Enterobacter cloacae* (Urmenyi and White-Franklin 1961). It was not until Farmer et al. in their seminal 1980 report provided critical microbiological and genomic characteristics of a collection of 57 strains (Farmer

Table 1. Typical phenotypic (biochemical) characteristics of *Cronobacter* and phylogenetically related species.[a]

Phenotype Reaction	Reaction For:									
	C. sakazakii	*C. malonaticus*	*C. dublinensis*	*C. muytjensii*	*C. turicensis*	*C. condimenti*	*C. universalis*	*S. turicensis*	*F. pulveris*	*F. helveticus*
Voges-Proskauer[b]	V	+	+	+	+	+	+	–	–	–
Methyl red[c]	V	–	–	–	–	–	–	+	+	+
Mucate	–	–	–	–	–	–	–	+	+	+
Dulcitol	V	–	–	+	+	–	+	+	V	+
D-Arabitol	–	–	–	–	–	–	–	+	+	+
5-Keto-D-gluconate	–	–	–	–	–	–	–	–	+	+
Sucrose	+	+	+	+	+	+	+	–	+	–
α-l-Rhamnose	+	+	+	+	+	+	+	+	+	+
Raffinose	+	+	+	+	+	+	+	–	+	–
Fumarate	+	+	+	+	+	ND	+	–	+	–
Quinate	+	+	+	+	+	ND	+	–	+	+
Ornithine decarboxylation[d]	V	V	+	+	+	+	V	–	–	–
Acid from[e]:										
D-Cellobiose	+	+	+	+	+	+	+	–	+	–
Palatinose	+	+	+	+	+	+	+	–	–	–

Table 1 contd. ...

...Table 1 contd.

Phenotype Reaction	Reaction For:									
	C. sakazakii	C. malonaticus	C. dublinensis	C. muytjensii	C. turicensis	C. condimenti	C. universalis	S. turicensis	F. pulveris	F. helveticus
Indole production[f]	–	–	+	+	–	+	–	–	–	–
Malonate utilization[g]	V[#]	+	V	+	+	+	V	+	–	+

[a] Data are derived from reports by Iversen et al. (2007a, b, 2008a), Joseph et al. (2012a), and Stephan et al. (2014). Negative tests should be incubated for seven days prior to being discarded unless otherwise indicated. Results for some of the phenotypic reactions, such as those for VP and ornithine decarboxylation, can be obtained directly by using the API 20E strip (bioMérieux, Inc., Hazelwood, Missouri) and by using reagents according to the manufacturer's instructions. +, 90 to 100% positive; V, 20 to 80% of members of the group are positive; –, 10 to 20% positive.

[b] The Voges-Proskauer test was performed by adding 40% potassium hydroxide in water and 5% 1-naphthol in 95% ethanol to cultures incubated for 24 h at 37°C in methyl red–Voges-Proskauer broth (CM0043, Oxoid, Inc., Thermo Fisher Scientific, Inc., Waltham, MA).

[c] The methyl red test was performed by adding the methyl red reagent (0.1 g per 300 mL of 95% ethanol) to cultures grown for 48 h at 37°C in 4 mL of methyl red–Voges-Proskauer broth (Thermo Fisher Scientific, Inc.).

[d] Ornithine decarboxylation is shown by an alkaline reaction according to the API 20E instructions.

[e] Acid production from carbohydrates was tested in phenol red broth base (BBL 221897, Thermo Fisher, Inc.) with the addition of filtered-sterilized carbohydrate solution (final concentration of 0.5% in cultures incubated for 24 h at 37°C).

[f] Determined using Kovac's reagent after growth in peptone broth (CM0009, Oxoid, Inc., Thermo Fisher Scientific, Inc.) of cultures incubated for 24 h at 37°C. The James reagent (Thermo Fisher Scientific, Inc.) used in conjunction with the API 20E strip is an alternative test.

[g] Determined using sodium malonate broth in cultures incubated for 24 h at 37°C.

[#] Gopinath et al. (2018) determined that ST64 *C. sakazakii* possesses the nine-gene malonate utilization operon that is found in the other *Cronobacter* species.

et al. 1980). Using a systematic scientific approach comprised of both phenotypic characterization and DNA-DNA hybridization (also known as DNA reassociation, a new molecular technique at the time; Brenner 1973), Farmer et al. (1980) were able to assign 15 different biogroup designations to these strains primarily based on malonate and dulcitol utilization, indole production, and a positive methyl red test. Later, Iversen et al. (2006) defined a 16th biogroup. Farmer et al. (1980) named the new species *Enterobacter sakazakii* in honor of the Japanese bacteriologist Riichi Sakazaki for his many contributions to our current understanding of various members of the families *Enterobacteriaceae* and *Vibrionaceae* and for recognition of his lifelong interest in enteric bacteriology. The proposed classification change was based on differences in biochemical reactions, pigment production, and antibiotic susceptibility between *E. cloacae* and *E. sakazakii*. Even though the expression of the yellow *Pantoea*-related pigment was a criterion that was used in later isolation schemes, it was a trait that Farmer had argued not be used as a determinative characteristic because of poor production among strains (Farmer et al. 1980; Lehner et al. 2006a; Tall et al. 2014). In support of Farmer's original comment (Farmer et al. 1980) about pigmentation, up to 21.0% of the *Cronobacter* strains studied lacked yellow pigmentation on TSA after incubation resulting in white colonies (Besse et al. 2006; Cawthorn et al. 2008).

The yellow pigment is encoded by a gene cluster with the organization *crtE-idi-XYIBZ* which was homologous to a similar gene cluster originally reported for *Pantoea agglomerans* (Lehner et al. 2006a). Additionally, Farmer's study was one of the first examples of using DNA-DNA hybridization for the taxonomic placement of clinical strains. Close or highly related strains (e.g., belonging to the same species) are usually taken to have close relative DNA reassociation values of 70% or more. Farmer et al. (1980) found that the *E. sakazakii* strains were 83% to 91% related to each other but were only 31% to 54% related to *E. cloacae*. The new species was placed in *Enterobacter* rather than *Citrobacter* because its closer phenotypic, and DNA reassociation values were more similar to *E. cloacae*, the type species of the genus *Enterobacter*, and because these strains were only 41% related by DNA hybridization analysis to *Citrobacter freundii*, the type species of *Citrobacter*.

During the initial research that led to the proposal for *Cronobacter*, Iversen et al. (2008a) thought that some strains did not fit the definition of *Cronobacter* and were thought to be more similar to *Enterobacter*. These strains were taxonomically placed into three novel species: *Enterobacter pulveris*, *Enterobacter helveticus*, and *Enterobacter turicensis*, which were characterized and described by Stephan et al. (2007; 2008). These organisms were isolated from dried fruit powders and other dried food ingredients, powdered infant formula (PIF) or infant milk formula (IMF), and several PIF-production environments. Iversen et al. (2008a) excluded them from the genus *Cronobacter* because these *Enterobacter* species more closely aligned with the description of *E. sakazakii* than that for *Cronobacter*. The decision to exclude them from *Cronobacter* was based on differences in their phenotypic characteristics, as well as data from DNA-DNA hybridization analysis and the phylogenetic analysis of the *rpoB* gene (Stephan et al. 2007; 2008). However, these novel species do share several phenotypic and metabolic characteristics with members of the genus *Cronobacter*, such as resistance to desiccation, production of a yellow *Pantoea*-like, carotenoid pigment (Lehner et al. 2006a), and constitutive metabolism of 5-bromo-4-chloro-3-indolyl-α-D-glucopyranoside, which is the substrate used in the differentiation of presumptive colonies of members of the genus *Cronobacter* when grown on the well-recognized chromogenic *Cronobacter* isolation agars, Druggan Forsythe and Iversen agar (DFI; Iversen et al. 2004), and *Enterobacter sakazakii* plating medium (ESPM; Restaino et al. 2006). Brady et al. (2013) re-evaluated the taxonomy of the genus *Enterobacter* and based primarily on multilocus sequence analysis (MLSA) of partial sequences of four housekeeping genes (*gyrB*, *rpoB*, *infB*, and *atpD*), these authors proposed that *E. helveticus*, *E. pulveris*, and *E. turicensis* be recognized as *Cronobacter* species. Stephan et al. (2014) subsequently presented new genome-scale data including average nucleotide identities, which is a sequence-based algorithm that has now replaced DNA-DNA hybridization (Arahal 2014). Genome-scale phylogeny and k-mer analyses, coupled with previously reported DNA-DNA reassociation values and phenotypic characterizations, strongly indicated that these three *Enterobacter* species are not members of the genus *Cronobacter*; nor did they belong to the re-evaluated genus *Enterobacter*. Furthermore, data from this polyphasic study indicated that all three species represented two new genera. *E. pulveris* and *E. helveticus* were taxonomically placed in the newly described genus *Franconibacter* as *Franconibacter pulveris* and *Franconibacter helveticus*, respectively, and *Enterobacter turicensis* was placed in the genus *Siccibacter* as *Siccibacter turicensis*. The genus *Franconibacter* was named after Augusto Franco for his contributions to the understanding of the various *Cronobacter* virulence plasmids and their associated virulence factors such as *Cronobacter* plasminogen activator, siderophore and iron acquisition systems, a *Bordetella*-like filamentous hemagglutinin, and a type 6 secretion system (T6SS) carried on them (Franco et al. 2011a, b; Grim et al. 2012).

Clinical Disease

Cronobacter species are a group of opportunistic pathogens that cause life-threatening infections in individuals of all age groups (Iversen and Forsythe 2003; Holý et al. 2014; Patrick et al. 2014; Alsonosi et al. 2015; Yong et al. 2018). *Cronobacter* can cause high fatality rates in immunocompromised individuals, such as the elderly, infants, and neonates (Lai 2001; Friedemann 2009). Except for *C. condimenti*, all species of *Cronobacter* have been isolated from clinical specimens. At present, there is not enough epidemiological information to tease apart adult infections originating nosocomially from those acquired through community-associated sources (Jang et al. 2020a) because most of the reports document infections of individuals already hospitalized (Holý et al. 2014; Patrick et al. 2014; Alsonosi et al. 2015) and identification of isolates have been difficult. Despite some advances, knowledge of pathogenesis or the nature and action of putative virulence factors is limited.

Infections Associated with Infants and Neonates

Premature (or preterm) infants, low birth-weight neonates, and infants with underlying medical conditions are at the highest risk for developing severe *Cronobacter* infections, such as septicemia, meningitis, and necrotizing enterocolitis (NEC) (Lai 2001; FAO/WHO 2004; Bowen and Braden 2006; Patrick et al. 2014; Elkhawaga et al. 2020). Mortality rates among this group at one time were reported to be as high as 80%, but now the rate has steadied to be ~ 27%. However, infants who survive, frequently suffer developmental delays, hydrocephaly, mental retardation, and other neurological sequelae (Lai 2001; Friedemann 2009; Holý et al. 2014; Patrick et al. 2014; Elkhawaga et al. 2020).

NEC is a very common inflammatory intestinal disorder seen in neonates, especially in preterm infants with very low birth weights (VLBW; defined as an individual with a birth weight of < 1.5 kg). Although the exact etiology of NEC remains to be fully understood, multiple risk factors have been connected to its pathogenesis. Besides prematurity (birth occurring at less than 36 weeks of gestation), a history of formula feeding, and underdeveloped microbial colonization of the gastrointestinal tract have been identified as risk factors (Holman et al. 1997; Hunter et al. 2008a, b; Grishin et al. 2016). The incidence rate of NEC in VLBW infants has somewhat stabilized over the years and is currently observed in approximately 5% to 7% of all babies admitted to neonatal intensive care units (NICUs) (Blackwood and Hunter 2016). Even though NEC has been associated with *Cronobacter* species, to date no single pathogen has been identified as being the sole etiologic agent responsible for the disease (Esposito et al. 2017).

Several new trends have come about since acknowledging *Cronobacter* as a neonatal pathogen. For example, prior to 2001, the majority of the *Cronobacter* cases seen by public health agencies were coming from infants who were hospitalized and admitted to NICUs. These individuals often have other underlying health issues, such as failure-to-thrive syndrome, hypoxia, hypothermia, and intestinal ischemia. Several outbreaks occurred in NICUs worldwide and until 2008, around 120 documented cases of *Cronobacter* spp. infection and at least 27 deaths had been reported (Lai 2001). From these studies, knowledge about infantile infections has shown that they have been epidemiologically linked to the consumption of intrinsically and extrinsically contaminated lots of reconstituted PIF (Noriega et al. 1990; Himelright et al. 2002; Henry and Fouladkhah 2019). Furthermore, Strysko et al. (2020) recently showed that within the United States, there are higher numbers of infections among full-term community-associated infants than hospitalized infants, which suggests that newborns are leaving hospitals after uncomplicated births only to return with severe onset disease. Another finding was that many of the cases were of infants who consumed PIF reconstituted from opened cans in a home setting (Strysko et al. 2020).

Retrospectively, Jason (2012) described another trend by evaluating historical surveillance data on 82 *Cronobacter* infant cases (between 1958 and 2010) where she showed that these infants became ill (defined here as a confirmed culture-positive case of septicemia or meningitis) after ingesting breast milk exclusively (without consumption of PIF, follow up the formula, or powdered human milk fortifiers) prior to illness onset. Friedemann (2007) also reported similar observations. To emphasize this new important epidemiological warning, Bowen et al. (2017) and McMullan et al. (2018) described several infantile cases of *C. sakazakii* septicemia/meningitis where these infants only consumed expressed maternal milk (EMM) during their first weeks after birth. Contaminated personal breast pumps were found to be the source of the contamination. Traceback investigations using molecular methods such as pulsed-field gel electrophoresis (PFGE) and whole genome sequencing (WGS) analyses of isolates concluded that the clinical isolates were indistinguishable from those cultured from frozen and stored contaminated expressed milk, a contaminated breast pump and home kitchen sink drain in the former case and the breast pump in the latter case. Together, these data suggest that there is a need to continue informing parents and infant care takers of the risks associated with breastfeeding and producing and storing EMM. Subsequent to these case reports, the Centers for Disease Control and Prevention (CDC) has provided breast pump cleaning instructions to infant care takers and nursing mothers (https://www.cdc.gov/hygiene/childcare/breast-pump.html, last accessed 5.20.2023).

There are numerous benefits of breastfeeding such as providing the infant with all the nutrition required for normal growth and development. Human breast milk is ideal for infant digestion and allows for effective psychological bonding to occur between infant and mother. Thus, breastfeeding is the first choice for providing the proper nutrients required for infant growth and development. When the option of breastfeeding is not available, IMF or PIF play essential roles; they are intended to be effective substitutes as they are formulated to mimic the nutritional components of breast milk. However, newborn infants are particularly vulnerable to infection due to the immaturity of their immune system (Lönnerdal 2012). Also, the establishment of the normal gut microflora, which acts as a barrier to infection may not be fully developed during the early weeks of neonatal life (Chap et al. 2009).

Infections Associated with Adults

Although *Cronobacter* species have been primarily associated with infections in infants, more than a few reports have emphasized the risk posed to adults, particularly the immunocompromised elderly (individuals with an age greater than 80 years of age) (Gosney et al. 2006; Holý et al. 2014; Patrick et al. 2014; Alsonosi et al. 2015). Various types of infections such as acute gastroenteritis, septicemia, urosepsis, osteomyelitis, wound infections, and pneumonia have been reported in adults (Holý et al. 2014; Patrick et al. 2014; Alsonosi et al. 2015; Yong et al. 2018). Additionally, the prevalence of *Cronobacter* infections in adults who have experienced strokes has been noted by Gosney et al. (2006) and because the stroke may have also diminished their ability to swallow (known as dysphagia), subsequently leading to a high risk of aspiration pneumonia in these individuals. This population may use rehydrated powder protein supplements as part of their diet and rehabilitation efforts (Gosney et al. 2006). This is a problem that is likely to become more common because of the aging of the world's population and an increase in the consumption of synthetic and dehydrated adult nutritional supplements in this population. Lastly, Yong et al. (2018) described an epidemiological investigation of an acute gastroenteritis outbreak in a high school, which was caused by a mixture of ST73 and ST4 *C. sakazakii* strains and ST567 *C. malonaticus* strains. The clinically implicated food and environmental samples included rectal swabs, a bean curd braised pork dish, and food delivery boxes (Yong et al. 2018). This is the first report of a foodborne gastroenteritis outbreak involving young teens and adults caused by *Cronobacter*.

There are few studies revealing that *Cronobacter* can be a member of the human intestinal tract microflora and may be carried in the intestinal tracts of healthy adults and neonates (Zogaj et al. 2003; Porres-Osante et al. 2015; Amaretti et al. 2020). However, additional epidemiologic evidence is needed to realize the role that *Cronobacter* may play as a member of the human gut microflora and whether its presence represents transient colonization or other related host-associations (Mao et al. 2015). Little is known about the perturbations associated with the gut microflora that may precede or develop during pregnancy and if there are any transient postpartum effects (Mutic et al. 2017). Lastly, surveillance studies using species-specific identification schemes that at least include results of species-specific PCR assays, or at best, species identities obtained from WGS or metagenomic studies are needed so that a clearer epidemiologic picture can be obtained as to which *Cronobacter* species are responsible for which infection (Savino et al. 2015; Chandrasekaran et al. 2018). Understanding the roles that the various members of the microflora of gut, vagina, and skin, both that of the mother and infant, are still in their early stages but hold promise to understanding postpartum recovery and newborn development.

Animal and Tissue Culture Models of Infection and Virulence Factors

It is commonly thought that for a pathogen to establish disease, successful colonization of target tissues must occur such as adherence to the upper respiratory mucosal epithelial layer in pneumonia, intestinal epithelium in necrotizing enterocolitis and gastroenteritis, or the cerebrospinal fluid-filled subarachnoid space in meningitis (Finlay and Falkow 1997). Pagotto et al. (2003) showed that some clinical and food isolates of *Cronobacter* were lethal when administered intraperitoneally in mice, but only two strains caused death by the oral route. They also reported that some, but not all strains, produced an enterotoxin that caused fluid accumulation in suckling mice, whereas other strains produced factors that lyzed or "rounded" tissue culture cells. A report by Raghav and Aggarwal (2007) described the purification and characterization of an *E. sakazakii* enterotoxin. However, at present, the linkage between enterotoxin production and pathogenesis is unclear because no gene (gene cluster) has been assigned. Additionally, Kothary et al. (2007) screened various clinical and environmental strains for the production of factors that affected Chinese hamster ovary (CHO) cells in tissue culture. A qualitative and preliminary study showed that many of the strains produced factors that caused the "rounding" of CHO cells. The rounding of tissue culture cells has been reported to be due to the action of various bacterial proteases and enterotoxins (Lockwood et al. 1982). The *Cronobacter* rounding factor was shown to be a zinc-containing metalloprotease which was cell-associated and poorly secreted into the culture supernatant (Kothary et al. 2007). Eshwar et al. (2018) used the zebrafish embryo infection model to study the interaction of the zinc metalloproteinase Zpx expressed by *Cronobacter turicensis* LMG 23827[T] with the eukaryotic MMP-9 proteinase that functions to cleave extracellular matrix gelatin and collagen. Cleavage and activation of the human recombinant pro-MMP-9 by Zpx-expressing *C. turicensis* cells were demonstrated *in vitro*, and the presence and increase of the processed, active form of zebrafish pro-MMP-9 were shown *in vivo*. This provides evidence that Zpx induces the expression of the MMP-9 but also increases the levels of processed MMP-9 during infection. Furthermore, this study provided evidence that MMP-9 is a substrate of Zpx and demonstrated a yet undescribed mutual cross-talk between these two proteases in infections mediated by *C. turicensis* LMG 23827[T]. According to NCBI, *zpx* is now annotated as a peptidase (Zpx peptidase) and is a member of the M4 protein family (https://www.ebi.ac.uk/merops/cgi-bin/pepsum?id=M04.023, last accessed 5.20.2023). Most members of this family have secreted enzymes that degrade extracellular proteins and peptides for use in bacterial metabolism.

The gene encoding a hemolysin (*hly*) was identified as a hemolysin III homolog (COG1272) by Cruz et al. (2011). Since then, using WGS, several investigators predicted that all *Cronobacter* species may possess a hemolysin III homolog (COG1272) gene (Kucerova et al. 2010; Grim et al. 2013; Moine et al. 2016). Additionally, Jang et al. (2020b) identified three other hemolysin-like genes by using a parallel next-generation DNA sequence-based approach utilizing a *Cronobacter* microarray and WGS and showed the presence of several hemolysin-like genes such as a cystathionine β-synthase (CBS) domain-containing hemolysin, a putative hemolysin, and a 21-kDa hemolysin. However, more in-depth genetic studies are needed to assign the functionality of these various hemolysin genes to a hemolytic phenotype.

Studies using animal infection and *in vitro* mammalian tissue culture models have shown that *Cronobacter* isolates demonstrate a variable virulence phenotype. However, several studies have demonstrated the role of the virulence plasmids (pESA3/pCTU1), several outer membrane proteins (OmpA and OmpX), exoproteins (Cpa, metalloprotease, Zpx, putative enterotoxin, and cis-trans prolyl isomerases (PPIases) such as Fkp), and a diffusible signal factor-type quorum sensing system in disease (Pagotto et al. 2003; Raghav and Aggarwal 2007; Nair et al. 2009; Franco et al. 2011b; Eshwar et al. 2015; 2016; Fehr et al. 2015; Suppiger et al. 2016). The authors refer readers to Jang et al. (2020a) for greater details of the role in the pathogenesis of these exoproteins.

Cronobacter *in Foods*

Cronobacter *Associated With Powdered Infant Formula (PIF) and Persistence Mechanisms*

Even though many members of the scientific community had suspicions that PIF was a source of infantile infections, epidemiological evidence was provided without a doubt by Himelright et al. (2002) that linked intrinsically contaminated PIF to neonatal disease. Subsequently, several outbreak investigations were reconstructively dissected which came to similar conclusions (Caubilla-Barron et al. 2007; Townsend et al. 2008; Masood et al. 2015). However, it is clear now that contamination of reconstituted PIF can occur intrinsically and extrinsically, although the main reservoir(s) and routes(s) of contamination have yet to be fully established.

Also, contamination can occur at any time point during the manufacture of PIF, its reconstitution, or storage of reconstituted PIF (Noriega et al. 1990; Friedemann 2007; Henry and Fouladkhah 2019; Strysko et al. 2020). Despite the best efforts of manufacturers, contamination of commercial PIF products still occurs, and *Cronobacter* spp. are routinely detected in microbiological analysis of products with a prevalence rate of 4.7% (Parra-Flores et al. 2020).

Cronobacter survives under extreme desiccation (high osmotic stress) growth conditions such as that present in low water activity (a_w) foods and this property is thought to influence its environmental persistence in PIF/FUF manufacturing facilities and other dried products or desiccated environments (Riedel and Lehner 2007; Feeney and Sleator 2011; Srikumar et al. 2019). In fact, there are examples of strains persisting within manufacturing facilities for long periods, up to five years (Reich et al. 2010; Jacobs et al. 2011; Yan et al. 2013; Yan et al. 2015a; Chase et al. 2017a). Extensive traceback studies showed that left over PIF residues (PIF dust in production areas) from operations can collect in air filters, on floors, in vacuum cleaners used to clean up dry spills, and in filtering (sieving) machines. Additionally, residual fluids from drains and swabs from other product contact surfaces and soil samples collected adjacent to the production facilities are primary positive areas for *Cronobacter* persistence (Mullane et al. 2008a; Reich et al. 2010; Fei et al. 2015; Yan et al. 2015a). This was supported by findings that the majority (78%) of *Cronobacter* isolates were recovered from surfaces associated with dryers and blenders located in various manufacturing facilities processing areas (Proudy 2009). Finding *Cronobacter* associated with residual powder in various vacuum cleaners located within PIF manufacturing facilities supports the outcomes reported by Srikumar et al. (2019), who showed that infant formula powder residue enhances the desiccation survival of persistent and xerotolerant strains such as *C. sakazakii* SP291 that was isolated from an Irish PIF manufacturing facility.

The bacterial primary response to high osmotic environments involves the rapid accumulation of the electrolytes potassium [K^+] and glutamate to increase the cell's internal osmotic pressure, thus counteracting the high external osmolar conditions (Srikumar et al. 2019). When a cell internalizes elevated potassium concentrations, glutamate also is co-transported (Srikumar et al. 2019). Eventually, the primary response transitions into a secondary response which is hallmarked by the synthesis of such osmoprotectants as glycine betaine, proline, carnitine, and trehalose (Srikumar et al. 2019). For more information about the osmotic response, see Feeney and Sleator (2011) and Srikumar et al. (2019). The widespread occurrence of *Cronobacter* and its resistance to desiccation are thought to be contributing factors to the commercial distribution of contaminated low a_w type foods, which poses a risk to susceptible consumers (Yan et al. 2012; Jaradat et al. 2014; Yan and Fanning 2015a, b). From an applied food safety research perspective, an improved understanding of *Cronobacter* osmotolerance response should lead to more effective control measures, which will limit the survival and persistence of *Cronobacter* in low a_w products and their manufacturing environments such as PIF/IMF production facilities.

The ability of *Cronobacter*, as is the case for many microorganisms, to withstand the harmful effects of exposure to different environmental stresses is crucial for their survival, persistence, and virulence (Hews et al. 2019). Results from additional studies conducted with several strains over a wide range of temperatures as well as survival in different foods suggest that most *Cronobacter* are thermotolerant in that strain-associated stress tolerance or prior heat adaptation can enhance thermal stability (Arroyo et al. 2009; Osaili et al. 2009; Arku et al. 2011; Huertas et al. 2015). Furthermore, besides the ability to tolerate osmotic stress or sudden changes in the solute concentrations surrounding a cell, *Cronobacter* can also adapt quite well to exposures of variable growth temperatures and this tolerance is beginning to be understood at the genomic level (Edelson-Mammel and Buchanan 2004; Arroyo et al. 2012). For example, Williams et al. (2005) identified a protein in heat-tolerant *Cronobacter* strains that Edelson-Mammel and Buchanan (2004) determined were highly heat-resistant. The protein was determined to be a hypothetical protein found in the thermal tolerant bacterium, *Methylobacillus flagellates* KT. Gajdosova et al. (2011) described an 18-kbp region that contained a cluster of genes, including the hypothetical protein found by Williams et al. (2005) that had significant homology with other known bacterial proteins involved in stress responses including heat, oxidation, and acid stress. However, not every strain that possesses a thermotolerant phenotype has this gene cluster (Yan et al. 2013), suggesting that other genomic mechanisms involving thermotolerance exist, quite possibly under the control of global regulators (Choi et al. 2012), of which little knowledge current exists.

Numerous studies have revealed that *Cronobacter* species can grow in environments of low pH levels (pH 3.9–4.5) and are more acid tolerant than most other phylogenetically related enteric pathogens (Alvarez-Ordóñez et al. 2014; Jaradat et al. 2014). However, such studies are in their infancy and more information is needed to fully understand how *Cronobacter* species can survive under these very different stressful growth conditions and if multiple commercially available treatment regimens can be used to successfully prevent growth in RTE foods and PIF. An interesting study by Jang and Rhee (2009), revealed the dynamics of combining caprylic acid and mild heat to reduce *Cronobacter* growth in reconstituted PIF. The combined treatment resulted in a "synergistic effect," in which *Cronobacter* populations were reduced much more rapidly by using increased temperatures and concentrations of caprylic acid compared to populations that were not treated. This study also pointed to how valuable the integrity of the Gram-negative outer membrane is in survival, and these data revealed that the addition of this natural GRAS antimicrobial agent to infant formula may have potential use for controlling *Cronobacter* prior to consumption. Understanding how *Cronobacter* persists under osmotic, thermal, and acidic stressful growth conditions is crucial for controlling this foodborne pathogen within the food manufacturing environment.

Cronobacter *Associated With Other Types of Foods*

Of equal significance is that *Cronobacter* species are largely thought now to be more ecologically widespread and have been found associated with many types of foods (including other low a_w foods) besides PIF. For example, *Cronobacter* species have been found associated with dried dairy protein products (milk and cheese protein powders), cereals, candies such as licorice and lemon-flavored cough drops, dried spices, teas, nuts, herbs, and pasta (Iversen and Forsythe 2003; Osaili et al. 2009; Sani and Odeyemi 2015; Hayman et al. 2020). It has also been found contaminating water and many different ready-to-eat and frozen vegetables, insect body surfaces and intestinal contents, and man-made environments such as PIF or dairy powder production facilities, and household sink drains (Iversen et al. 2006; Pava-Ripoll et al. 2012; Müller et al. 2013; Sani and Odeyemi 2015; Chase et al. 2017a, b; Jang et al. 2018a, b; 2020b; 2022; Hayman et al. 2020). Isolation from livestock has been uncommon and limited survival in the animal gut is indicative of a non-zoonotic nature, although contamination of meat and milk has been reported (Molloy et al. 2010; Joseph and Forsythe 2012; Zeng et al. 2020).

Cronobacter *Isolation and Detection Methods*

Classical microbiology methods using both pre- and selective-enrichment broths, and differential and selective plating media are the mainstay procedures used to isolate *Cronobacter* from foods. For proper identification to be accomplished, phenotypic analysis needs to be combined with molecular methods, including polymerase chain reaction (PCR) amplification or PCR coupled with immunomagnetic separation (IMS) or whole genome sequencing (WGS) of genes, such as *fusA* or entire genomes. Eventually, culture-independent molecular methods such as metagenomic analyses will replace culture methods, but presently there are still many regulatory challenges that need to be considered (Dwivedi and Jaykus 2011; Carleton et al. 2019).

Initially, the FDA described a protocol for the isolation of *Cronobacter* from PIF and closely related products, such as powdered substitutes for breast milk, which depended on standard methods used for the detection of all members of the *Enterobacteriaceae* (Muytjens et al. 1988; FAO/WHO 2004; 2008). Briefly, sterile water was added to PIF (1:10 dilution) followed by another 1:10 dilution in *Enterobacteriaceae* enrichment (EE) broth. Both broths were incubated at 36°C for 18 hours. The EE broth is then streaked onto Violet Red Bile Glucose Agar (VRBGA) and incubated overnight at 36°C. Presumptive colonies are picked and plated on Trypticase Soy Agar (TSA) for at least 72 hours at 25°C. Yellow-pigmented colonies that are oxidase negative are confirmed using a biochemical test strip (such as API 20E, BioMerieux, Hazelwood, MO). This protocol required 5 to 7 days for completion and was very cumbersome and time-consuming. As mentioned earlier, the selection of yellow-pigmented colonies and their passage onto TSA lacks specificity for isolating *Cronobacter* spp. because other genera such as *Proteus, Morganella, Enterobacter, Franconibacter, Siccibacter* species, *Erwinia*, and *Pantoea* can be considered presumptive positives until the final biochemical/molecular confirmation steps are completed. Besides the lack of specificity in the above-mentioned method, other shortcomings exist such as the brilliant green component in EE broth has been found to inhibit some strains of *Cronobacter* (Fox and Jordan 2008) and as reported by Iversen and Forsythe (2007) it was also found that ~ 4% of the strains tested failed to grow in Brilliant Green Bile Broth (BGBB) incubated at 37°C and 44°C for 48 hours. EE agar also performed poorly with spiral plating for recovering normal and stressed *Cronobacter* spp. cells

(Gurtler and Beuchat 2005). For VRBGA, the crystal violet ingredient can inhibit certain strains of *Cronobacter* (Druggan and Iversen 2009).

In 2002, when this FDA method was developed, procedures for isolating *Cronobacter* spp. from PIF were lacking and since then improvements in culture techniques have occurred. In 2009 and 2010, a revised method for the isolation and detection of *Cronobacter* from PIF with greater sensitivity and specificity was developed, validated, and currently replaces the VRBGA method in the Bacteriological Analytical Manual (BAM) (Chen et al. 2009; Chen et al. 2010). The revised FDA method shown in Figure 2 includes a pre-enrichment step in Buffered Peptone Water (BPW) incubated at 36°C. After 4–6 h, 40 mL of the enrichment broth is centrifuged for 10 min at 3,000 × g, the supernatant is discarded, and the pellet is resuspended in 200 μL of Phosphate Buffered Saline (PBS). A real-time PCR procedure targeting the *dnaG* region first described by Seo and Brackett (2005) was developed that included primers and internal control was combined with the method and performed using a boiled DNA template preparation from the resuspended enrichment culture pellets (Chen et al. 2009; 2010; 2012). The real-time PCR method of Seo and Brackett (2005) with the modification of 40 total cycles was the basis of the improved BAM qPCR method. If the results from the PCR are positive, the centrifuge step is repeated on the 24 hours pre-enrichment broth and 100 μL of the PBS is spread-plated/streaked onto DFI and *Enterobacter sakazakii* Chromogenic Plating Medium (ESPM). The plates are incubated at 36°C for 24 hours. Typical colonies are confirmed using Rapid ID 32E or VITEK Compact 2.0 kits and with the genus-level PCR assay.

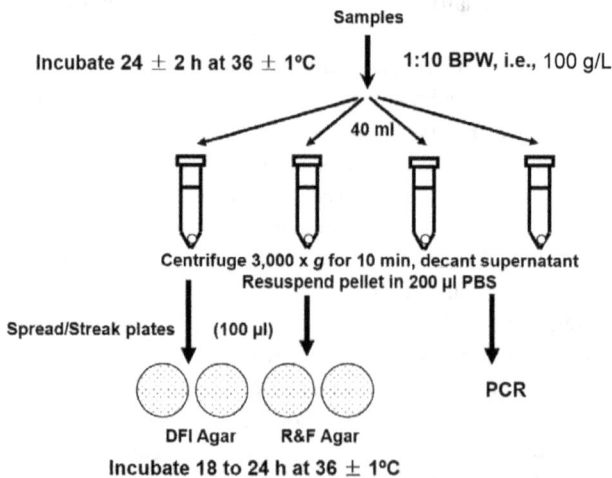

Figure 2. Schematic workflow describing the revised FDA *Cronobacter* method for isolation and detection of *Cronobacter* from PIF that is illustrated in Chapter 29 of the BAM. For details, please see https://www.fda.gov/food/laboratory-methods-food/bam-chapter-29-cronobacter (last accessed 5.20.2023). Take note that two colonies from the Chromogenic plates are processed for the qPCR confirmation assay and either VITEK 2.0 or Rapid ID 32E as that described by Chen et al. (2012).

1. Pre-enrichment
- Sample (10 g or 10 mL) + Buffered peptone water (90 mL)
- Incubation at 34-38°C for 18 ± 2 h

2. Selective enrichment
- Culture (0.1 mL) + CSB (*Cronobacter* Selective Broth, 10 mL)
- Incubation at 41.5 ± 1°C for 24 ± 2 h

3. Plating-out
- Streaking onto Chromogenic *Cronobacter* Isolation (CCI) agar
- Incubation at 41.5 ± 1°C for 24 ± 2 h

4. Confirmation
- A typical colony (blue-green) selected
- Streaking onto non-selective agar (e.g. TSA)
- Incubation at 34-38°C for 21 ± 3 h
- Biochemical characterization

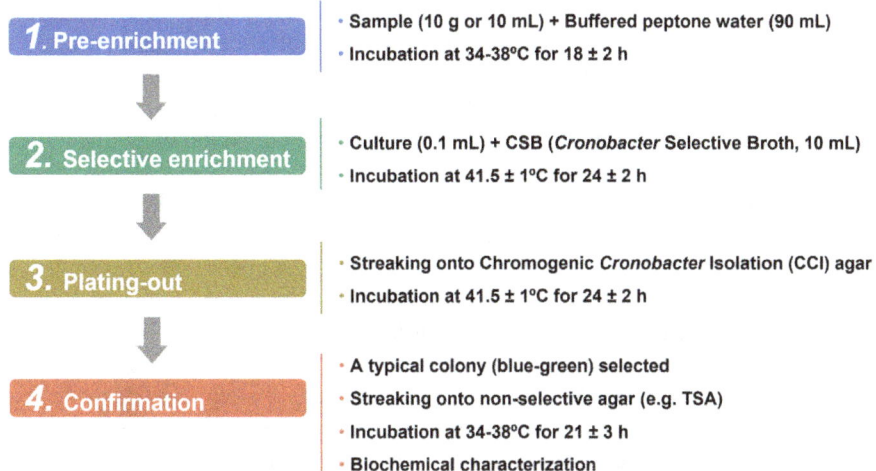

Figure 3. Procedure steps of ISO 22964:2017 standard for *Cronobacter* detection, which describes a horizontal method for the detection of *Cronobacter* spp. in food, animal feed, and environmental samples (ISO 2017; de Benito et al. 2018).

A horizontal International Organization for Standardization (ISO) method was published and has undergone validation as a technical standard protocol, known as ISO 22964:2017. As shown in Figure 3, it can be used for the detection of *Cronobacter* species in food products for humans, animal feeds, and environmental samples (de Benito et al. 2019).

This method, which technically revises an earlier protocol, ISO/TS 22964:2006, in addition to PIF now includes all food products for humans and feeds for animals as well as environmental samples. Other changes from the earlier ISO method include replacing the selective-enrichment broth, modified lauryl sulfate tryptose broth (mLST) with *Cronobacter* selective broth (CSB; Iversen et al. 2008b) and the isolation agar, *E. sakazakii* isolation agar (which is a formulation of DFI agar), with chromogenic *Cronobacter* isolation agar (CCI). ESPM agar can also be substituted or used in conjugation with CCI agar. The goal of CSB is to screen negative *Cronobacter* samples without any further evaluation or testing after 48 hours. CSB yielded more positive *Cronobacter* samples than mLST in artificially inoculated PIF at low levels, and compared favorably with the ISO/TS 22964 method for isolating *Cronobacter* from naturally contaminated PIF, PIF ingredients, and environmental samples (Iversen et al. 2008b). This updated ISO method has also been used by a variety of research groups to isolate *Cronobacter* from vegetables and RTE foods (Berthold-Pluta et al. 2017; Brandão et al. 2017; Ueda 2017; Moravkova et al. 2018; Vasconcellos et al. 2018).

Since α-glucosidase activity is quite inclusive and unique to *Cronobacter,* in contrast to *Enterobacter* (Muytjens et al. 1984) and the *Enterobacteriaceae* family (Kämpfer et al. 1991), various plating media have used this characteristic for differentiation of *Cronobacter* from other isolates. The fluorogenic substrate

4-MU-α-Glc was added to nonselective nutrient agar (Leuschner and Bew 2004) and *Cronobacter* colonies displayed a blue/violet fluorescence on this agar using a longwave emitting ultraviolet light. For low-level contamination of *Cronobacter* in PIF, the sensitivity of this agar was like that of VRBGA (Lehner et al. 2006b) so its usefulness as a selective medium may be limited to confirmation of α-glucosidase activity.

Various chromogenic plating media have incorporated the chromogen 5-bromo-4-chloro-3-indoxyl-α-D-glucopyranoside (X-α-Glc) for the sole detection of *Cronobacter's* α-glucosidase activity. Since the chromophore dimer complex is water-insoluble compared to the water-soluble 4-MU, the blue indigo color stays within the colony, displaying a distinct intrinsic improvement in an agar system versus the fluorogenic reaction. The various chromogenic plating media (including typical colonial morphologies and temperature parameters) using X-α-Glc to identify α-glucosidase activity in *Cronobacter* strains are presented in Table 2.

A fast, simple, and reliable method for differentiating *Franconibacter* and *Siccibacter* from *Cronobacter* based on intact-cell matrix-assisted laser desorption ionization-time of flight mass spectrometry (MALDI-TOF MS) has been developed by Svobodová et al. (2017). Stephan et al. (2010) was the first group to apply matrix-assisted laser desorption ionization-time of flight mass spectrometry (MALDI-TOF MS) for species identification of *Cronobacter* species. Since then, other groups have applied and developed this technology for rapid species identification (Wang et al. 2017; Bastin et al. 2018).

Table 2. Growth characteristics of *Cronobacter* on chromogenic plating media incorporating 5-bromo-4-choro-3-indoxyl-α-D-glucopyranoside for α-glucosidase detection.

Chromogenic Plating Medium	Presumptive Positive Colonial Morphologies	Incubation Parameters (Temp./Time)
DFI/CCI	Entirely blue-green color	37°C/24 hours
mDFI	Entirely blue-green color	42°C/24 hours
COMPASS agar	Blue-green to pale green color	44°C/24 hours
ESIA	Blue to turquoise color	44°C/24 hours
Chromocult® *Enterobacter sakazakii* agar (CES)	Entirely turquoise color	44°C/24 hours
Rapid' Sakazakii Medium (RSM)	Turquoise color	44°C/24 hours
ESPM (chromID™ Sakazakii)	Blue-black to blue-gray	37°C as well as at 41.5°C/24 hours

Immunomagnetic Separation

Cronobacter Detection Using Magnetic Bead Capture

Using a cationic–magnetic capture procedure, Mullane et al. (2006) showed that 1 to 5 *Cronobacter* cells artificially inoculated per 500 g of PIF could be reliably detected after enrichment. This protocol reliably detected *Cronobacter* in less than

24 hours (Mullane et al. 2006). The incorporation of a magnetic bead capture step showed promise, especially in reducing the time for a negative result, but needs to be further evaluated with respect to the duration of incubation of the sample, its feasibility of getting a negative result in 24 hours, and how it performs in other naturally contaminated food products.

Lateral Flow Devices

Several immunoassays have been developed to rapidly detect *Cronobacter* from cultures and PIF samples, for example using liposomes as the capture particle and enzyme-linked immunosorbent technology (Song et al. 2015; 2016). Other techniques such as enhanced lateral flow immunoassay (LFA) using nanogold particles conjugated with capture antibodies have been used to capture *Cronobacter* cells from food products (Pan et al. 2018), lateral flow devices that could rapidly detect and simultaneously serotype cells contaminating PIF (Scharinger et al. 2017), and lateral flow devices that combined technologies such as impedance and RNA hybridization (Zhu et al. 2012; Tomas et al. 2018). Other technologies such as loop-mediated isothermal amplification-based lateral flow dipsticks that could simultaneously detect *Cronobacter* (and other pathogens such as *Salmonella* and *Staphylococcus aureus*) in PIF have also been developed (Fan et al. 2012; Liu et al. 2012; Jiang et al. 2020).

Molecular Genus and Species Identification and Subtyping Characterization Methods

The use of the PCR as a detection tool has become indispensable in the field of infectious disease diagnostics and its similar use in food safety is no exception. Primarily, PCR has been used to connect the gap between presumptive identification and confirmation and has replaced the batteries of biochemical and susceptibility tests, which were laborious and time-consuming. Several PCR assays have been developed for the detection and identification of *Cronobacter.*

Detection of Cronobacter *spp. Using PCR*

Not surprisingly, the 16S rRNA gene was among the first gene targets for use in *E. sakazakii*-specific PCR detection assays (Hassan et al. 2007; Jaradat et al. 2009; Kang et al. 2007). Additionally, PCR assays targeting one of two genes encoding α-glucosidase activity, *gluA*, were also developed (Iversen and Forsythe 2007; Lehner et al. 2006b), as well as a PCR used to detect the outer membrane protein, *ompA* (Nair et al. 2006), and zinc metalloprotease, *zpx* (Kothary et al. 2007). Another genus-specific PCR assay targeting the *dnaG* gene was developed by Seo and Brackett (Seo and Brackett 2005). However, as mentioned earlier it has been reported that some of the *E. sakazakii*-like organisms (e.g., *F. helveticus, S. turicensis*) may be falsely identified (Chen et al. 2009; 2010). As noted, most of these earlier PCR assays were developed for *E. sakazakii* and that, given the heterogeneity of this group of organisms (e.g., the description of seven species of *Cronobacter*), it is expected that most of them will yield a number of false negatives. This does in fact appear to be the case, as reported by Jaradat et al. (2009), who compared several of the published end-point PCR assays described at the time. Indeed, Lerner et al. (2012) reported that 16S rRNA genes varied significantly (97.9%) between *E. sakazakii* (*C. sakazakii*) ATCC 29544[T] and (*C. muytjensii*) ATCC 51329[T]. Identification of *Cronobacter* by using 16S rDNA gene sequence information is further complicated by the fact that 16S rDNA sequence analysis has failed to accurately provide species-level characterization because the sequence divergence is not great enough to sufficiently distinguish between all the *Cronobacter* species, specifically between *C. sakazakii* and *C. malonaticus* (Joseph et al. 2012b; Strydom et al. 2012). Additionally, the sequence diversity between multiple copies of the 16S rDNA operon within *Cronobacter* can also introduce discrepancies (Acinas et al. 2004). Another target that was reported useful in detecting *Cronobacter* species was *gyrB* gene that encodes the B subunit of DNA gyrase (topoisomerase type II), which was found to be suitable for the identification of *Cronobacter* species (Chen et al. 2013). Lastly, several researchers have reported that the genus-level PCR assay targeting the *zpx* gene misses some *C. muytjensii* strains (B.D. Tall, personal communication).

Species-Specific PCR Assays

Following the change in taxonomic standing from *E. sakazakii* to *Cronobacter* spp., Stoop et al. (2009) and Lehner et al. (2012) described two *Cronobacter* species-specific PCR assays based on species-specific single nucleotide polymorphisms (SNPs) associated with the *rpoB* gene. This was followed by a description of a multiplex PCR assay targeting the *cgcA* gene (Carter et al. 2013; Hu et al. 2016). This PCR assay is best run as two duplex reactions using the DNA template sample with primers for *C. sakazakii*, *C. malonaticus*, *C. turicensis*, and *C. universalis* in a first reaction and then a second PCR assay with the same template with the *C. dublinensis* and *C. muytjensii* primers (B.D. Tall, personal communication).

In addition to traditional end-point PCR assays, several real-time PCR assays have also been reported, including commercial products. Seo and Brackett first reported an *E. sakazakii*-specific real-time PCR assay targeting the macromolecular synthesis operon, in this case, the 39 bp region of *rpsU* to the 59 bp region of *dnaG* (Seo and Brackett 2005). Subsequently, Drudy et al. (2006) modified the Taqman probe sequence to improve the specificity of the assay. Liu et al. (2006) developed two real-time PCR assays based on TaqMan and SYBR green technology. Both assays used primers to target the 16–23S rRNA spacer region and could detect 1.1 CFU of *E. sakazakii* per 100 g of infant formula after a 25-h enrichment. Several real-time assays followed, including those targeting 16S rRNA (Kang et al. 2007) and 16S–23S rRNA internal transcribed spacer (ITS) sequences (Liu et al. 2006; Chen et al. 2013).

A number of commercially available kits for the detection of *Cronobacter* spp. (not inclusive) include the 3M™ Molecular Detection Assay 2 – *Cronobacter* (https://www.3m.com/3M/en_US/p/d/v000269173, last accessed 5.20.2023), the BAX® system PCR assay for *Enterobacter sakazakii* (https://www.hygiena.com/bax-esak-std.html, last accessed 10.14.2020), the Assurance GDS™ *Enterobacter sakazakii* (https://www.thermofisher.com/order/catalog/product/4485034?SID=srch-srp-4485034, last accessed 5.20.2023), and the foodproof® *Enterobacteriaceae* plus *E. sakazakii* Detection Kit (BIOTECON Diagnostics, Potsdam, Germany; https://www.hygiena.com/wp-content/uploads/2023/02/Product-Sheet-foodproof-Enterobacteriaceae-and-Cronobacter-Detection-Kits.pdf, last accessed 5.20.2023), and Shukla et al. (2016) discusses the usefulness of several commercial products that detect *Cronobacter*.

The FDA BAM *Cronobacter* chapter (https://www.fda.gov/food/laboratory-methods-food/bam-chapter-29-cronobacter, last accessed 5.20.2023) has been updated to incorporate the real-time PCR method of Seo and Brackett (2005) with the modification of 40 total cycles (Chen et al. 2009; Chen et al. 2010; Chen et al. 2012). However, the 1,680 bp *gluA* PCR reported by Lerner et al. (2006b) has also been used by many researchers (Hogue et al. 2010; Giammanco et al. 2011; Ye et al. 2009; 2010).

Molecular Subtyping of Cronobacter

Molecular subtyping is a useful approach to help investigate outbreaks and connect contaminating pathogens in food sources with organisms present in clinical samples, and it clarifies the biology of bacteria colonizing other microbial habitats (Swaminathan et al. 2001). Mullane et al. (2007) applied PFGE to characterize and follow *Cronobacter* species colonizing econiches of a PIF processing facility. It was not long afterward that the PFGE protocol was validated and added to CDC's PulseNet international epidemiological laboratory network system, which compares the DNA fingerprints of bacteria from patients to find clusters of disease that might represent unrecognized outbreaks (Brengi et al. 2012; Yan et al. 2012). PFGE was commonly considered the gold standard for epidemiologic studies, but now PulseNet has added WGS to the surveillance network. Using WGS for molecular subtyping enhances, PulseNet's ability to detect and resolve outbreaks faster with more accuracy (Yachison et al. 2017) and allows for better source attribution.

Multilocus Sequence Typing (MLST)

Multilocus sequence typing (MLST) is a technique used in molecular biology for the typing of multiple loci. The procedure characterizes isolates of a microbial species using the DNA sequences of internal fragments (approximately 450–500 bp internal fragments of each gene) of multiple housekeeping genes, usually between five to seven alleles are used. For each housekeeping gene, the technique involves PCR amplification followed by DNA sequencing, the different sequences present within a bacterial species are assigned as distinct alleles; for each isolate, the alleles at each of the loci define the allelic profile or sequence type (ST). The *Cronobacter* MLST was initially applied to distinguish between *C. sakazakii* and *C. malonaticus* because the lack of resolution with 16S rDNA sequencing is not always accurate and biotyping is too subjective (Baldwin et al. 2009). The *Cronobacter* MLST scheme uses 7 alleles: *atpD*, *fusA*, *glnS*, *gltB*, *gyrB*, *infB*, and *ppsA* which ultimately gives a concatenated sequence of 3,036 bp for phylogenetic analysis (MLSA; multilocus sequence analysis) and comparative genomics. MLST has also been used in the formal recognition of new *Cronobacter* species (Joseph et al. 2012b). The method has revealed a strong association between one genetic *C. sakazakii* lineage, sequence type 4 (ST4), and cases of neonatal meningitis (Joseph and Forsythe 2011). The *Cronobacter* MLST site is at http://www.pubMLST.org/cronobacter (last accessed 5.20.2023).

CRISPR-Cas Subtyping System

In general, CRISPR-Cas systems contain three genetic elements: a Cas gene cluster, an AT-rich leader sequence, followed by a CRISPR spacer array composed of short (~ 24- to 48-nucleotide) direct repeat sequences separated by similarly sized, unique spacers which are usually derived from mobile genetic elements such as bacteriophage and plasmids (Ogrodzki and Forsythe 2016). It is thought that CRISPR-Cas systems provide a type of adaptive immunity from invasive genetic elements (phages and plasmids), functioning to regulate lysogeny and biofilm formation. In 2016, Ogrodzki and Forsythe (2016) described a clustered regularly interspaced short palindromic repeat (CRISPR)-Cas profiling system for *C. sakazakii*. It has been determined that the variability of a CRISPR array can provide a greater power of differentiation for genotyping within clonal lineages (Ogrodzki and Forsythe 2016).

Zeng et al. (2017) found that genome sizes of *C. sakazakii* strains had a significant correlation with the total genome sizes of lysogenized prophages. The percentage that prophages contribute to the genetic diversity (pan-genome) of *C. sakazakii* was calculated to be 16.57%. They further described subtypes I-E CRISPR-Cas system and five types of CRISPR arrays which were in a conserved site in *C. sakazakii* strains. CRISPR1 and CRISPR2 loci with highly variable spacers were active and predicted to protect against unwanted phage attacks (Zeng et al. 2017). Zeng et al. in other studies (2019; 2020) further developed this molecular typing procedure by developing a routine PCR method to establish the CRISPR-Cas system. This group evaluated 257 isolates of *C. sakazakii*, *C. malonaticus*, and *C. dublinensis* to verify the feasibility of the method. Results showed that 161 *C. sakazakii* strains could be divided into 129 CRISPR types (CTs), among which CT15 (n = 7) was the most prevalent CT followed by CT6 (n = 4). Further, 65 *C. malonaticus* strains were divided into 42 CTs and CT23 (n = 8) was the most prevalent followed by CT2, CT3, and CT13 (n = 4). Finally, 31 *C. dublinensis* strains belonged to 31 CTs. They also found a relationship among CT, ST, food types, and serotypes. They proposed that the PCR-based CRISPR-Cas method could identify sources in *Cronobacter* foodborne outbreaks and source-attribute clinical cases to food sources as well as production sites.

Molecular Serotyping

For members of *Salmonella*, the Kauffmann and White classification scheme, was proposed in 1934 and listed 44 serovars and classifies the genus *Salmonella* into serotypes based on surface antigens which are encoded in lipopolysaccharide (LPS) O antigen biosynthesis, flagellar, and capsular gene clusters (*Salmonella*

Subcommittee 1934). The *Cronobacter* LPS biosynthetic gene cluster like many members of the *Enterobacteriaceae* is flanked by the *galF* and *gnd* genes, contains 6 to 19 genes, and is generally conserved within a species (Yan et al. 2015b). For many of the organisms found in *Enterobacteriaceae*, similar serological-based schemes have been established. However, there is no established serological classification system for *Cronobacter* or for that matter *Enterobacter*. To aid in the characterization of *Cronobacter* strains isolated from foods, PIF manufacturing environments, and clinical samples, a harmonized molecular serotyping (LPS) identification scheme was suggested by Yan et al. (2015b), which is based on information obtained from studies by Mullane et al. (2008b), Jarvis et al. (2011; 2013), and Sun et al. (2011). Yan et al. (2015b) characterized 409 *Cronobacter* isolates representing the seven *Cronobacter* species using the Mullane-Jarvis and Sun molecular serotyping schemes. PCR analysis revealed many overlapping results obtained by the two serotyping schemes. Interestingly, many of the LPS biosynthetic gene loci contained species-specific SNPs which could be used to identify these O antigens (Mullane et al. 2008b).

Virulence Plasmid Typing

WGS of *C. sakazakii* BAA-894 and *C. turicensis* LMG 23827[T] (synonym = z3032) revealed that they harbor homologous plasmids identified as pESA3 (131 kb) and pCTU1 (138 kb), respectively (Kucerova et al. 2010; Stephan et al. 2011). *In silico* analysis showed that both plasmids have a single RepFIB-like origin of replication gene, *repA*, and encode both common and plasmid-specific virulence factors (Figure 4; Table 3) (Franco et al. 2011a).

Shared gene attributes associated with both plasmids include the presence of two iron acquisition systems (*eitCBAD* and *iucABCD/iutA*). The *iucABCD/iutA* gene

Table 3. Comparison of prevalence and distribution of pESA3/pCTU1 (incFIB), pESA2/pCTU2 (incF2), and pCTU3 (incH1) plasmids among 570 *Cronobacter* isolates.

Species	No. of Isolates	No. of Isolates with the Indicated Plasmid Incompatibility Class (%)[a]		
		pESA3/pCTU1 (incFIB)	pESA2/pCTU2 (incF2)	pCTU3 (incH1)
C. sakazakii	507	493 (97)	20 (4)	142 (28)
C. malonaticus	30	30 (100)	3 (10)	12 (40)
C. turicensis	13	13 (100)	1 (8)	8 (62)
C. muytjensii	12	9 (75)	0 (0)	1 (8)
C. dublinensis	5	4 (80)	0 (0)	0 (0)
C. universalis	2	2 (100)	0 (0)	1 (50)
C. condimenti	1	1 (100)	0 (0)	0 (0)
Total	570	552 (97)	24 (4)	164 (28)

[a] Numbers within parentheses are the percentage PCR-positive for each plasmid replicon gene locus (*repA*) as described by Franco et al. (2011a). The prevalence of pESA3/pCTU1 (incFIB), pESA2/ pCTU2 (incF2), and pCTU3 (incH1) plasmids among the strains were calculated using the total number of strains tested. The table was adapted from Jang et al. (2020a).

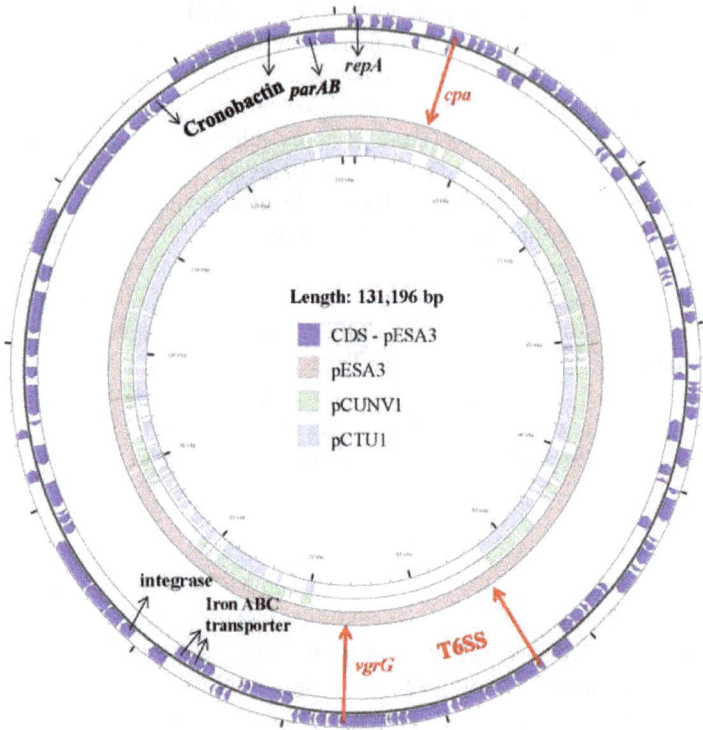

Figure 4. Sequence alignment of pESA3, pCUNV1, and pCTU1 produced on the CGView Server from the Stothard Research Group that uses BLAST analysis to illustrate conserved and missing genomic sequences (Available online: https://github.com/paulstothard/cgview, last accessed 5.20.2023). Two circular plasmid genomes, pCUNV1 (NZ_CP012258) and pCTU1 (NC_013283) were compared against the reference pESA3 (NC_009780). GenBank annotations of the reference pESA3 (CDS in blue arranged in two outside rings) were downloaded as a GFF file for analysis using the default configuration on the CGView server. Select genes or loci of interest are shown across the circular genomes as follows: Siderophore loci with Cronobactin gene, Iron ABC transporter genes, type 6 secretion system (T6SS), *parAB* genes and the toxin *cpa* gene are adapted from Franco et al. (2011a, b). Missing regions identified by the BLAST analysis on the CGView server are shown as 'gaps' on each of the two circular genomes. Genes and loci missing in pCUNV1 or pCTU1 plasmids are in red. As expected, T6SS is seen only on the reference pESA3 from *C. sakazakii* while the toxin encoding *cpa* gene is absent in the plasmid pCTU1 from *C. turicensis*. The figure was adapted from Jang et al. (2020a).

cluster is to date the only siderophore gene cluster, thus far identified in *Cronobacter*. Additionally, pESA3 contains a *Yersinia pestis* plasminogen activator-like (omptin) gene homolog named *cpa* (*Cronobacter* plasminogen activator) and a 17-kb type 6 secretion system (T6SS) locus, while pCTU1 contains a 27-kb region encoding a *Bordetella pertussis*-like filamentous hemagglutinin gene (*fhaB*), its specific transporter gene (*fhaC*), and associated putative adhesins (FHA locus). Taken together these results suggest that these are virulence plasmids. The role of pESA3/ pCTU1 plasmids that were described by Franco et al. (2011a) and Stephan et al. (2010) as virulence plasmids were confirmed by Eshwar et al. (2016), who showed that strains harboring plasmids pESA3 and pCTU1 exhibited twice the mortality rate than isogenically plasmid-cured strains or naturally occurring plasmid-free strains using the zebrafish infection model. These data suggest that these plasmids are virulence-associated but may not represent the entire virulence factor gene repertoire of *Cronobacter*. Jang et al. (2020a) extended these studies by analyzing 570 *Cronobacter* strains by the PCR primers described by Franco et al. (2011a) and showed that 97% of the strains possessed a pESA3/pCTU1-like virulence plasmid, as shown in Table 3.

Next Generation Sequencing (NGS) Approaches

DNA Microarray

Soon after the publication of the genome for *C. sakazakii* strain BAA-894, the first *Cronobacter* genome sequenced (Kucerova et al. 2010), DNA microarrays were developed to study the genomic diversity of members of this newly classified genus. Healy et al. (2009) developed a DNA microarray with 276 selected genes from the genome of *C. sakazakii* BAA-894 to compare it with other strains representing six of the seven *Cronobacter* species (*C. condimenti* had not been described at that time).

Their results supported the reclassification scheme proposed by Iversen et al. (2008a) in which each species was grouped into clusters, except for some of the *C. malonaticus* strains grouped within a larger cluster of *C. sakazakii* strains. Other strains were grouped within a large cluster that also contained *C. turicensis*, *C. universalis*, and an unknown *Cronobacter* species. The DNA microarray developed by Kucerova et al. (2010) featured the entire genome (387,000-probe oligonucleotide tiling DNA microarray) of *C. sakazakii* BAA-894 for comparison with strains representing five of the seven *Cronobacter* species. Among the 4,382 annotated genes identified in *C. sakazakii* BAA-894, approximately 55% were common to all *C. sakazakii* isolates and 43% were common to all *Cronobacter* strains. Many of the genes absent in more than half of the strains evaluated were identified as phage genes, which emphasizes the importance of the accessary genome or mobilome in driving the genomic diversity of *Cronobacter* and supports the findings reported by Zeng et al. (2017). Both studies used a common method of analysis which incorporated the use of direct comparisons of gene probe intensities obtained for the different strains to probe intensities of similar genes from the reference strain, *C. sakazakii* BAA-894.

In contrast, to the two microarray approaches described above, a more general multi-genome array approach containing the annotated pangenomes from 15 sequenced strains of *Cronobacter*, representative of all seven species, was developed by Tall et al. (2015) and was based on the design of Jackson et al. (2011) for *Escherichia coli*. The sequencing of these genomes was accomplished by a five-member international consortium in 2008–2009 that used complete and draft genomes from various sequencing efforts (Stephan et al. 2011; 2014; Grim et al. 2013; Yan et al. 2013). The microarray was a custom-designed Affymetrix array that offered an opportunity to survey species-specific and unique genes from different *Cronobacter* species. Gene orthologs across the seven species were represented by the smallest number of probesets that theoretically covered all alleles of a gene target. The microarray consisted of 19,287 independent genomic targets and genes

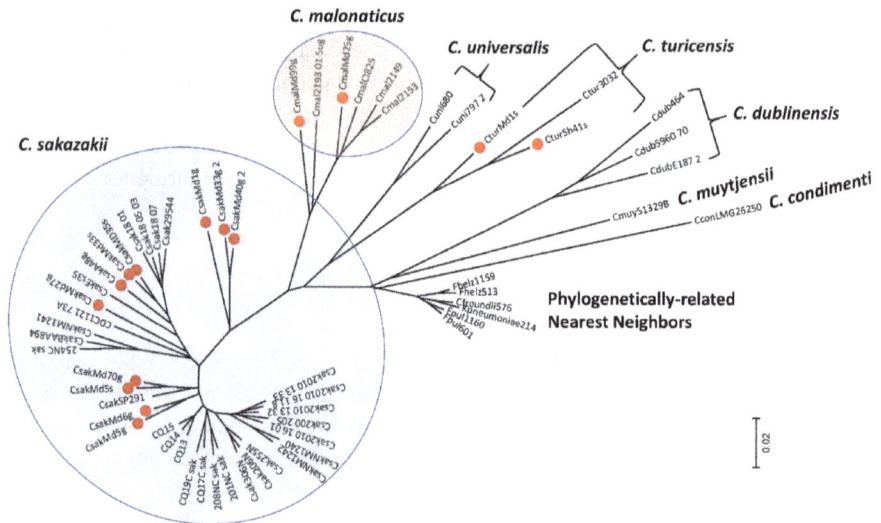

Figure 5. Phylogenetic analysis among 15 *Cronobacter* isolates obtained from filth flies (identified with a red dot) and 45 other *Cronobacter* and phylogenetically related strains. The tree was inferred using the Neighbor-Joining method based on the 19,287 *Cronobacter* gene targets (Saitou and Nei 1987) of the FDACRONOa520845F microarray. The optimal tree with the sum of branch length = 1.86949337 is shown. The tree is drawn to scale, with branch lengths in the same units as those of the evolutionary distances used to infer the phylogenetic tree. The evolutionary distances were computed using the p-distance method (Nei and Kumar 2000) and are in the units of the number of base differences per site. The analysis involved 60 strains evaluated by using the FDA *Cronobacter* microarray. All positions containing gaps and missing data were eliminated. There was a total of 21,402 positions in the final dataset. Evolutionary analyses were conducted in MEGA7 (Kumar et al. 2016). These *Cronobacter* and phylogenetically related strains were analyzed from the presence-absence gene matrix (data not shown). The microarray experimental protocol as described by Jackson et al. (2011) and as modified by Tall et al. (2015) was used for the interrogation of the strains for this analysis. The phylogenetic tree illustrates that the *Cronobacter* microarray could clearly separate the seven species of *Cronobacter* with each species forming its distinct cluster, and that representative fly strains clustered (identified with red dots) according to their species taxon. The scale bar represents a 0.02 base substitution per site. The figure was adapted from Jang et al. (2020b).

of 18 *Cronobacter* plasmids and 2,371 virulence factor genes of phylogenetically related Gram-negative bacteria (Tall et al. 2015). It was determined that the *Cronobacter* microarray could distinguish the seven *Cronobacter* species from one another and non-*Cronobacter* species (Tall et al. 2015). Additionally, within each species, strains are grouped into distinct clusters according to ST and clonal complexes. These results also supported the phylogenic divergence of the genus and clearly highlighted the genomic diversity in each member of the genus as shown in Figure 5. Other studies showed that microarray analysis could correctly assess the phylogenetic relatedness among persistent *Cronobacter* strains obtained from PIF manufacturing facilities, among strains isolated from plant-origin foods and flies, which showed that clinically relevant strains were phylogenetically related to these strains and helped to understand the nucleotide divergence of outer membrane proteins captured within outer membrane vesicles (Tall et al. 2015; Yan et al. 2015a; Chase et al. 2017a; Kothary et al. 2017; Jang et al. 2018a, b, 2020a). This led to the finding that microarray and BLAST analyses of *Cronobacter* fly sequence datasets were corroborative and showed that the presence and absence of virulence factors followed species and ST evolutionary lines even though such genes were orthologous. The microarray was also able to accurately assess each strain's identity, could differentiate *Cronobacter* species from phylogenetically related species such as *Franconibacter* and *Siccibacter*, and has been a useful tool to assess phylogenetic relatedness and gene content among strains.

Genomic Features Obtained From BLAST Analyses of Genomes

Malonate Utilization Metabolic Island

Gopinath et al. (2018) determined that ST64 *C. sakazakii* possesses the nine-gene malonate utilization operon like that found in the other *Cronobacter* species (Table 1). Prior to this study, it was thought that *C. sakazakii* was generally unable to utilize malonate. Parallel WGS and MA were useful in characterizing the operon, which found that the malonate utilization gene cluster-flanking regions were flanked upstream by *gyrB*; this encodes a topoisomerase IV subunit B gene (EC 5.99.1), and downstream by *katG*, which encodes for a catalase/peroxidase gene (EC 1.11.1.6; EC 1.11.1.7). Interestingly, these two genes were found to be conserved among all *Cronobacter* species even malonate-negative *C. sakazakii* strains; however, instead of the 7.7 kbp malonate utilization gene cluster, there is an undescribed 323–325 bp nucleotide region as shown in Figure 6. Also, within the gene cluster is a gene that encodes for an auxin efflux carrier. Auxin is a universal hormone in plants, which participates in many aspects of plant developmental and growth processes. Auxin is usually synthesized in the shoot apex, as well as the developing leaf primordia, and transported to the targeted tissues by bulk flow via vascular tissues or direct polar transport. Presently the role of the auxin efflux carrier protein in plant associations by *Cronobacter* and why it is within the malonate utilization operon is not known.

Xylose Utilization Metabolic Island

Xylose and arabinose combined make up more than 30% of the total sugars in agricultural by-products and xylose is the second most abundant sugar in nature besides glucose and primarily exists as D-xylose (Olofsson et al. 2008). However, it is usually found as a polymeric component of plant cell wall polysaccharides called xylans, e.g., arabinoxylans, hemicellulose (xylan and glucuronoxylan), and xyloglucan (Olofsson et al. 2008). A xylose utilization operon (average size of ~ 16,771 bp; 11 genes) which possessed a G+C content of 54.9%, was found among spice-associated *C. sakazakii* strains (Jang et al. 2018a). A map of the operon for *C. sakazakii* strain MOD1_AS15 is shown in Figure 7.

The genomic structure of the *Cronobacter* xylose utilization operon was similar to that found in *E. coli* strain K-12 (strain MG1655; GenBank assembly accession: GCA_000005845; RefSeq assembly accession: GCF_000005845) except that two genes present in the *Cronobacter* xylose operon, *xylS,* and α-*xynT* are missing

Figure 6A

Figure 6B

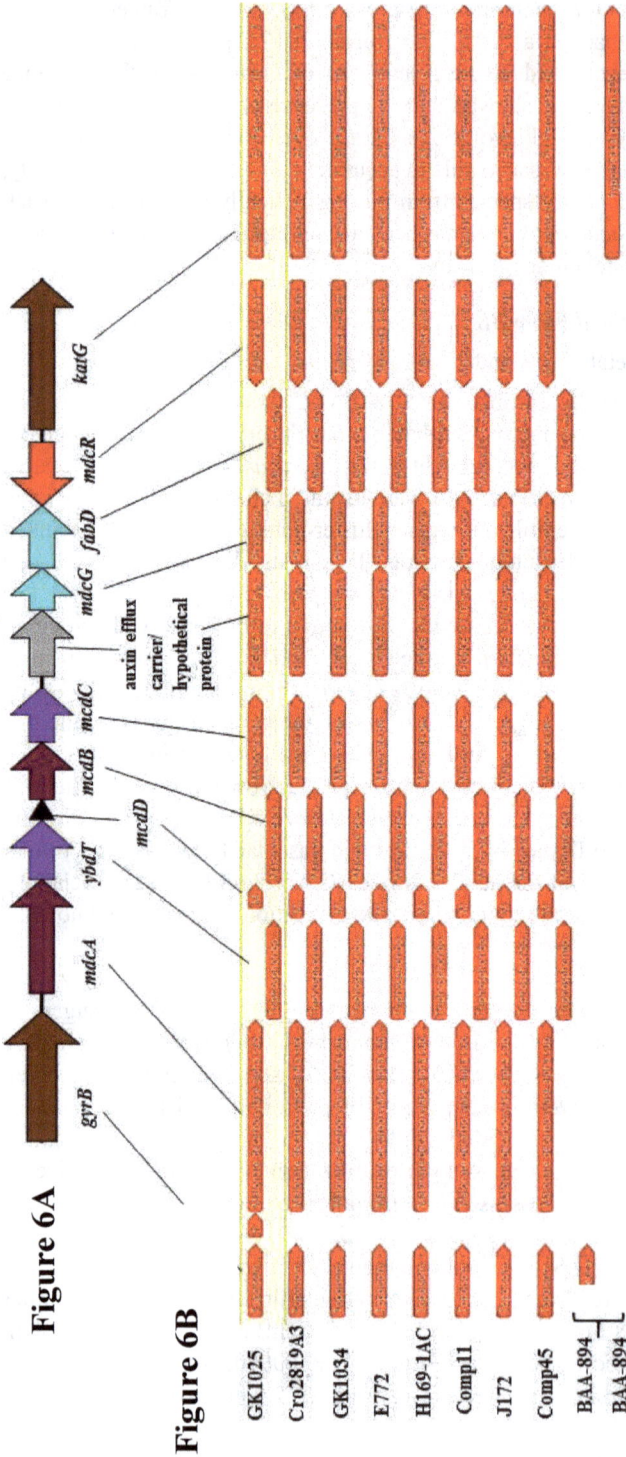

Figure 6. (A) Map of the 7.7 kbp malonate utilization gene cluster consisting of nine genes flanked by *gyrB* and *katG* and as annotated from ST64 *C. sakazakii* strain GK1025. (B) Comparison of the malonate operons from different ST64 *C. sakazakii* strains: The reference *C. sakazakii* strains GK1025B and Cro2819A3 were previously reported by Gopinath et al. (2018). Only the partial sequence of *gyrB* was included for illustration. The two flanking genes, *gyrB* and *katG* are shown in the last two tracks which in *C. sakazakii* BAA-894 and other malonate-negative strains are separated by a 323–325 bp intergenic region. The figure was adapted from Gopinath et al. (2018).

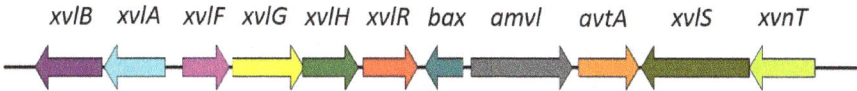

Figure 7. Xylose utilization operon. Schematic map made using XPlas, Map DNA for Mac OS XAp showing the annotated xylose utilization operon from *C. sakazakii* MOD1_AS-15. Figure was adapted from Jang et al. (2018a).

from within the operon in *E. coli* strain MG1655, which resulted in ~ 13,041 bp sized operon. Additionally, there was a genomic size variation noted for these *C. sakazakii* strains (size ranged from 16,340 to 16,790 bp). Previously the presence of a xylose utilization operon in *C. sakazakii* strain GP1999, which was isolated from a tomato's rhizoplane/rhizosphere continuum (Chase et al. 2017b), further supports the hypothesis that plants may be the ancestral econiche for *Cronobacter* spp. as posited by Schmid et al. (2009) and Joseph et al. (2012b). The G+C content of a 17,970 bp region upstream and a 17,422 bp region downstream of the *C. sakazakii* xylose utilization operon possessed G+C contents of 58.1% and 59.6%, respectively (Jang et al. 2018a). This change in G+C content suggests that the *Cronobacter* xylose utilization operon may be a predicted genomic or metabolic island (Soares et al. 2016). However, such G+C content change was not seen in the genomes of *E. coli* and the *X. axonopodis* pv. *citri* strains. Xylose and arabinose utilization by *Cronobacter* and its role in a plant association lifestyle are currently not well understood.

Plant Associations-Indole Production

Indole-3-acetic acid (IAA), the most common naturally occurring auxin, is a hormone produced by plants, fungi, and bacteria (Woodward and Bartel 2005; Duca and Glick 2020). IAA and its related metabolic intermediates [indole acetonitrile (IAN), indole acetamide (IAM), and indole pyruvic acid (IPA)] play central roles in modulating plant growth and development. In addition to being synthesized by plants, IAA is also produced by some bacteria in the rhizosphere, where it acts as a signaling molecule that has significant effects on the communication between plants and microorganisms promoting plant growth (i.e., IAA in microbial and microorganism-plant signaling) (Spaepen et al. 2007). In microbes, tryptophan is a main precursor for IAA biosynthesis. IAA, which is produced by plant growth-promoting bacteria (PGPB) is thought to also improve bacterial stress tolerance and microbe-microbe communication. Interestingly, two separate reports by Afridi et al. (2019) and Eida et al. (2020) revealed that the inoculation of plants by ACC (1-Aminocyclopropane-1-carboxylic acid, a precursor of plant hormone ethylene) deaminase producing PGPBs is a potential tool for the enhancement of plant growth and stress tolerances. The presence of genes in *Cronobacter* for plant nutrient acquisition and phytohormone production could explain the ability of *Cronobacter* to colonize plants and sustain plant growth under stress conditions as well as control the growth of phytopathogens. However, understanding how *Cronobacter* maintains its plant-associated lifestyle is only in its infancy.

Antibiotic Resistance

Muytjens et al. (1986) reported the antibiotic susceptibility of 195 *Cronobacter* (reported as *E. sakazakii*) strains compared with seven other *Enterobacter* species and showed that *E. sakazakii* was the most susceptible species but was resistant to cephalothin and sulfamethoxazole. This was followed up by a study by Pitout et al. (1997), which identified both Bush group 1 and Group 2 cephalosporinases which were thought to be expressed constitutively at high levels. Lai (2001) showed that increasing antibiotic resistance among *E. sakazakii* strains was occurring and proposed the judicious use of carbapenems or newer cephalosporins in combination with a second agent such as an aminoglycoside to treat infections, especially in view of the increased incidence of extended-spectrum β-lactamases capable of inactivating the cephalosporins and extended-spectrum penicillin. Clinically, Block et al. (2002) reported initial clinical descriptions of several cases of *E. sakazakii* neonatal infections along with three asymptomatic fecal carrier cases which were susceptible to ampicillin. However, these isolates were resistant to cefazolin but susceptible to all other agents tested, including the more advanced penicillins and cephalosporins, carbapenems, fluoroquinolones, aminoglycosides, tetracycline, trimethoprim-sulfamethoxazole, and chloramphenicol. All isolates tested were also β-lactamase positive, probably representing the Bush group 1 β-lactamase (cephalosporinase) produced constitutively at low levels as reported earlier by Pitout et al. (1997). In 2012, Kilonzo-Nthenge et al. (2012) determined the prevalence of *C. sakazakii* which was present in domestic kitchens in middle Tennessee. The authors observed multidrug resistance in *Cronobacter* strains with the highest resistance found to be against penicillin (76.1% of isolates) followed by tetracycline (66.6%), ciprofloxacin (57.1%), and nalidixic acid (47.6%). None of the *C. sakazakii* isolates were resistant to gentamicin. These results suggest that multidrug-resistant *C. sakazakii* strains could also be present at various sites in domestic kitchens.

The increased use of WGS technology to genotype foodborne pathogens (Allard et al. 2016) has led to a simple approach to classifying the strain as either susceptible or resistant to specific antibiotics. This straightforward approach is to use a "rules-based" classification based on the presence of one or more known antimicrobial resistance (AMR) genes or mutations (Su et al. 2019). This requires cross-referencing the genome sequence against databases of antibiotic resistance determinants. Databases have been developed mostly from the curation of the literature on molecular genetic studies that link antibiotic resistance phenotypes to genes (Su et al. 2019) in a lot but not all pathogens. However, as more and more pathogens are sequenced these databases will only be further enriched.

A report by Liu et al. (2017) showed the presence of a colistin resistance gene, *mcr-1*, and a Delhi metallo-β-lactamase gene, bla_{NDM-9}, respectively in two carbapenem-resistant *C. sakazakii* strains. This is the first time that a *mcr-1* resistance gene in *C. sakazakii* was found. This report was followed by another very alarming report by Shi et al. (2018) who described *C. sakazakii* strain 505108, which was isolated from a neonatal sputum specimen with severe pneumonia and showed that it co-harbored two plasmids that carried a large number of antimicrobial drug resistance genes (total of 20 AMR genes), such as on p505108-MDR (NCBI accession #: KY978628) the following antimicrobial resistance/tolerance genes were found: bla_{TEM-1}, *dfrA19*, *aph(3")-Ib*, *aph(6)-Id*, *mcr-9.1*, *aph(3')-Ia*, *catA2*, *tet(D)*, *aac(6')-Ib3*, bla_{SHV-12}, *sul1*, bla_{DHA-1}, *qnrB4*, *qacEdelta1*, *arr*, *aac(3)-II*, and *aac(6')-IIc* with predicted resistances/tolerances to β-Lactam based compounds, Trimethoprim, Aminoglycoside, Colistin, Phenicol, Tetracycline, Sulfonamide, Quinolone, Quaternary Ammonium, and Rifamycin antibiotics, respectively. The antimicrobial resistance genes found on plasmid p505108-NDM (KY978629) were identified as *ble*, bla_{NDM-1}, and bla_{SHV-12} with predicted resistances to bleomycin and NDM-1 β-Lactam antibiotics. These authors showed that the plasmid-borne antibiotic resistance genes were associated with numerous insertion sequences, integrons, and transposons, indicating that their assembly and mobilization were facilitated by transposition and/or homologous recombination. This report also provided a deeper understanding of plasmid-mediated multidrug resistance in *Cronobacter* that may be occurring nosocomially.

Müller et al. (2014) described a family of class C β-lactamase (*bla*) resistance genes, $bla_{CSA/CMA}$, carried by *Cronobacter* which were noninducible and were cephalosporinases. Using the AMRFinderPlus tool within the Center for Food Safety and Applied Nutrition's GalaxyTrakR instance, Jang et al. (2020b) demonstrated that *Cronobacter* strains isolated from filth flies all possessed these β-lactamase genes. The *C. turicensis* and *C. malonaticus* strains carried a *C. malonaticus* (CMA) class C bla_{CMA} resistance gene and the filth fly *C. sakazakii* strains carried a *C. sakazakii* CSA class C bla_{CSA} resistance gene. Six of these CSA class C *bla* resistance genes were only identified at the family level, whereas the remaining class C *bla* resistance genes were identified as either bla_{CSA-2} or bla_{CSA-1} variants. Additionally, the *C. turicensis* strains also possessed a *fosA* family fosfomycin resistance glutathione transferase gene (Jang et al. 2020b).

WGS Analysis Using Core Genes

Previously, whole genome SNP-based clustering or phylogenetic analysis with conserved homologs of *C. sakazakii* BAA894 and other species had been carried out by Stephan et al. (2014), Tall et al. (2015), Chase et al. (2016; 2017a, b), Gopinath et al. (2018), and Jang et al. (2020b). The core gene reference loci data set (2,000 core genes) that are developed by spine analysis can be used as a powerful tool to accurately capture subtle differences in strains belonging to the same ST or ecological niche (as demonstrated in Figure 8). This standardized genome-wide SNP finding tool thus provides the community with a method to query an ever-expanding repertoire of *Cronobacter* genome draft assemblies from different geographical areas and/or sources without any sample-bias, allowing the least ambiguity in SNP calls. It was noted that several of the ST4 *C. sakazakii* fly strains clustered indistinguishable from *C. sakazakii* strains 8155 and SP291 which are strains that were isolated from a can of dried milk in the 1950s and the environment of an Irish PIF manufacturing facility in 2014, respectively (Yan et al. 2015a). These results suggest that the genomic backbone of these isolates from very disparate sample

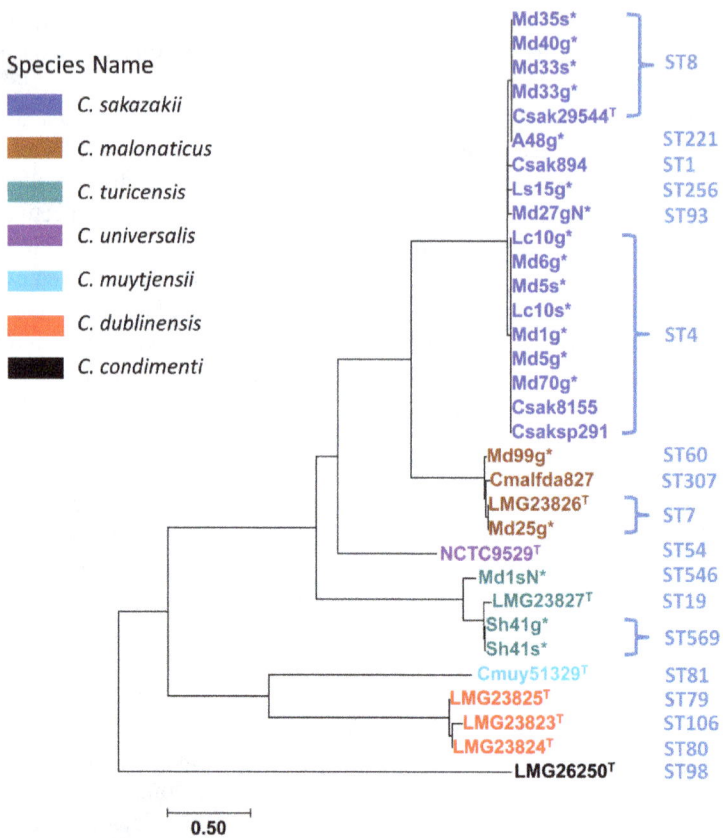

Figure 8. WGS analysis of *Cronobacter* strains using core genes. The phylogenetic analysis was inferred using the Neighbor-Joining method (Saitou and Nei 1987). The optimal tree with the sum of branch length = 10.02357736 is shown. The tree is drawn to scale, with branch lengths in the same units as those of the evolutionary distances used to infer the phylogenetic tree. The evolutionary distances were computed using the Maximum Composite Likelihood method (Tamura et al. 2004) and are in the units of the number of base substitutions per site. The analysis involved 32 nucleotide sequences. All positions containing gaps and missing data were eliminated. There was a total of 308,989 positions in the final dataset. Evolutionary analyses were conducted in MEGA7 representing 2,000 core genes obtained from Spine analysis (Ozer et al. 2014; Kumar et al. 2016). The sequenced fly strains are labeled with an asterisk and the type strains are labeled with an uppercase T. Figure was adapted from Jang et al. (2020b).

sources (fly strains, dried milk, and PIF manufacturing environment) appear to be similar and highly conserved. Nevertheless, using the resulting data matrix from applying the 2,000 wg-core gene analysis, it is possible to extract SNP profiles in the wg-core loci shared by these two sets of isolates from exclusive sample sources for further analysis and methods development. Moreover, a robust dataset is generally needed for rigorous statistical analysis with bootstrap values to increase confidence in any phylogenetic analyses. Unlike traditional MLST methods that are susceptible to loss of resolution due to missing allelic information in the query dataset, the spine analysis approach uses hundreds of conserved core genes for detecting phylogenetic features in a collection of isolates (Ozer et al. 2014). As the populations of pathogens like *Cronobacter* are subject to varied evolutionary forces in different environments impacting genome sequences differentially, it will result in the emergence of smaller, but distinct clades. Results from surveillance studies by Chase et al. (2017a) and Jang et al. (2018a, b) reported emerging diversity of *Cronobacter* isolates. Recently, a comprehensive and deep analysis of plant-origin and low-moisture foods using WGS datasets identified major and minor 'lineage groups' representing sub-clades within *C. sakazakii* (Jang et al. 2022). Studies like these using sequence data from strains isolated from new and less-sampled food and environmental niches suggest a genus with broader sequence variations than currently recognized. A standardized genome-wide tool that combines SNP data points with phylogenetic tree topologies and bootstrap support enhances the better interpretation of WGS data (Pightling et al. 2018). The 2,000 wg-core gene schema presented here fills a critical gap for such a genome-wide analysis. This genome-wide approach was also applied to corroborate the overall similarity in the pattern of clustering as seen from pan-genomic microarrays.

Future Considerations

Cronobacter are opportunistic pathogens with remarkable adaptability in interacting with their environment whether during colonization of the gastrointestinal tract, breast pump internal tubing, or a PIF manufacturing facility surface. Outbreaks of disease due to *Cronobacter* can affect susceptible individuals, including neonates, infants, and elderly individuals alike, and continues to galvanize national and international media headlines. Epidemiologically, there are more adult infections seen than there are infant infections. However, neonates and immunocompromised elderly individuals are most at higher risk for severe disease. Though extrinsically and intrinsically contaminated PIF has historically been an important known vehicle of infant disease, less information has been found to understand the principal sources leading to adults becoming infected. Although *Cronobacter* species have been found associated with dried dairy protein products (milk and cheese protein powders), cereals, candies such as licorice and lemon-flavored cough drops, dried spices, teas, nuts, herbs, and pasta, there is no epidemiological evidence linking these sources to adult infections. These organisms can persist for long periods within PIF manufacturing facilities. However, little knowledge is known about its presence and persistence in other food manufacturing facilities. Species-specific virulence factors have been found that adversely affect a wide range of host processes including protein synthesis, cell division, and pro-inflammatory host responses. These factors are encoded on a variety of mobile genetic elements such as plasmids, transposons, and pathogenicity islands. This genomic plasticity denotes ongoing re-assortment and possibly acquisition of future virulence factors that will undoubtedly complicate our efforts to categorize these organisms into sharply delineated genomo-pathotypes. NGS technologies such as WGS analysis have the potential to be used as a powerful tool to accurately capture subtle differences in strains belonging to the same ST or ecological niche. The recent finding of multidrug antibiotic resistance in strains from a variety of sources advocates for continued surveillance of this foodborne pathogen. This dynamic nature will only present additional challenges for the global food safety and public health communities in the diagnosis, treatment, and prevention of infections.

Acknowledgments

The authors would like to thank the Joint Institute for Food Safety and Applied Nutrition (JIFSAN) for supporting JIFSAN interns: Flavia Negrete, Katie Ko, and Leah Weinstein through a cooperative agreement with the FDA (no. FDU001418). We thank the Oak Ridge Institute for Science and Education of Oak Ridge, Tennessee for sponsoring research fellow Hyein Jang.

References

Acinas, S.G., L.A. Marcelino, V. Klepac-Ceraj and M.F. Polz. 2004. Divergence and redundancy of 16S rRNA sequences in genomes with multiple *rrn* operons. J. Bacteriol. 186: 2629–2635.

Afridi, M.S., Amna, Sumaira, T. Mahmood, A. Salam, T. Mukhtar et al. 2019. Induction of tolerance to salinity in wheat genotypes by plant growth promoting endophytes: Involvement of ACC deaminase and antioxidant enzymes. Plant Physiol. Biochem. 139: 569–577.

Allard, M.W., E. Strain, D. Melka, K. Bunning, S.M. Musser, E.W. Brown et al. 2016. Practical value of food pathogen traceability through building a whole-genome sequencing network and database. J. Clin. Microbiol. 54: 1975–1983.

Alsonosi, A., S. Hariri, M. Kajsík, M. Oriešková, V. Hanulík, M. Röderová et al. 2015. The speciation and genotyping of *Cronobacter* isolates from hospitalised patients. Eur. J. Clin. Microbiol. Infect. Dis. 34: 1979–1988.

Alvarez-Ordóñez, A., C. Cummins, T. Deasy, T. Clifford, M. Begley and C. Hill. 2014. Acid stress management by *Cronobacter sakazakii*. Int. J. Food Microbiol. 178: 21–28.

Amaretti, A., L. Righini, F. Candeliere, E. Musmeci, F. Bonvicini, G.A. Gentilomi et al. 2020. Antibiotic resistance, virulence factors, phenotyping, and genotyping of non-*Escherichia coli* Enterobacterales from the gut microbiota of healthy subjects. Int. J. Mol. Sci. 21: 1847.

Arahal, D.R. 2014. Chapter 6, Whole-genome analyses: average nucleotide identity. pp. 103–122. *In*: Goodfellow, M., I. Sutcliffe and J. Chun (eds.). Methods in Microbiology, Volume 41. Academic Press, Waltham, MA 02451, USA.

Arku, B., S. Fanning and K. Jordan. 2011. Heat adaptation and survival of *Cronobacter* spp. (formerly *Enterobacter sakazakii*). Foodborne Pathog. Dis. 8: 975–981.

Arroyo, C., S. Condón and R. Pagán. 2009. Thermobacteriological characterization of *Enterobacter sakazakii*. Int. J. Food Microbiol. 136: 110–118.

Arroyo, C., G. Cebrián, S. Condón and R. Pagán. 2012. Development of resistance in *Cronobacter sakazakii* ATCC 29544 to thermal and nonthermal processes after exposure to stressing environmental conditions. J. Appl. Microbiol. 112: 561–570.

Baldwin, A., M. Loughlin, J. Caubilla-Barron, E. Kucerova, G. Manning, C. Dowson et al. 2009. Multilocus sequence typing of *Cronobacter sakazakii* and *Cronobacter malonaticus* reveals stable clonal structures with clinical significance which do not correlate with biotypes. BMC Microbiol. 9: 223.

Bastin, B., P. Bird, M.J. Benzinger, E. Crowley, J. Agin, D. Goins et al. 2018. Confirmation and identification of *Salmonella* spp., *Cronobacter* spp., and other Gram-negative organisms by the Bruker MALDI Biotyper method: Collaborative Study, First Action 2017.09. J. AOAC Int. 101: 1593–1609.

Berthold-Pluta, A., M. Garbowska, I. Stefańska and A. Pluta. 2017. Microbiological quality of selected ready-to-eat leaf vegetables, sprouts and non-pasteurized fresh fruit-vegetable juices including the presence of *Cronobacter* spp. Food Microbiol. 65: 221–230.

Besse, N.G., A. Leclercq, V. Maladen, C. Tyburski and L. Bertrand. 2006. Evaluation of the International Organization for Standardization-International Dairy Federation (ISO-IDF) draft standard method for detection of *Enterobacter sakazakii* in powdered infant food formulas. J. AOAC Int. 89: 1309–1316.

Blackwood, B.P. and C.J. Hunter. 2016. *Cronobacter* spp. Microbiol. Spectr. 4: EI10-0002-2015.

Block, C., O. Peleg, N. Minster, B. Bar-Oz, A. Simhon, I. Arad et al. 2002. Cluster of neonatal infections in Jerusalem due to unusual biochemical variant of *Enterobacter sakazakii*. Eur. J. Clin. Microbiol. Infect. Dis. 21: 613–616.

Bowen, A.B. and C.R. Braden. 2006. Invasive *Enterobacter sakazakii* disease in infants. Emerg. Infect. Dis. 12: 1185–1189.

Bowen, A., H.C. Wiesenfeld, J.L. Kloesz, A.W. Pasculle, A.J. Nowalk, L. Brink et al. 2017. Notes from the field: *Cronobacter sakazakii* infection associated with feeding extrinsically contaminated expressed human milk to a premature infant—Pennsylvania. Morb. Mortal. Wkly. Rep. 66: 761–762.

Brady, C., I. Cleenwerck, S. Venter, T. Coutinho and P. De Vos. 2013. Taxonomic evaluation of the genus *Enterobacter* based on multilocus sequence analysis (MLSA): proposal to reclassify *E. nimipressuralis* and *E. amnigenus* into *Lelliottia* gen. nov. as *Lelliottia nimipressuralis* comb. nov. and *Lelliottia amnigena* comb. nov., respectively, *E. gergoviae* and *E. pyrinus* into *Pluralibacter* gen. nov. as *Pluralibacter gergoviae* comb. nov. and *Pluralibacter pyrinus* comb. nov., respectively, *E. cowanii*, *E. radicincitans*, *E. oryzae* and *E. arachidis* into *Kosakonia* gen. nov. as *Kosakonia cowanii* comb. nov., *Kosakonia radicincitans* comb. nov., *Kosakonia oryzae* comb. nov. and *Kosakonia arachidis* comb. nov., respectively, and *E. turicensis*, *E. helveticus* and *E. pulveris* into *Cronobacter* as *Cronobacter zurichensis* nom. nov., *Cronobacter helveticus* comb. nov. and *Cronobacter pulveris* comb. nov., respectively, and emended description of the genera *Enterobacter* and *Cronobacter*. Syst. Appl. Microbiol. 36: 309–319.

Brandão, M.L.L., N.S. Umeda, E. Jackson, S.J. Forsythe and I. de Filippis. 2017. Isolation, molecular and phenotypic characterization, and antibiotic susceptibility of *Cronobacter* spp. from Brazilian retail foods. Food Microbiol. 63: 129–138.

Brengi, S.P., S.B. O'Brien, M. Pichel, C. Iversen, M. Arduino, N. Binsztein et al. 2012. Development and validation of a PulseNet standardized protocol for subtyping isolates of *Cronobacter* species. Foodborne Pathog. Dis. 9: 861–867.

Brenner, D.J. 1973. Deoxyribonucleic acid reassociation in the taxonomy of enteric bacteria. Int. J. Sys. Bacteriol. 23: 298–307.

Carleton, H.A., J. Besser, A.J. Williams-Newkirk, A. Huang, E. Trees and P. Gerner-Smidt. 2019. Metagenomic approaches for public health surveillance of foodborne infections: Opportunities and challenges. Foodborne Pathog. Dis. 16: 474–479.

Carter, L., L.A. Lindsey, C.J. Grim, V. Sathyamoorthy, K.G. Jarvis, G. Gopinath et al. 2013. Multiplex PCR assay targeting a diguanylate cyclase-encoding gene, *cgcA*, to differentiate species within the genus *Cronobacter*. Appl. Environ. Microbiol. 79: 734–737.

Caubilla-Barron, J., E. Hurrell, S. Townsend, P. Cheetham, C. Loc-Carrillo, O. Fayet et al. 2007. Genotypic and phenotypic analysis of *Enterobacter sakazakii* strains from an outbreak resulting in fatalities in a neonatal intensive care unit in France. J. Clin. Microbiol. 45: 3979–3985.

Cawthorn, D.M., S. Botha and R.C. Witthuhn. 2008. Evaluation of different methods for the detection and identification of *Enterobacter sakazakii* isolated from South African infant formula milks and the processing environment. Int. J. Food Microbiol. 127: 129–138.

Chandrasekaran, S., C.D. Burnham, B.B. Warner, P.I. Tarr and T.N. Wylie. 2018. Carriage of *Cronobacter sakazakii* in the very preterm infant gut. Clin. Infect. Dis. 67: 269–274.

Chap, J., P. Jackson, R. Siqueira, N. Gaspar, C. Quintas, J. Park et al. 2009. International survey of *Cronobacter sakazakii* and other *Cronobacter* spp. in follow up formulas and infant foods. Int. J. Food Microbiol. 136: 185–188.

Chase, H.R., G.R. Gopinath, A.K. Eshwar, A. Stoller, C. Fricker-Feer, J. Gangiredla et al. 2017a. Comparative genomic characterization of the highly persistent and potentially virulent *Cronobacter sakazakii* ST83, CC65 strain H322 and other ST83 strains. Front. Microbiol. 8: 1136.

Chase, H.R., L. Eberl, R. Stephan, H. Jeong, C. Lee, S. Finkelstein et al. 2017b. Draft genome sequence of *Cronobacter sakazakii* GP1999, sequence type 145, an epiphytic isolate obtained from the tomato's rhizoplane/rhizosphere continuum. Genome Announc 5: e00723-17.

Chen, W., L. Ai, J. Yang, J. Ren, Y. Li and B. Guo. 2013. Development of a PCR assay for rapid detection of *Cronobacter* spp. from food. Can. J. Microbiol. 59: 656–661.

Chen, Y., T.S. Hammack, K.Y. Song and K.A. Lampel. 2009. Evaluation of a revised U.S. Food and Drug Administration method for the detection and isolation of *Enterobacter sakazakii* in powdered infant formula: precollaborative study. J. AOAC Int. 92: 862–872.

Chen, Y., K.Y. Song, E.W. Brown and K.A. Lampel. 2010. Development of an improved protocol for the isolation and detection of *Enterobacter sakazakii* (*Cronobacter*) from powdered infant formula. J. Food Prot. 73: 1016–1022.

Chen, Y., K.E. Noe, S. Thompson, C.A. Elems, E.W. Brown, K.A. Lampel et al. 2012. Evaluation of a revised U.S. Food and Drug Administration method for the detection of *Cronobacter* in powdered infant formula: a collaborative study. J. Food Prot. 75: 1144–1147.

Choi, Y., K.P. Kim, K. Kim, J. Choi, H. Shin, D.H. Kang et al. 2012. Possible roles of LysR-type transcriptional regulator (LTTR) homolog as a global regulator in *Cronobacter sakazakii* ATCC 29544. Int. J. Med. Microbiol. 302: 270–275.

Cruz, A., J. Xicohtencatl-Cortes, B. González-Pedrajo, M. Bobadilla, C. Eslava and I. Rosas. 2011. Virulence traits in *Cronobacter* species isolated from different sources. Can. J. Microbiol. 57: 735–744.

de Benito, A., N. Gnanou Besse, I. Desforges, B. Gerten, B. Ruiz and D. Tomás. 2019. Validation of standard method EN ISO 22964:2017 - Microbiology of the food chain - Horizontal method for the detection of *Cronobacter* spp. Int. J. Food Microbiol. 288: 47–52.

Drudy, D., M. O'Rourke, M. Murphy, N.R. Mullane, R. O'Mahony, L. Kelly et al. 2006. Characterization of a collection of *Enterobacter sakazakii* isolates from environmental and food sources. Int. J. Food Microbiol. 110: 127–134.

Druggan, P. and C. Iversen. 2009. Culture media for the isolation of *Cronobacter* spp. Int. J. Food Microbiol. 136: 169–178.

Duca, D.R. and B.R. Glick. 2020. Indole-3-acetic acid biosynthesis and its regulation in plant-associated bacteria. Appl. Microbiol. Biotechnol. 104: 8607–8619.

Dwivedi, H.P. and L.A. Jaykus. 2011. Detection of pathogens in foods: the current state-of-the-art and future directions. Crit. Rev. Microbiol. 37: 40–63.

Edelson-Mammel, S.G. and R.L. Buchanan. 2004. Thermal inactivation of *Enterobacter sakazakii* in rehydrated infant formula. J. Food Prot. 67: 60–63.

Eida, A.A., S. Bougouffa, F. L'Haridon, I. Alam, L. Weisskopf, V.B. Bajic et al. 2020. Genome insights of the plant-growth promoting bacterium *Cronobacter muytjensii* JZ38 with volatile-mediated antagonistic activity against phytophthora infestans. Front. Microbiol. 11: 369.

Elkhawaga, A.A., H.F. Hetta, N.S. Osman, A. Hosni and M.A. El-Mokhtar. 2020. Emergence of *Cronobacter sakazakii* in cases of neonatal sepsis in upper Egypt: First report in North Africa. Front. Microbiol. 11: 215.

Eshwar, A.K., T. Tasara, R. Stephan and A. Lehner. 2015. Influence of FkpA variants on survival and replication of *Cronobacter* spp. in human macrophages. Res. Microbiol. 166: 186–195.

Eshwar, A.K., B.D. Tall, J. Gangiredla, G.R. Gopinath, I.R. Patel, S.C. Neuhauss et al. 2016. Linking Genomo- and Pathotype: Exploiting the zebrafish embryo model to investigate the divergent virulence potential among *Cronobacter* spp. PLoS One 11: e0158428.

Eshwar, A.K., N. Wolfrum, R. Stephan, S. Fanning and A. Lehner. 2018. Interaction of matrix metalloproteinase-9 and Zpx in *Cronobacter turicensis* LMG 23827[T] mediated infections in the zebrafish model. Cell. Microbiol. 20: e12888.

Esposito, F., R. Mamone, M. Di Serafino, C. Mercogliano, V. Vitale, G. Vallone et al. 2017. Diagnostic imaging features of necrotizing enterocolitis: a narrative review. Quant. Imaging Med. Surg. 7: 336–344.

Fan, H., B. Long, X. Wu and Y. Bai. 2012. Development of a loop-mediated isothermal amplification assay for sensitive and rapid detection of *Cronobacter sakazakii*. Foodborne Pathog. Dis. 9: 1111–1118.

FAO/WHO. *Enterobacter sakazakii* and other microorganisms in powdered infant formula: Meeting report; Microbiological Risk Assessment Series, No. 6; FAO/WHO: Rome, Italy. 2004. Available online: http://www.fao.org/3/a-y5502e.pdf (last accessed 10.14.2020).

FAO/WHO. *Enterobacter sakazakii* (*Cronobacter* spp.) in powdered follow-up formulae; Microbiol Risk Assess Series No. 15; FAO/WHO: Rome, Italy. 2008. Available online: https://www.who.int/foodsafety/publications/micro/MRA_followup.pdf (last accessed 10.14.2020).

Farmer, J.J., M.A. Asbury, F.W. Hickman, D.J. Brenner and the *Enterobacteriaceae* Study Group. 1980. *Enterobacter sakazakii*: a new species of *"Enterobacteriaceae"* isolated from clinical specimens. Int. J. Syst. Bacteriol. 30: 569–584.

Farmer, J.J., 3rd. 2015. My 40-year history with *Cronobacter/Enterobacter sakazakii* - Lessons learned, myths debunked, and recommendations. Front. Pediatr. 3: 84.

Feeney, A. and R.D. Sleator. 2011. An *in silico* analysis of osmotolerance in the emerging gastrointestinal pathogen *Cronobacter sakazakii*. Bioeng. Bugs 2: 260–270.

Fehr, A., A.K. Eshwar, S.C. Neuhauss, M. Ruetten, A. Lehner and L. Vaughan. 2015. Evaluation of zebrafish as a model to study the pathogenesis of the opportunistic pathogen *Cronobacter turicensis*. Emerg. Microbes Infect. 4: e29.

Fei, P., C. Man, B. Lou, S.J. Forsythe, Y. Chai, R. Li et al. 2015. Genotyping and source tracking of *Cronobacter sakazakii* and *C. malonaticus* isolates from powdered infant formula and an infant formula production factory in China. Appl. Environ. Microbiol. 81: 5430–5439.

Finlay, B.B. and S. Falkow. 1997. Common themes in microbial pathogenicity revisited. Microbiol. Mol. Biol. Rev. 61: 136–169.

Fox, E.M. and K.N. Jordan. 2008. Towards a one-step *Enterobacter sakazakii* enrichment. J. Appl. Microbiol. 105: 1091–1097.

Franco, A.A., L. Hu, C.J. Grim, G. Gopinath, V. Sathyamoorthy, K.G. Jarvis et al. 2011a. Characterization of putative virulence genes on the related RepFIB plasmids harbored by *Cronobacter* spp. Appl. Environ. Microbiol. 77: 3255–3267.

Franco, A.A., M.H. Kothary, G. Gopinath, K.G. Jarvis, C.J. Grim, L. Hu et al. 2011b. Cpa, the outer membrane protease of *Cronobacter sakazakii*, activates plasminogen and mediates resistance to serum bactericidal activity. Infect. Immun. 79: 1578–1587.

Friedemann, M. 2007. *Enterobacter sakazakii* in food and beverages (other than infant formula and milk powder). Int. J. Food Microbiol. 116: 1–10.

Friedemann, M. 2009. Epidemiology of invasive neonatal *Cronobacter* (*Enterobacter sakazakii*) infections. Eur. J. Clin. Microbiol. Infect. Dis. 28: 1297–1304.

Gajdosova, J., K. Benedikovicova, N. Kamodyova, L. Tothova, E. Kaclikova, S. Stuchlik et al. 2011. Analysis of the DNA region mediating increased thermotolerance at 58 degrees C in *Cronobacter* sp. and other enterobacterial strains. Antonie Van Leeuwenhoek 100: 279–289.

Giammanco, G.M., A. Aleo, I. Guida and C. Mammina. 2011. Molecular epidemiological survey of *Citrobacter freundii* misidentified as *Cronobacter* spp. (*Enterobacter sakazakii*) and *Enterobacter hormaechei* isolated from powdered infant milk formula. Foodborne Pathog. Dis. 8: 517–525.

Gopinath, G.R., H.R. Chase, J. Gangiredla, A. Eshwar, H. Jang, I. Patel et al. 2018. Genomic characterization of malonate positive *Cronobacter sakazakii* serotype O:2, sequence type 64 strains, isolated from clinical, food, and environment samples. Gut Pathog. 10: 11.

Gosney, M.A., M.V. Martin, A.E. Wright and M. Gallagher. 2006. *Enterobacter sakazakii* in the mouths of stroke patients and its association with aspiration pneumonia. Eur. J. Intern. Med. 17: 185–188.

Grim, C.J., M.H. Kothary, G. Gopinath, K.G. Jarvis, J.J. Beaubrun, M. McClelland et al. 2012. Identification and characterization of *Cronobacter* iron acquisition systems. Appl. Environ. Microbiol. 78: 6035–6050.

Grim, C.J., M.L. Kotewicz, K.A. Power, G. Gopinath, A.A. Franco, K.G. Jarvis et al. 2013. Pan-genome analysis of the emerging foodborne pathogen *Cronobacter* spp. suggests a species-level bidirectional divergence driven by niche adaptation. BMC Genomics 14: 366.

Grishin, A., J. Bowling, B. Bell, J. Wang and H.R. Ford. 2016. Roles of nitric oxide and intestinal microbiota in the pathogenesis of necrotizing enterocolitis. J. Pediatr. Surg. 51: 13–17.

Gurtler, J.B. and L.R. Beuchat. 2005. Performance of media for recovering stressed cells of *Enterobacter sakazakii* as determined using spiral plating and ecometric techniques. Appl. Environ. Microbiol. 71: 7661–7669.

Hassan, A.A., O. Akineden, C. Kress, S. Estuningsih, E. Schneider and E. Usleber. 2007. Characterization of the gene encoding the 16S rRNA of *Enterobacter sakazakii* and development of a species-specific PCR method. Int. J. Food Microbiol. 116: 214–220.

Hayman, M.M., S.G. Edelson-Mammel, P.J. Carter, Y.I. Chen, M. Metz, J.F. Sheehan et al. 2020. Prevalence of *Cronobacter* spp. and *Salmonella* in milk powder manufacturing facilities in the United States. J. Food Prot. 83: 1685–1692.

Healy, B., S. Huynh, N. Mullane, S. O'Brien, C. Iversen, A. Lehner et al. 2009. Microarray-based comparative genomic indexing of the *Cronobacter* genus (*Enterobacter sakazakii*). Int. J. Food Microbiol. 136: 159–164.

Henry, M. and A. Fouladkhah. 2019. Outbreak history, biofilm formation, and preventive measures for control of *Cronobacter sakazakii* in infant formula and infant care settings. Microorganisms 7: 77.

Hews, C.L., T. Cho, G. Rowley and T.L. Raivio. 2019. Maintaining integrity under stress: envelope stress response regulation of pathogenesis in Gram-megative bacteria. Front. Cell. Infect. Microbiol. 9: 313.

Himelright, I., E. Harris, V. Lorch, M. Anderson, T. Jones, A. Craig et al. 2002. *Enterobacter sakazakii* infections associated with the use of powdered infant formula—Tennessee, 2001. Morb. Mortal. Wkly. Rep. 51: 297–300.

Holman, R.C., B.J. Stoll, M.J. Clarke and R.I. Glass. 1997. The epidemiology of necrotizing enterocolitis infant mortality in the United States. Am. J. Public Health 87: 2026–2031.

Holý, O., J. Petrželová, V. Hanulík, M. Chromá, I. Matoušková and S.J. Forsythe. 2014. Epidemiology of *Cronobacter* spp. isolates from patients admitted to the Olomouc University Hospital (Czech Republic). Epidemiol. Mikrobiol. Imunol. 63: 69–72.

Hoque, A., T. Ahmed, M. Shahidullah, A. Hossain, A. Mannan, K. Noor et al. 2010. Isolation and molecular identification of *Cronobacter* spp. from powdered infant formula (PIF) in Bangladesh. Int. J. Food Microbiol. 142: 375–378.

Hu, S., Y. Yu, R. Li, X. Wu, X. Xiao and H. Wu. 2016. Rapid detection of *Cronobacter sakazakii* by real-time PCR based on the *cgcA* gene and TaqMan probe with internal amplification control. Can. J. Microbiol. 62: 191–200.

Huertas, J.P., A. Álvarez-Ordóñez, R. Morrissey, M. Ros-Chumillas, M.D. Esteban, J. Maté et al. 2015. Heat resistance of *Cronobacter sakazakii* DPC 6529 and its behavior in reconstituted powdered infant formula. Food Res. Int. 69: 401–409.

Hunter, C.J., B. Podd, H.R. Ford and V. Camerini. 2008a. Evidence vs experience in neonatal practices in necrotizing enterocolitis. J. Perinatol. 28 Suppl. 1: S9–S13.

Hunter, C.J., M. Petrosyan, H.R. Ford and N.V. Prasadarao. 2008b. *Enterobacter sakazakii*: an emerging pathogen in infants and neonates. Surg. Infect. (Larchmt) 9: 533–539.

ISO. 2017. ISO EN 22964:2017 Microbiology of the food chain—Horizontal method for the detection of *Cronobacter* spp. International Organization for Standardization, Geneva.

Iversen, C. and S.J. Forsythe. 2003. Risk profile of *Enterobacter sakazakii*, an emergent pathogen associated with infant milk formula. Trends Food Sci. Technol. 14: 443–454.

Iversen, C., P. Druggan and S. Forsythe. 2004. A selective differential medium for *Enterobacter sakazakii*, a preliminary study. Int. J. Food Microbiol. 96: 133–139.

Iversen, C., M. Waddington, J.J. Farmer, 3rd and S.J. Forsythe. 2006. The biochemical differentiation of *Enterobacter sakazakii* genotypes. BMC Microbiol. 6: 94.

Iversen, C. and S.J. Forsythe. 2007. Comparison of media for the isolation of *Enterobacter sakazakii*. Appl. Environ. Microbiol. 73: 48–52.

Iversen, C., A. Lehner, N. Mullane, E. Bidlas, I. Cleenwerck, J. Marugg et al. 2007a. The taxonomy of *Enterobacter sakazakii*: proposal of a new genus *Cronobacter* gen. nov. and descriptions of *Cronobacter sakazakii* comb. nov. *Cronobacter sakazakii* subsp. *sakazakii*, comb. nov., *Cronobacter sakazakii* subsp. *malonaticus* subsp. nov., *Cronobacter turicensis* sp. nov., *Cronobacter muytjensii* sp. nov., *Cronobacter dublinensis* sp. nov. and *Cronobacter* genomospecies 1. BMC Evol. Biol. 7: 64.

Iversen, C., A. Lehner, N. Mullane, J. Marugg, S. Fanning, R. Stephan et al. 2007b. Identification of "*Cronobacter*" spp. (*Enterobacter sakazakii*). J. Clin. Microbiol. 45: 3814–3816.

Iversen, C., N. Mullane, B. McCardell, B.D. Tall, A. Lehner, S. Fanning et al. 2008a. *Cronobacter* gen. nov., a new genus to accommodate the biogroups of *Enterobacter sakazakii*, and proposal of *Cronobacter sakazakii* gen. nov., comb. nov., *Cronobacter malonaticus* sp. nov., *Cronobacter turicensis* sp. nov., *Cronobacter muytjensii* sp. nov., *Cronobacter dublinensis* sp. nov., *Cronobacter* genomospecies 1, and of three subspecies, *Cronobacter dublinensis* subsp. *dublinensis* subsp. nov.,

Cronobacter dublinensis subsp. *lausannensis* subsp. nov. and *Cronobacter dublinensis* subsp. *lactaridi* subsp. nov. Int. J. Syst. Evol. Microbiol. 58: 1442–1447.

Iversen, C., P. Druggan, S. Schumacher, A. Lehner, C. Feer, K. Gschwend et al. 2008b. Development of a novel screening method for the isolation of "*Cronobacter*" spp. (*Enterobacter sakazakii*). Appl. Environ. Microbiol. 74: 2550–2553.

Jackson, S.A., I.R. Patel, T. Barnaba, J.E. LeClerc and T.A. Cebula. 2011. Investigating the global genomic diversity of *Escherichia coli* using a multi-genome DNA microarray platform with novel gene prediction strategies. BMC Genomics 12: 349.

Jacobs, C., P. Braun and P. Hammer. 2011. Reservoir and routes of transmission of *Enterobacter sakazakii* (*Cronobacter* spp.) in a milk powder-producing plant. J. Dairy Sci. 94: 3801–3810.

Jang, H.I. and M.S. Rhee. 2009. Inhibitory effect of caprylic acid and mild heat on *Cronobacter* spp. (*Enterobacter sakazakii*) in reconstituted infant formula and determination of injury by flow cytometry. Int. J. Food Microbiol. 133: 113–120.

Jang, H., A. Eshwar, A. Lehner, J. Gangiredla, I.R. Patel, J.J. Beaubrun et al. 2022. Characterization of *Cronobacter sakazakii* strains originating from plant-origin foods using comparative genomic analyses and zebrafish infectivity studies. Microorganisms 10: 1396.

Jang, H., J. Woo, Y. Lee, F. Negrete, S. Finkelstein, H.R. Chase et al. 2018a. Draft genomes of *Cronobacter sakazakii* strains isolated from dried spices bring unique insights into the diversity of plant-associated strains. Stand. Genomic Sci. 13: 35.

Jang, H., N. Addy, L. Ewing, J. Jean-Gilles Beaubrun, Y. Lee, J. Woo et al. 2018b. Whole-genome sequences of *Cronobacter sakazakii* isolates obtained from foods of plant origin and dried-food manufacturing environments. Genome Announc. 6: e00223-18.

Jang, H., G.R. Gopinath, A. Eshwar, S. Srikumar, S. Nguyen, J. Gangiredla et al. 2020a. The secretion of toxins and other exoproteins of *Cronobacter*: Role in virulence, adaption, and persistence. Microorganisms 8: 229.

Jang, H., H.R. Chase, J. Gangiredla, C.J. Grim, I.R. Patel, M.H. Kothary et al. 2020b. Analysis of the molecular diversity among *Cronobacter* species isolated from filth flies using targeted PCR, pan genomic DNA microarray, and whole genome sequencing analyses. Front. Microbiol. 11: 561204.

Jaradat, Z.W., Q.O. Ababneh, I.M. Saadoun, N.A. Samara and A.M. Rashdan. 2009. Isolation of *Cronobacter* spp. (formerly *Enterobacter sakazakii*) from infant food, herbs and environmental samples and the subsequent identification and confirmation of the isolates using biochemical, chromogenic assays, PCR and 16S rRNA sequencing. BMC Microbiol. 9: 225.

Jaradat, Z.W., W. Al Mousa, A. Elbetieha, A. Al Nabulsi and B.D. Tall. 2014. *Cronobacter* spp.—opportunistic food-borne pathogens. A review of their virulence and environmental-adaptive traits. J. Med. Microbiol. 63: 1023–1037.

Jarvis, K.G., C.J. Grim, A.A. Franco, G. Gopinath, V. Sathyamoorthy, L. Hu et al. 2011. Molecular characterization of *Cronobacter* lipopolysaccharide O-antigen gene clusters and development of serotype-specific PCR assays. Appl. Environ. Microbiol. 77: 4017–4026.

Jarvis, K.G., Q.Q. Yan, C.J. Grim, K.A. Power, A.A. Franco, L. Hu et al. 2013. Identification and characterization of five new molecular serogroups of *Cronobacter* spp. Foodborne Pathog. Dis. 10: 343–352.

Jason, J. 2012. Prevention of invasive *Cronobacter* infections in young infants fed powdered infant formulas. Pediatrics 130: e1076–1084.

Jiang, Y., S. Chen, Y. Zhao, X. Yang, S. Fu, J.L. McKillip et al. 2020. Multiplex loop-mediated isothermal amplification-based lateral flow dipstick for simultaneous detection of 3 food-borne pathogens in powdered infant formula. J. Dairy Sci. 103: 4002–4012.

Joseph, S. and S.J. Forsythe. 2011. Predominance of *Cronobacter sakazakii* sequence type 4 in neonatal infections. Emerg. Infect. Dis. 17: 1713–1715.

Joseph, S. and S.J. Forsythe. 2012. Insights into the emergent bacterial pathogen *Cronobacter* spp., generated by multilocus sequence typing and analysis. Front. Microbiol. 3: 397.

Joseph, S., E. Cetinkaya, H. Drahovska, A. Levican, M.J. Figueras and S.J. Forsythe. 2012a. *Cronobacter condimenti* sp. nov., isolated from spiced meat, and *Cronobacter universalis* sp. nov., a species designation for *Cronobacter* sp. genomospecies 1, recovered from a leg infection, water and food ingredients. Int. J. Syst. Evol. Microbiol. 62: 1277–1283.

Joseph, S., H. Sonbol, S. Hariri, P. Desai, M. McClelland and S.J. Forsythe. 2012b. Diversity of the *Cronobacter* genus as revealed by multilocus sequence typing. J. Clin. Microbiol. 50: 3031–3039.

Kämpfer, P., O. Rauhoff and W. Dott. 1991. Glycosidase profiles of members of the family *Enterobacteriaceae*. J. Clin. Microbiol. 29: 2877–2879.

Kang, S.E., Y.S. Nam and K.W. Hong. 2007. Rapid detection of *Enterobacter sakazakii* using TaqMan real-time PCR assay. J. Microbiol. Biotechnol. 17: 516–519.

Kilonzo-Nthenge, A., E. Rotich, S. Godwin, S. Nahashon and F. Chen. 2012. Prevalence and antimicrobial resistance of *Cronobacter sakazakii* isolated from domestic kitchens in middle Tennessee, United States. J. Food Prot. 75: 1512–1517.

Kothary, M.H., B.A. McCardell, C.D. Frazar, D. Deer and B.D. Tall. 2007. Characterization of the zinc-containing metalloprotease encoded by *zpx* and development of a species-specific detection method for *Enterobacter sakazakii*. Appl. Environ. Microbiol. 73: 4142–4151.

Kucerova, E., S.W. Clifton, X.Q. Xia, F. Long, S. Porwollik, L. Fulton et al. 2010. Genome sequence of *Cronobacter sakazakii* BAA-894 and comparative genomic hybridization analysis with other *Cronobacter* species. PLoS One 5: e9556.

Kumar, S., G. Stecher and K. Tamura. 2016. MEGA7: Molecular evolutionary genetics analysis version 7.0 for bigger datasets. Mol. Biol. Evol. 33: 1870–1874.

Lai, K.K. 2001. *Enterobacter sakazakii* infections among neonates, infants, children, and adults. Case reports and a review of the literature. Medicine (Baltimore) 80: 113–122.

Lehner, A., M. Grimm, T. Rattei, A. Ruepp, D. Frishman, G.G. Manzardo et al. 2006a. Cloning and characterization of *Enterobacter sakazakii* pigment genes and in situ spectroscopic analysis of the pigment. FEMS Microbiol. Lett. 265: 244–248.

Lehner, A., S. Nitzsche, P. Breeuwer, B. Diep, K. Thelen and R. Stephan. 2006b. Comparison of two chromogenic media and evaluation of two molecular based identification systems for *Enterobacter sakazakii* detection. BMC Microbiol. 6: 15.

Lehner, A., C. Fricker-Feer and R. Stephan. 2012. Identification of the recently described *Cronobacter condimenti* by an *rpoB*-gene-based PCR system. J. Med. Microbiol. 61: 1034–1035.

Leuschner, R.G. and J. Bew. 2004. A medium for the presumptive detection of Enterobacter sakazakii in infant formula: interlaboratory study. J. AOAC Int. 87: 604–613.

Liu, B.T., F.J. Song, M. Zou, Z.H. Hao and H. Shan. 2017. Emergence of colistin resistance gene *mcr-1* in *Cronobacter sakazakii* producing NDM-9 and in *Escherichia coli* from the same animal. Antimicrob. Agents Chemother. 61: e01444-16.

Liu, X., J. Fang, M. Zhang, X. Wang, W. Wang, Y. Gong et al. 2012. Development of a loop-mediated isothermal amplification assay for detection of *Cronobacter* spp. (*Enterobacter sakazakii*). World J. Microbiol. Biotechnol. 28: 1013–1020.

Liu, Y., X. Cai, X. Zhang, Q. Gao, X. Yang, Z. Zheng et al. 2006. Real time PCR using TaqMan and SYBR Green for detection of *Enterobacter sakazakii* in infant formula. J. Microbiol. Methods 65: 21–31.

Lockwood, D.E., A.S. Kreger and S.H. Richardson. 1982. Detection of toxins produced by *Vibrio fluvialis*. Infect. Immun. 35: 702–708.

Lönnerdal, B. 2012. Preclinical assessment of infant formula. Ann. Nutr. Metab. 60: 196–199.

Mao, B., D. Li, J. Zhao, X. Liu, Z. Gu, Y.Q. Chen et al. 2015. *In vitro* fermentation of fructooligosaccharides with human gut bacteria. Food Funct. 6: 947–954.

Masood, N., K. Moore, A. Farbos, K. Paszkiewicz, B. Dickins, A. McNally et al. 2015. Genomic dissection of the 1994 *Cronobacter sakazakii* outbreak in a French neonatal intensive care unit. BMC Genomics 16: 750.

McMullan, R., V. Menon, A.G. Beukers, S.O. Jensen, S.J. van Hal and R. Davis. 2018. *Cronobacter sakazakii* infection from expressed breast milk, Australia. Emerg. Infect. Dis. 24: 393–394.

Moine, D., M. Kassam, L. Baert, Y. Tang, C. Barretto, C. Ngom Bru et al. 2016. Fully closed genome sequences of five type strains of the Genus *Cronobacter* and one *Cronobacter sakazakii* strain. Genome Announc. 4: e00142-16.

Molloy, C., C. Cagney, S. Fanning and G. Duffy. 2010. Survival characteristics of *Cronobacter* spp. in model bovine gut and in the environment. Foodborne Pathog. Dis. 7: 671–675.

Moravkova, M., V. Verbikova, V. Huvarova, V. Babak, H. Cahlikova, R. Karpiskova et al. 2018. Occurrence of *Cronobacter* spp. in ready-to-eat vegetable products, frozen vegetables, and sprouts examined using cultivation and real-time PCR methods. J. Food Sci. 83: 3054–3058.

Mullane, N., J. Murray, D. Drudy, N. Prentice, P. Whyte, P.G. Wall et al. 2006. Detection of *Enterobacter sakazakii* in dried infant milk formula by cationic-magnetic-bead capture. Appl. Environ. Microbiol. 72: 6325–6330.

Mullane, N., P. Whyte, P.G. Wall, T. Quinn and S. Fanning. 2007. Application of pulsed-field gel electrophoresis to characterise and trace the prevalence of *Enterobacter sakazakii* in an infant formula processing facility. Int. J. Food Microbiol. 116: 73–81.

Mullane, N., B. Healy, J. Meade, P. Whyte, P.G. Wall and S. Fanning. 2008a. Dissemination of *Cronobacter* spp. (*Enterobacter sakazakii*) in a powdered milk protein manufacturing facility. Appl. Environ. Microbiol. 74: 5913–5917.

Mullane, N., P. O'Gaora, J.E. Nally, C. Iversen, P. Whyte, P.G. Wall et al. 2008b. Molecular analysis of the *Enterobacter sakazakii* O-antigen gene locus. Appl. Environ. Microbiol. 74: 3783–3794.

Muller, A., R. Stephan, C. Fricker-Feer and A. Lehner. 2013. Genetic diversity of *Cronobacter sakazakii* isolates collected from a Swiss infant formula production facility. J. Food Prot. 76: 883–887.

Müller, A., H. Hächler, R. Stephan and A. Lehner. 2014. Presence of AmpC beta-lactamases, CSA-1, CSA-2, CMA-1, and CMA-2 conferring an unusual resistance phenotype in *Cronobacter sakazakii* and *Cronobacter malonaticus*. Microb. Drug. Resist. 20: 275–280.

Mutic, A.D., S. Jordan, S.M. Edwards, E.P. Ferranti, T.A. Thul and I. Yang. 2017. The postpartum maternal and newborn microbiomes. MCN Am. J. Matern. Child Nurs. 42: 326–331.

Muytjens, H.L., J. van der Ros-van de Repe and H.A. van Druten. 1984. Enzymatic profiles of *Enterobacter sakazakii* and related species with special reference to the alpha-glucosidase reaction and reproducibility of the test system. J. Clin. Microbiol. 20: 684–686.

Muytjens, H.L. and J. van der Ros-van de Repe. 1986. Comparative *in vitro* susceptibilities of eight *Enterobacter* species, with special reference to *Enterobacter sakazakii*. Antimicrob. Agents Chemother. 29: 367–370.

Muytjens, H.L., H. Roelofs-Willemse and G.H. Jaspar. 1988. Quality of powdered substitutes for breast milk with regard to members of the family *Enterobacteriaceae*. J. Clin. Microbiol. 26: 743–746.

Nair, M.K. and K.S. Venkitanarayanan. 2006. Cloning and sequencing of the *ompA* gene of *Enterobacter sakazakii* and development of an *ompA*-targeted PCR for rapid detection of *Enterobacter sakazakii* in infant formula. Appl. Environ. Microbiol. 72: 2539–2546.

Nair, M.K., K. Venkitanarayanan, L.K. Silbart and K.S. Kim. 2009. Outer membrane protein A (OmpA) of *Cronobacter sakazakii* binds fibronectin and contributes to invasion of human brain microvascular endothelial cells. Foodborne Pathog. Dis. 6: 495–501.

Nei, M. and S. Kumar. 2000. Molecular Evolution and Phylogenetics. Oxford University Press, Inc., New York, NY 10016, USA.

Noriega, F.R., K.L. Kotloff, M.A. Martin and R.S. Schwalbe. 1990. Nosocomial bacteremia caused by *Enterobacter sakazakiki* and *Leuconostoc mesenteroides* resulting from extrinsic contamination of infant formula. Pediatr. Infect. Dis. J. 9: 447–449.

Ogrodzki, P. and S.J. Forsythe. 2016. CRISPR-cas loci profiling of *Cronobacter sakazakii* pathovars. Future Microbiol. 11: 1507–1519.

Olofsson, K., M. Bertilsson and G. Lidén. 2008. A short review on SSF—an interesting process option for ethanol production from lignocellulosic feedstocks. Biotechnol. Biofuels 1: 7.

Osaili, T.M., R.R. Shaker, M.S. Al-Haddaq, A.A. Al-Nabulsi and R.A. Holley. 2009. Heat resistance of *Cronobacter* species (*Enterobacter sakazakii*) in milk and special feeding formula. J. Appl. Microbiol. 107: 928–935.

Ozer, E.A., J.P. Allen and A.R. Hauser. 2014. Characterization of the core and accessory genomes of *Pseudomonas aeruginosa* using bioinformatic tools Spine and AGEnt. BMC Genomics 15: 737.

Pagotto, F.J., M. Nazarowec-White, S. Bidawid and J.M. Farber. 2003. *Enterobacter sakazakii*: infectivity and enterotoxin production *in vitro* and *in vivo*. J. Food Prot. 66: 370–375.

Pan, R., Y. Jiang, L. Sun, R. Wang, K. Zhuang, Y. Zhao et al. 2018. Gold nanoparticle-based enhanced lateral flow immunoassay for detection of *Cronobacter sakazakii* in powdered infant formula. J. Dairy Sci. 101: 3835–3843.

Parra-Flores, J., E. Maury-Sintjago, A. Rodriguez-Fernández, S. Acuña, F. Cerda, J. Aguirre et al. 2020. Microbiological quality of powdered infant formula in Latin America. J. Food Prot. 83: 534–541.

Patrick, M.E., B.E. Mahon, S.A. Greene, J. Rounds, A. Cronquist, K. Wymore et al. 2014. Incidence of *Cronobacter* spp. infections, United States, 2003–2009. Emerg. Infect. Dis. 20: 1520–1523.

Pava-Ripoll, M., R.E. Pearson, A.K. Miller and G.C. Ziobro. 2012. Prevalence and relative risk of *Cronobacter* spp., *Salmonella* spp., and *Listeria monocytogenes* associated with the body surfaces and guts of individual filth flies. Appl. Environ. Microbiol. 78: 7891–7902.

Pightling, A.W., J.B. Pettengill, Y. Luo, J.D. Baugher, H. Rand and E. Strain. 2018. Interpreting whole-genome sequence analyses of foodborne bacteria for regulatory applications and outbreak investigations. Front. Microbiol. 9: 1482.

Pitout, J.D., E.S. Moland, C.C. Sanders, K.S. Thomson and S.R. Fitzsimmons. 1997. Beta-lactamases and detection of beta-lactam resistance in *Enterobacter* spp. Antimicrob. Agents Chemother. 41: 35–39.

Porres-Osante, N., Y. Sáenz, S. Somalo and C. Torres. 2015. Characterization of beta-lactamases in faecal *Enterobacteriaceae* recovered from healthy humans in Spain: Focusing on AmpC polymorphisms. Microb. Ecol. 70: 132–140.

Proudy, I. 2009. *Enterobacter sakazakii* in powdered infant food formulas. Can. J. Microbiol. 55: 473–500.

Raghav, M. and P.K. Aggarwal. 2007. Purification and characterization of *Enterobacter sakazakii* enterotoxin. Can. J. Microbiol. 53: 750–755.

Reich, F., R. König, W. von Wiese and G. Klein. 2010. Prevalence of *Cronobacter* spp. in a powdered infant formula processing environment. Int. J. Food Microbiol. 140: 214–217.

Restaino, L., E.W. Frampton, W.C. Lionberg and R.J. Becker. 2006. A chromogenic plating medium for the isolation and identification of *Enterobacter sakazakii* from foods, food ingredients, and environmental sources. J. Food Prot. 69: 315–322.

Riedel, K. and A. Lehner. 2007. Identification of proteins involved in osmotic stress response in *Enterobacter sakazakii* by proteomics. Proteomics 7: 1217–1231.

Saitou, N. and M. Nei. 1987. The neighbor-joining method: a new method for reconstructing phylogenetic trees. Mol. Biol. Evol. 4: 406–425.

Salmonella Subcommittee of the Nomenclature Committee of the International Society for Microbiology. 1934. The Genus *Salmonella* Lignières, 1900. J. Hyg. (Lond) 34: 333–350.

Sani, N.A. and O.A. Odeyemi. 2015. Occurrence and prevalence of *Cronobacter* spp. in plant and animal derived food sources: a systematic review and meta-analysis. Springerplus 4: 545.

Savino, F., S. Fornasero, S. Ceratto, A. De Marco, N. Mandras, J. Roana et al. 2015. Probiotics and gut health in infants: A preliminary case-control observational study about early treatment with *Lactobacillus reuteri* DSM 17938. Clin. Chim. Acta. 451: 82–87.

Scharinger, E.J., R. Dietrich, T. Wittwer, E. Märtlbauer and K. Schauer. 2017. Multiplexed lateral flow test for detection and differentiation of *Cronobacter sakazakii* serotypes O1 and O2. Front. Microbiol. 8: 1826.

Schmid, M., C. Iversen, I. Gontia, R. Stephan, A. Hofmann, A. Hartmann et al. 2009. Evidence for a plant-associated natural habitat for *Cronobacter* spp. Res. Microbiol. 160: 608–614.

Seo, K.H. and R.E. Brackett. 2005. Rapid, specific detection of *Enterobacter sakazakii* in infant formula using a real-time PCR assay. J. Food Prot. 68: 59–63.

Shi, L., Q. Liang, Z. Zhan, J. Feng, Y. Zhao, Y. Chen et al. 2018. Co-occurrence of 3 different resistance plasmids in a multi-drug resistant *Cronobacter sakazakii* isolate causing neonatal infections. Virulence 9: 110–120.

Shukla, S., G. Lee, X. Song, J.H. Park, H. Cho, E.J. Lee et al. 2016. Detection of *Cronobacter sakazakii* in powdered infant formula using an immunoliposome-based immunomagnetic concentration and separation assay. Sci. Rep. 6: 34721.

Soares, S.C., L.C. Oliveira, A.K. Jaiswal and V. Azevedo. 2016. Genomic islands: an overview of current software and future improvements. J. Integr. Bioinform. 13: 301.

Song, X., S. Shukla, S. Oh, Y. Kim and M. Kim. 2015. Development of fluorescence-based liposome immunoassay for detection of *Cronobacter muytjensii* in pure culture. Curr. Microbiol. 70: 246–252.

Song, X., S. Shukla, G. Lee, S. Park and M. Kim. 2016. Detection of *Cronobacter* genus in powdered infant formula by enzyme-linked immunosorbent assay using anti-*Cronobacter* antibody. Front. Microbiol. 7: 1124.

Spaepen, S., J. Vanderleyden and R. Remans. 2007. Indole-3-acetic acid in microbial and microorganism-plant signaling. FEMS Microbiol. Rev. 31: 425–448.

Srikumar, S., Y. Cao, Q. Yan, K. Van Hoorde, S. Nguyen, S. Cooney et al. 2019. RNA sequencing-based transcriptional overview of xerotolerance in *Cronobacter sakazakii* SP291. Appl. Environ. Microbiol. 85: e01993-18.

Stephan, R., S. Van Trappen, I. Cleenwerck, M. Vancanneyt, P. De Vos and A. Lehner. 2007. *Enterobacter turicensis* sp. nov. and *Enterobacter helveticus* sp. nov., isolated from fruit powder. Int. J. Syst. Evol. Microbiol. 57: 820–826.

Stephan, R., S. Van Trappen, I. Cleenwerck, C. Iversen, H. Joosten, P. De Vos et al. 2008. *Enterobacter pulveris* sp. nov., isolated from fruit powder, infant formula and an infant formula production environment. Int. J. Syst. Evol. Microbiol. 58: 237–241.

Stephan, R., D. Ziegler, V. Pflüger, G. Vogel and A. Lehner. 2010. Rapid genus- and species-specific identification of *Cronobacter* spp. by matrix-assisted laser desorption ionization-time of flight mass spectrometry. J. Clin. Microbiol. 48: 2846–2851.

Stephan, R., A. Lehner, P. Tischler and T. Rattei. 2011. Complete genome sequence of *Cronobacter turicensis* LMG 23827, a food-borne pathogen causing deaths in neonates. J. Bacteriol. 193: 309–310.

Stephan, R., C.J. Grim, G.R. Gopinath, M.K. Mammel, V. Sathyamoorthy, L.H. Trach et al. 2014. Re-examination of the taxonomic status of *Enterobacter helveticus*, *Enterobacter pulveris* and *Enterobacter turicensis* as members of the genus *Cronobacter* and their reclassification in the genera *Franconibacter* gen. nov. and *Siccibacter* gen. nov. as *Franconibacter helveticus* comb. nov., *Franconibacter pulveris* comb. nov. and *Siccibacter turicensis* comb. nov., respectively. Int. J. Syst. Evol. Microbiol. 64: 3402–3410.

Stoop, B., A. Lehner, C. Iversen, S. Fanning and R. Stephan. 2009. Development and evaluation of *rpoB* based PCR systems to differentiate the six proposed species within the genus *Cronobacter*. Int. J. Food Microbiol. 136: 165–168.

Strydom, A., M. Cameron and R.C. Witthuhn. 2012. Phylogenetic analysis of *Cronobacter* isolates based on the rpoA and 16S rRNA genes. Curr. Microbiol. 64: 251–258.

Strysko, J., J.R. Cope, H. Martin, C. Tarr, K. Hise, S. Collier et al. 2020. Food safety and invasive *Cronobacter* infections during early infancy, 1961–2018. Emerg. Infect. Dis. 26.

Su, M., S.W. Satola and T.D. Read. 2019. Genome-based prediction of bacterial antibiotic resistance. J. Clin. Microbiol. 57: e01405-18.

Sun, Y., M. Wang, H. Liu, J. Wang, X. He, J. Zeng et al. 2011. Development of an O-antigen serotyping scheme for *Cronobacter sakazakii*. Appl. Environ. Microbiol. 77: 2209–2214.

Suppiger, A., A.K. Eshwar, R. Stephan, V. Kaever, L. Eberl and A. Lehner. 2016. The DSF type quorum sensing signalling system RpfF/R regulates diverse phenotypes in the opportunistic pathogen *Cronobacter*. Sci. Rep. 6: 18753.

Svobodová, B., J. Vlach, P. Junková, L. Karamonová, M. Blažková and L. Fukal. 2017. Novel method for reliable identification of *Siccibacter* and *Franconibacter* Strains: from "Pseudo-*Cronobacter*" to new *Enterobacteriaceae* genera. Appl. Environ. Microbiol. 83: e00234-17.

Swaminathan, B., T.J. Barrett, S.B. Hunter, R.V. Tauxe and C.D.C.P.T. Force. 2001. PulseNet: the molecular subtyping network for foodborne bacterial disease surveillance, United States. Emerg. Infect. Dis. 7: 382–389.

Tall, B.D., Y. Chen, Q. Yan, G.R. Gopinath, C.J. Grim, K.G. Jarvis et al. 2014. *Cronobacter*: an emergent pathogen causing meningitis to neonates through their feeds. Sci. Prog. 97: 154–172.

Tall, B.D., J. Gangiredla, G.R. Gopinath, Q. Yan, H.R. Chase, B. Lee et al. 2015. Development of a custom-designed, pan genomic DNA microarray to characterize strain-level diversity among *Cronobacter* spp. Front. Pediatr. 3: 36.

Tamura, K., M. Nei and S. Kumar. 2004. Prospects for inferring very large phylogenies by using the neighbor-joining method. Proc. Natl. Acad. Sci. U.S.A. 101: 11030–11035.

Tomas, D., M. Fan, S. Zhu and A. Klijn. 2018. Use of biochemical miniaturized galleries, rRNA based lateral flow assay and Real Time PCR for *Cronobacter* spp. confirmation. Food Microbiol. 76: 189–195.

Townsend, S., E. Hurrell and S. Forsythe. 2008. Virulence studies of *Enterobacter sakazakii* isolates associated with a neonatal intensive care unit outbreak. BMC Microbiol. 8: 64.

Ueda, S. 2017. Occurrence of *Cronobacter* spp. in dried foods, fresh vegetables and soil. Biocontrol Sci. 22: 55–59.

Urmenyi, A.M. and A. White-Franklin. 1961. Neonatal death from pigmented coliform infection. Lancet 1: 313–315.

Vasconcellos, L., C.T. Carvalho, R.O. Tavares, V. de Mello Medeiros, C. de Oliveira Rosas, J.N. Silva et al. 2018. Isolation, molecular and phenotypic characterization of *Cronobacter* spp. in ready-to-eat salads and foods from Japanese cuisine commercialized in Brazil. Food Res. Int. 107: 353–359.

Wang, Q., X.J. Zhao, Z.W. Wang, L. Liu, Y.X. Wei, X. Han et al. 2017. Identification of *Cronobacter* species by matrix-assisted laser desorption/ionization time-of-flight mass spectrometry with an optimized analysis method. J. Microbiol. Methods 139: 172–180.

Williams, T.L., S.R. Monday, S. Edelson-Mammel, R. Buchanan and S.M. Musser. 2005. A top-down proteomics approach for differentiating thermal resistant strains of *Enterobacter sakazakii*. Proteomics 5: 4161–4169.

Woodward, A.W. and B. Bartel. 2005. Auxin: regulation, action, and interaction. Ann. Bot. 95: 707–735.

Yachison, C.A., C. Yoshida, J. Robertson, J.H.E. Nash, P. Kruczkiewicz, E.N. Taboada et al. 2017. The validation and implications of using whole genome sequencing as a replacement for traditional serotyping for a national *Salmonella* reference laboratory. Front. Microbiol. 8: 1044.

Yan, Q., O. Condell, K. Power, F. Butler, B.D. Tall and S. Fanning. 2012. *Cronobacter* species (formerly known as *Enterobacter sakazakii*) in powdered infant formula: a review of our current understanding of the biology of this bacterium. J. Appl. Microbiol. 113: 1–15.

Yan, Q., K.A. Power, S. Cooney, E. Fox, G.R. Gopinath, C.J. Grim et al. 2013. Complete genome sequence and phenotype microarray analysis of *Cronobacter sakazakii* SP291: a persistent isolate cultured from a powdered infant formula production facility. Front. Microbiol. 4: 256.

Yan, Q., J. Wang, J. Gangiredla, Y. Cao, M. Martins, G.R. Gopinath et al. 2015a. Comparative genotypic and phenotypic analysis of *Cronobacter* species cultured from four powdered infant formula production facilities: Indication of pathoadaptation along the food chain. Appl. Environ. Microbiol. 81: 4388–4402.

Yan, Q., K.G. Jarvis, H.R. Chase, K. Hebert, L.H. Trach, C. Lee et al. 2015b. A proposed harmonized LPS molecular-subtyping scheme for *Cronobacter* species. Food Microbiol. 50: 38–43.

Yan, Q. and S. Fanning. 2015a. Pulsed-field gel electrophoresis (PFGE) for pathogenic *Cronobacter* species. Methods Mol. Biol. 1301: 55–69.

Yan, Q. and S. Fanning. 2015b. Strategies for the identification and tracking of *Cronobacter* species: an opportunistic pathogen of concern to neonatal health. Front. Pediatr. 3: 38.

Ye, Y., Q. Wu, L. Yao, X. Dong, K. Wu and J. Zhang. 2009. A comparison of polymerase chain reaction and international organization for standardization methods for determination of *Enterobacter sakazakii* contamination of infant formulas from Chinese mainland markets. Foodborne Pathog. Dis. 6: 1229–1234.

Ye, Y., Q. Wu, X. Xu, X. Yang, X. Dong and J. Zhang. 2010. The phenotypic and genotypic characterization of *Enterobacter sakazakii* strains from infant formula milk. J. Dairy Sci. 93: 2315–2320.

Yong, W., B. Guo, X. Shi, T. Cheng, M. Chen, X. Jiang et al. 2018. An investigation of an acute gastroenteritis outbreak: *Cronobacter sakazakii*, a potential cause of food-borne illness. Front. Microbiol. 9: 2549.

Zeng, H., J. Zhang, C. Li, T. Xie, N. Ling, Q. Wu et al. 2017. The driving force of prophages and CRISPR-cas system in the evolution of *Cronobacter sakazakii*. Sci. Rep. 7: 40206.

Zeng, H., C. Li, W. He, J. Zhang, M. Chen, T. Lei et al. 2019. *Cronobacter sakazakii, Cronobacter malonaticus*, and *Cronobacter dublinensis* genotyping based on CRISPR locus diversity. Front. Microbiol. 10: 1989.

Zeng, H., C. Li, N. Ling, J. Zhang, M. Chen, T. Lei et al. 2020. Prevalence, genetic analysis and CRISPR typing of *Cronobacter* spp. isolated from meat and meat products in China. Int. J. Food Microbiol. 321: 108549.

Zhu, S., S. Schnell and M. Fischer. 2012. Rapid detection of *Cronobacter* spp. with a method combining impedance technology and rRNA based lateral flow assay. Int. J. Food Microbiol. 159: 54–58.

Zogaj, X., W. Bokranz, M. Nimtz and U. Römling. 2003. Production of cellulose and curli fimbriae by members of the family *Enterobacteriaceae* isolated from the human gastrointestinal tract. Infect. Immun. 71: 4151–4158.

Chapter 2

Current Practices for Isolation, Identification and Characterization of *Campylobacter* in Foods and Recommendations for Future

Zajeba Tabashsum,[1] *Mayur Krishna,*[1] *Mia Lang,*[2] *Syed Hussain,*[1]
Nishi Shah,[1] *Paulina B. Romo*[2] *and Debabrata Biswas*[1,2,3,]*

Introduction

Campylobacter is one of the leading causative agents of diarrheal diseases and bacterial gastroenteritis worldwide (World Health Organization 2018). As it is a self-limited disease, often the patients do not require visiting the hospital or doctors. Though in many cases the infection caused by *Campylobacter*, known as campylobacteriosis, can go unreported; the Center for Disease Control (CDC) estimates that over 1.3 million people suffer from campylobacteriosis each year in the US (Centers for Disease Control and Prevention 2017). Symptoms of *Campylobacter* infection include a range of gastrointestinal issues, such as diarrhea, severe abdominal cramps, vomiting, and fever (Food Safety 2019). These symptoms usually resolve within 2–5 days for a healthy individual, though serious complications or consequences can occur in those individuals with pre-existing health conditions or compromised immune systems, and in such cases, the infected person may require antibiotic

[1] Biological Sciences Program, University of Maryland, College Park, MD 20742.
[2] Department of Animal and Avian Sciences, University of Maryland, College Park, MD 20742.
[3] Center for Food Safety and Security Systems, University of Maryland, College Park, MD 20742.
* Corresponding author: dbiswas@umd.edu

treatment or even hospitalization (Galanis 2007). The most common sources of *Campylobacter* are common food products, particularly poultry meat and milk.

Campylobacters are unable to grow in atmospheric oxygen concentration and rather need a microaerophilic condition (5% O_2, 10% CO_2, and 85% N_2) for growth in laboratory conditions (Garénaux et al. 2008). *Campylobacter* grows at a wide variety of temperature ranges, though at a lower temperature; *Campylobacter* goes into a viable but non-culturable (VBNC) state (Federighi et al. 1988; Garénaux et al. 2008). Due to these special growth requirements, it is very difficult to detect and grow *Campylobacter* in laboratory conditions, and special growth conditions are required. *Campylobacter* has 31 species reported but only a few are important with respect to pathogenicity to humans, including *C. jejuni*, *C. coli*, and *C. fetus*. Therefore, in this chapter, we talk about various sources of *Campylobacter*, current practices of *Campylobacter* detection and characterization, and recommendations for better practices mostly focusing on *C. jejuni*, *C. coli*, and *C. fetus*.

Potential Food Sources of *Campylobacter* and Current Microbiological Methods of Analysis

Common Food Sources of Campylobacter

The most common sources of *Campylobacter* include raw poultry meats, unpasteurized milk, and contaminated water (World Health Organization 2018). Consumption of raw fruits and vegetables has also been linked to increased incidence of campylobacteriosis in humans. Fruits and vegetables can become contaminated throughout many steps of the retail process. For example, contact with contaminated raw poultry meats or water in a packaging facility can be the source of cross-contamination of produce products. Levels of *Campylobacter* are much lower in produce products than in raw poultry meats and meat products. The presence of bacteria in or on produce products can be largely eliminated by careful processing of produces for sale at the facility and then washing the produces before eating on the consumers' end (Mohammadpour et al. 2018). In this section, we will focus on the more common breeding grounds for *Campylobacter* and sources of *Campylobacter* infection, like poultry, meat, milk, and sea-foods.

Meat and Meat Products as Sources of Campylobacter

Campylobacter sometimes exhibits commensal relationships with their animal hosts (Silva et al. 2011). *Campylobacter* often resides in the intestinal tract of warm-blooded animals, such as chicken and cattle, and only causes gastrointestinal problems when it transfers to humans. Animals infected with *Campylobacter* are typically asymptomatic, making it difficult to prevent the spread of the bacteria to humans through contact with contaminated raw meat or other contaminated sources (Mezher et al. 2016). Processing and handling of raw animal products coupled with the unique survival ability of *Campylobacter* allow for cross-contamination to a variety of foods (Silva et al. 2011). Raw poultry meat and other poultry-related products are the most common sources of *Campylobacter* and play a huge role in the infection and cross-contamination of foods, surfaces, and subsequently humans.

Poultry

Among other meat and avian sources of *Campylobacter*, raw poultry constitutes the leading cause of campylobacteriosis in humans (Sahin et al. 2015). Not only poultry is the most widely consumed meat product worldwide, but also poultry gut provides ideal conditions for the growth of *Campylobacter*. *Campylobacter* grows in the warm intestinal tract of poultry with most of the bacteria residing in the colon and the cecum (Silva et al. 2011). When the poultry carcass is processed, the intestinal walls may rupture, allowing the *Campylobacter* to spread to other parts of the processed poultry, including channels in the skin. This migration of *Campylobacter* to the poultry skin increases the possibility of cross-contamination with other foods, which are also processed in the same facility (Berrang et al. 2001). Additionally, while they survive best under warm conditions, *Campylobacter* can survive within the poultry carcass under colder temperatures, allowing chances of further cross-contamination even after the poultry is processed, refrigerated, and distributed (Silva et al. 2011). So, improper storage of raw poultry can lead to the further spread of the bacteria (Scherer et al. 2006).

Beef

While *Campylobacter* is the most prevalent in poultry, contamination with *Campylobacter* can also occur in red meat, such as cows (Silva et al. 2011). Like in poultry, the bacteria inhabit the digestive tracts of these animals and can be transferred to humans via improper handling of raw or unprocessed meats and cross-contamination of surfaces, and human contact with infected animal feces (Humphrey et al. 2007). While studies are still being done to better understand the optimal environment for *Campylobacter* survival, studies have shown that contamination of red meat and beef products may be less common than in poultry/poultry products in part due to the different processing procedures. In a study, Humphrey et al. (2007) found that the drying aspect of processing meats helps to reduce relative amounts of *Campylobacter* growth in red meats. The lower prevalence of *Campylobacter* in red meat sources can also be due to the overall lower body temperature of cows in comparison to chickens (Lopes et al. 2018).

Milk and Other Food Products as Sources of Campylobacter

Due to the high prevalence of cross-contamination involving *Campylobacter*, the bacteria are not only limited to raw poultry and meat products, but some other key sources of *Campylobacter* include raw (unpasteurized) milk, contaminated water, and shellfish. As mentioned earlier, contamination at the retail processing is a crucial point and improper handling during food preparation can be another major point at the consumers' end. For example, not washing hands between preparation of dishes like chicken and fish and as a result, contaminating the fish dish. Though the bacterial load will be less in fish, still it can be a source of *Campylobacter*.

Milk

Raw milk can also be a source of *Campylobacter* due to cross-contamination of the milk during the milking process by feces from an infected dairy cow. Raw milk can also contain *Campylobacter* due to the infection of the udders of the dairy cow by the bacteria (Silva et al. 2011). Maintaining milking hygiene by washing the udders of the cows and keeping the milking facility clean can help to reduce *Campylobacter* levels in raw milk, but the most effective and safe way is done during post-processing, i.e., exterminating the bacteria via pasteurization (Humphrey and Beckett 1987). Based on multiple case studies of campylobacteriosis related to raw milk consumption in the USA, CDC recommends the consumption of pasteurized milk only as current testing procedures of raw milk may not be sensitive enough to detect *Campylobacter* (Davis et al. 2016). Fortunately, cheeses are not considered a major source of *Campylobacter* as the bacteria can only live-in fresh cheeses for much shorter amounts of time than they can in raw milk (Butzler and Oosterom 1991).

Water

Water supplies can be contaminated by animal feces containing *Campylobacter*. While *Campylobacter* live and reproduce primarily in the intestinal tract of animals, they are capable of completely changing their survival abilities once they enter a water source. For this reason, it is very difficult to isolate and culture *Campylobacter* from contaminated water sources (Schallenberg et al. 2005). Interestingly, studies done by Korhonen and Martikalnon (1991) have shown that *Campylobacter* survive better in filtered water than in untreated water, suggesting that *Campylobacter* survive best without competition with other organisms for nutrients in the water. But they do not survive when the water is treated to remove most of the potential nutrients for the bacteria (Korhonen and Martikalnon 1991). Certain strains of the bacteria can survive for months in various water sources, causing the contaminated water to be a major source of infection (Schallenberg et al. 2005). Human consumption of contaminated water, or the use of contaminated water to wash produce, can directly lead to campylobacteriosis (World Health Organization 2018; Centers for Disease Control and Prevention 2017).

Shellfish and Seafood

Studies have shown that consumption of various types of infected shellfish can be linked to increased occurrence of campylobacteriosis in humans. Both WHO and CDC warn against the consumption of raw shellfish due to the potential for infections by different enteric pathogens including campylobacteriosis (Centers for Disease Control and Prevention 2019). Shellfish can be infected by the bacteria through contaminated waters (Teunis 1997) and improper handling during retail processing and/or food preparation.

Microbiological Methods Used for Campylobacter Detection in Food Industries

The genus *Campylobacter* are mostly small ($0.2–0.8$ μm × $0.5–5$ μm), Gram-negative, slender, and are spirally curved rods, where most species have an unsheathed polar flagellum at one or both ends (Silva et al. 2011). Under unfavorable conditions, *Campylobacter* can form VBNC cells (Portner et al. 2007). The optimum temperatures for the growth of *Campylobacter* are between 37°C and 42°C. *Campylobacter* grow optimally at pH 6.5–7.5 and will not survive above pH 9.0 or below pH 4.9. They are microaerophilic and non-spore-forming fastidious bacteria growing in lower oxygen atmospheric conditions (5% O_2, 10% CO_2, and 85% N_2) (Garénaux et al. 2008). This selective spectrum for the growth of *Campylobacter* has enabled it to grow on very particular media, such as blood, ferrous iron, pyruvates, etc., and with selective agents like antibiotics in laboratory conditions (Corry et al. 1995). Diverse types of broths, like Bolton broth, Preston broth, etc., are used to enrich and grow *Campylobacter* (Baylis et al. 2000). The presence of the enzyme oxyrase is known to be effective in reducing oxygen levels and isolating *Campylobacter* from various sources (Abeyta et al. 1997). The most common plating method for *Campylobacter* growth and isolation is to use mCCDA (modified Charcoal Cefoperazone Deoxycholate Agar) as the selective agar after enrichment in Bolton broth at 37°C in a microaerophilic atmosphere for 4–6 hours, followed by incubation at 41.5°C for 40–48 hours. Another agar medium named Karmali can also be used to increase the efficacy of the detection procedure. A study by Federighi et al. (1990) showed that Park and Sanders' broth followed by the isolation on Karmali agar is a very effective combination for the isolation of *Campylobacter*.

Campylobacter is one of the main causes of bacterial gastroenteritis in developed countries around the world and one of the most common organisms that cause bacterial infections which make their way through basic diet. These bacteria follow a commensal lifestyle with various livestock like poultry and dairy sources as mentioned above and ultimately behaves like a pathogen after being ingested by humans. Hendrixson et al. (2004) used signature-tagged transposon mutagenesis to identify various *Campylobacter* to understand precisely how *C. jejuni* bacterium promotes commensalism by identifying different mutants, each representing different genes involved in the chicken gastrointestinal tract. *Campylobacter* naturally colonize the gastrointestinal tract of birds and animals, but it is a pathogen for humans and results in productive gastroenteritis, which causes mild to bloody diarrhea—one of the main symptoms that can help identify *Campylobacter* as the likely pathogen (Hendrixson et al. 2004). *Campylobacter* isolates can mainly be found in the caeca and the large intestine of poultry birds (Beery et al. 1988). Colonization of the chicken gastrointestinal tract usually occurs in chickens at an early age and can last up to the time of the slaughter (Lindblom et al. 1986). Different *Campylobacter* can be identified according to the region of colonization. Pathogenic strains of *C. jejuni* have a commensal relationship with chickens, favoring the caeca, and this was confirmed by studies done by Beery et al. (1988). The identification of *Campylobacter* was done by performing a competition assay between different strains of *Campylobacter* mutants. It has been shown that, for successful colonization by *Campylobacter*, motility is very important (Nachamkin et al. 1993; Wassenaar et al. 1993). The mutations in the genes that cause flagellar motility usually result in a 10 to 1,000-fold reduction of the bacterial loads in the chick caeca (Hendrixson et al. 2004). Although the role of motility is currently unknown, *Campylobacter* with flagellar apparatus can colonize a more substantial area for a longer time, providing a distinction between the strains that contain flagella from the ones that do not contain apparatus for motility. One of the genes that can help identify *C. jejuni* specifically is *cj0019c*, which encodes a methyl-accepting chemotaxis protein, which signals proteins that help transduce signals to the chemotaxis regulatory proteins (Hendrixson et al. 2004).

Another way to identify different species of *Campylobacter* from animals and humans is with the help of bacterial restriction endonuclease DNA analysis (BRENDA) (Kakoyiannis et al. 1984). The inability of *C. jejuni* to hydrolyze hippurate in the BRENDA test is the key feature that distinguishes *C. jejuni* from *C. coli* (Skirrow et al. 1980). The intestinal tract of pigs is usually colonized by *C. coli* (Munroe et al. 1983), which is the most commonly found *Campylobacter* species in pigs, while the other predominant species, *C. jejuni* is frequently isolated from poultry (Rossef et al. 1982). Presently, there are two serotyping techniques to identify *Campylobacter* species developed by Lior et al. (1982). One technique deals with *C. jejuni* where 14% to 17% of strains are untypable, while another technique is able to identify both *C. coli* and *C. jejuni* (Penner et al. 1980). It is also worthwhile to mention that plasmid typing is necessary as an additional means of differentiation (Bradbury et al. 1983). BRENDA patterns of the strains that are isolated from pigs are usually uncommon and are typically not found in the isolates from humans. But the BRENDA patterns that are found in the isolates from chickens are remarkably similar to the patterns found in the isolates from humans. One of the advantages of using BRENDA to identify different *Campylobacter* species is that the patterns remain stable for a very long time (Kakoyiannis et al. 1984). These BRENDA patterns can be used as reference standards to be compared with other patterns to identify a specific type of strain. BRENDA can also be used to identify strains that are very similar and only differ from each other in degrees of pathogenicity. To make it more precise, the BRENDA patterns can be applied to restriction enzymes of two organisms which show identical patterns after digestion with the initial enzyme (Kakoyianis et al. 1984). However, it should be noted that minor differences in the patterns of the same organism as *C. coli* exists, but these differences are only due to a change in the number of plasmids which vary between each individual organism/strain. BRENDA can also be used to identify *C. pyloridis* isolates. Just like different species of *Campylobacter* colonizing different areas of the gastrointestinal tract, *C. pyloridis* usually colonize the gastric mucosa of the upper abdominal region in human (Langenberg et al. 1986). No biotyping or serotyping schemes exist for *C. pyloridis*. This strain is only identifiable by the analysis of its proteins by sodium dodecyl sulfate-polyacrylamide (SDS) gel electrophoresis, which results in rather uniform patterns (Langenberg et al. 1986). However, this analysis has limitations to identify *C. pyloridis* after the restriction endonuclease identification because other species like *C. jejuni* and *C. coli* can also be identified by the method showing a similar pattern. An exclusive characteristic of *C. pyloridis* is that they can hydrolyze urea rapidly. This production of urease excludes *C. jejuni*-like organisms which are also able to inhabit the gastric mucosa occasionally. Langenberg et al. (1986) demonstrate that restriction endonuclease analysis with HindIII is a very sensitive method for differentiating *C. pyloridis* isolates. However, there is a possibility that the colonization of the gastric mucosa is possible by more than one strain of

C. pyloridis isolates. This is evident from the identity of the DNA digest patterns which are produced by different colony-type isolates recovered from one patient on various occasions (Langenberg et al. 1986). This method of identification of *C. pyloridis* isolates is more accurate and can be used for epidemiologic surveillance, like the investigation of hospital infections, or to study hypochlorhydria, a disease attributed to the infection of *C. pyloridis*.

Another method used to identify *Campylobacter* is multiplex-polymerase chain reaction (PCR). Multiplex PCR is a common laboratory technique used for the detection of *C. jejuni* and *C. coli*. In most diagnostic laboratories, at least 95% of human *Campylobacter* isolates belong to either *C. coli* or *C. jejuni* when a selective medium is applied (Endtz et al. 1991). Primers used in PCR operate on different loci in these two organisms. For *C. jejuni*, the hippuricase gene (*hipO*) primer and for *C. coli*, an aspartokinase gene (*lysC*) primer is used and for the positive control of the PCR, a universal 16S rDNA gene primer is used (Persson et al. 2005). Previously, the only method used for distinguishing between *C. coli* and *C. jejuni* was hippurate hydrolysis, but this method was less accurate and more time-consuming (Rautelin et al. 1999). In the study by Persson et al. (2005), a three-gene multiplex-PCR-based method was used when *Campylobacter* is exposed to room temperature and atmospheric air, their survival rate decreases dramatically (Holler et al. 1998), thus making PCR-based detection methods more accurate.

Characterization of Campylobacter Isolated From Food Products

Campylobacter is microaerophilic meaning they "require O_2 for growth but are sensitive to high oxygen tensions" (Kaakoush et al. 2007). Due to their sensitivity to high oxygen tensions, *Campylobacter* grow best in atmospheres with low oxygen tensions (5% O_2, 10% CO_2, and 85% N_2) (Silva et al. 2011). Although *Campylobacter* utilizes oxygen as its terminal electron acceptor for respiration, numerous alternative electron acceptors can also be utilized (Sellars et al. 2002). For example, in reduced oxygen conditions growth of *C. jejuni* is possible when fumarate, nitrite, and TMAO are utilized as electron acceptors for respiration (Sellars et al. 2002). However, the growth of *C. jejuni* is not supported under strictly anaerobic conditions (Sellars et al. 2002).

The growth of *Campylobacter* is also highly dependent on the temperature of the environment. Different *Campylobacter* isolated from food products are thermotolerant, growing between 37°C and 42°C with optimal growth at 41.5°C (Silva et al. 2011). As the temperature approaches the lower temperature limit, the growth of *Campylobacter* drops dramatically unlike other microorganisms where growth decreases gradually (Park 2002). By 30°C, there is a complete absence of growth in *Campylobacter* (Park 2002). This dramatic decline in growth can be explained by the absence of cold shock proteins in *Campylobacter* (Park 2002). The presence of cold shock proteins in other microorganisms allows them to adapt and grow at lower temperatures (Park 2002; Levin 2007). Therefore, the lack of cold shock proteins in *Campylobacter* does not allow the species to grow at lower temperatures (Park 2012; Levin 2007). Although unable to grow at lower temperatures, *Campylobacter* can still be metabolically active below their optimal growth temperature (Park 2002). Respiration and ATP generation are supported at temperatures as low as 4°C in *C. jejuni*, the primary cause of enteric illness in the US (Park 2002). However, although unable to replicate, the motility of *C. jejuni* remains unaffected at lower temperatures (Park 2002). The ability of *C. jejuni* to survive and retain motility allows the species to migrate to more favorable environments for survival (Park 2002). The ability of *C. jejuni* to survive and be motile at temperatures associated with food storage allows for *C. jejuni* to be successful in human infection (Park 2002). Although *Campylobacter* grows at higher temperatures, they are better classified as "thermotolerant" due to their inability to grow at temperatures above 55°C (Levin 2007; Silva et al. 2011). Their sensitivity to higher heat makes cooking and pasteurization successful tools in killing *Campylobacter* (Park 2002).

Unlike other gastrointestinal bacteria, the nutritional needs of *Campylobacter* are unique requiring a diet low in carbohydrates and high in proteins (Hofreuter 2014). This diet is highly dependent on the inability of *Campylobacter* to utilize many common carbohydrates as carbon sources (Hofreuter 2014; Epps et al. 2013; Stahl et al. 2012). The absence of metabolic pathways necessary for the breakdown of glucose, galactose, and many other carbohydrates results in the non-saccharolytic nature of these bacteria (Hofreuter 2014; Stahl et al. 2012). Due to the absence of glucokinase, the phosphorylation of extracellular glucose is not possible by *Campylobacter* (Stahl et al. 2012). *Campylobacter* also lack the key enzyme 6-phosphofructokinase which is responsible for "the irreversible phosphorylation of fructose-6-phosphate to fructose-1,6-diphosphate during glycolysis" (Stahl et al. 2012). Although glycolysis is not possible, the enzyme, pyruvate kinase which is necessary for a different irreversible step of glycolysis is present (Stahl et al. 2012). Similarly, even though the enzymes for the non-oxidative portion of the pentose phosphate pathway are present, the pentose phosphate pathway cannot be performed because the oxidative portion is absent (Hofreuter 2014). Although unable to utilize extracellular glucose, *Campylobacter* possess the enzymes required to convert pyruvate to glucose-6-phosphate and therefore can perform gluconeogenesis (Stahl et al. 2012).

Because *Campylobacter* is unable to utilize many carbohydrates as carbon sources, they rely on amino acids for their primary nutrient sources (Stahl et al. 2012; Weingarten et al. 2009). Both amino acids and the intermediates of the citric acid cycle are important carbon sources necessary for the growth and survival of *C. jejuni* (Epps et al. 2013). *C. jejuni* can utilize amino acids in the following sequential order: serine, aspartate, asparagine, and glutamate (Epps et al. 2013; Stahl et al. 2012). However, certain *C. jejuni* strains can also metabolize proline after the other amino acids have been completely utilized (Epps et al. 2013; Stahl et al. 2012). The most common amino acids found in chick excreta are serine, aspartate, glutamate, and proline, therefore, explaining the colonization of *C. jejuni* in the intestines of chickens (Stahl et al. 2012).

Flagella-mediated motility is an essential characteristic of the virulence of *Campylobacter* (Silva et al. 2011). *Campylobacters* have single polar flagella that mediate rapid, darting motility (Guerry 2007). This flagella-mediated motility is essential for the colonization of the small intestine by *Campylobacter* (Silva et al.

2011). The spiral shape along with the polar flagellum of *Campylobacter,* allows for motility in highly viscous environments such as in mucous which typically causes paralysis in other motile rod-shaped bacteria (Park 2002). *Campylobacter* flagellin are made up of FlaA and FlaB, the major and minor flagella protein respectively (Guerry 2007). The classic flagellin promoter σ^{28} regulates the *flaA* gene while the *flaB* gene is regulated by a σ^{54}-dependent promoter (Guerry 2007). When the *flaA* gene is mutated a severe reduction in motility has been reported causing the production of a truncated flagellar filament made up of *flaB* protein (Silva et al. 2011; Guerry 2007). However, mutations in *flaB* do not result in structural and motility differences (Silva et al. 2011; Guerry 2007). Adherence, gastrointestinal colonization, and invasion of host cells are dependent on the *flaA* gene, therefore mutations of *flaA* genes affect virulence properties (Silva et al. 2011). Many reports have also described that *Campylobacter* flagella can secrete non-flagellar proteins that may regulate virulence (Silva et al. 2011; Guerry 2007).

The ability of *Campylobacter* to produce a cytolethal-distending toxin (CDT) is also thought to be an important mechanism regulating virulence (Silva et al. 2011; Park 2002). CDT is responsible for ceasing the division of eukaryotic cells in the G_2 phase of the cell cycle (Silva et al. 2011; Park 2002). This leads to cell death by preventing the cells from entering mitosis (Silva et al. 2011). It is hypothesized that when this occurs in the intestine, it can cause loss of function or erosion of the epithelial layer triggering diarrhea (Park 2002). CDT is made up of three subunits encoded by the *cdtA, cdtB,* and *cdtC* genes (Silva et al. 2011). For CDT to be functionally active, *cdtA, cdtB,* and *cdtC* genes all must be present (Silva et al. 2011). It is thought that *cdtA* and *cdtC* are necessary to transport *cdtB* into the host cell (Silva et al. 2011). Then this active *cdtB* subunit can break the double strand of the host DNA inflicting damage on the host (Silva et al. 2011).

Treatment of infections caused by *Campylobacter* is becoming increasingly difficult with antibiotics. Antibiotic treatment with macrolides, tetracycline, and fluoroquinolones is generally only used in severe cases of infection (Silva et al. 2011). However, due to the rise in antibiotic resistance of *Campylobacter*, treatment with fluoroquinolones, tetracycline, and erythromycin has become increasingly difficult (Silva et al. 2011). It is hypothesized that the unregulated use of antibiotics in food animal production has caused this increasing trend of antibiotic resistance (Silva et al. 2011). For example, following the approval of the use of fluoroquinolones in poultry in Europe and the US, an increase in fluoroquinolone-resistant *Campylobacter* was found in human and animal samples (Silva et al. 2011). Genetic mutations in *Campylobacter* are largely attributed to antibiotic resistance (Reddy and Zishiri 2017). Amino acid substitutions in the quinolone resistance-determining region (QRDR) regulate antibiotic resistance to fluoroquinolones (Reddy and Zishiri 2017). Point mutations in *gyrA* genes are responsible for high-level resistance to nalidixic acid as well as low-level resistance to ciprofloxacin (Reddy and Zishiri 2017). Unlike resistance in fluoroquinolones and macrolides, tetracycline resistance is plasmid-mediated (Reddy and Zishiri 2017). The *tetO* gene, which is responsible for encoding ribosomal protection proteins (RPPs), is present in both *C. jejuni* and *C. coli* (Reddy and Zishiri 2017). The presence of the *tetO* gene leads to extremely elevated levels of tetracycline resistance (Reddy and Zishiri 2017). Due to the increase in antibiotic resistance among species of *Campylobacter*, treatment of infections in humans are becoming more complicated to treat using antibiotics.

Limitation of Currently Used Detection Techniques

Many methods and procedures have been developed and used for the detection of *Campylobacter*, as described in previous sections of this chapter. The standard for the detection of *Campylobacter* is culturing and enrichment, which can be a time-consuming process. Other methods have been explored such as real-time PCR detection, and they are equally reliable but faster than culturing and enrichment. In one study conducted by De Boer et al. (2015), real-time PCR far outpaced culturing, taking 4 hours compared to the 2 days expended for culturing. Additionally, dead-end ultrafiltration (DEUF) has been utilized to detect low concentrations of *Campylobacter* in contaminated water sources (Ferrari et al. 2019). Another technique used to characterize *Campylobacter* is whole-genome sequencing, which analyzes genes, markers, and phylogeny for the identification of *Campylobacter* (Redondo et al. 2019). Other methods have been validated and more are currently being investigated.

Methodological Difficulties Due To Complex Cultural Requirements

There are many complications and shortcomings associated with the various methods of detection of *Campylobacter*. These often have to do with the rapidity and accuracy of *Campylobacter* identification by the method in question. As already mentioned, the standard *Campylobacter* detection method is enrichment and culturing, which lacks speed. It can take up to 2–3 days for results to be collected, which is extremely slow in most laboratory settings (De Boer et al. 2015). Although this method has been in use far longer than any other method and has proven itself to be accurate, it is inefficient and laborious. Additionally, culturing only allows the identification of bacteria, in this case, *Campylobacter*, to the genus level (Kulkarni et al. 2002) and so different species may not be differentiated through the culturing method. Therefore, to increase efficiency, other methods have been developed but also come with their limitations.

Studies conducted using real-time PCRs for the detection of *Campylobacter* have shown that the method can be reliable and much faster than enrichment followed by culturing. However, in a study comparing the two methods, there were a small number of *Campylobacter*-containing samples for which only the classic culturing worked (the real-time PCRs returned false negatives) (Lund et al. 2004). This defection could be attributed to the lack of expertise in sample collection and processing. Additionally, there can be certain chemicals, termed inhibitory substances in the samples that alter the efficacy of real-time PCR (Lund et al. 2004). These inhibitory chemicals inhibit the real-time PCR from functioning correctly, which can lower or nullify the yield of DNA amplification and result in false-negative results.

Although real-time PCRs have been widely discussed as a faster alternative to enrichment and culturing method, conventional PCR has also been used in the detection of *Campylobacter*. However, more shortcomings come with this method as it is not nearly as specific as real-time PCR. While real-time PCR allows for quantification of the DNA that has been amplified, conventional PCR techniques involve the separation of the DNA strands and visualization of the bands by gel electrophoresis as the only indicator of amplification and specificity. This is problematic since the method for detection must be specific enough to precisely determine the presence or absence of *Campylobacter* in each sample. Additionally, it is difficult to set apart the dead and viable bacterial cells through PCR analysis since both are usually detected by this method (Giesendorf et al. 1992). Another setback in conventional PCR is the issue of contamination, as a carryover from previous PCRs can introduce new nucleic acids that can be amplified and result in false positives (Lund et al. 2004).

Dead-end ultrafiltration is another technique that is used for the detection of *Campylobacter*, specifically in the detection of small numbers in large quantities of water. Although it is effective in some cases, its use also carries various weaknesses. For example, if *Campylobacter* is in a non-culturable state, but is still viable and can reproduce but this method may return false negatives (Ferrari et al. 2019). Additionally, false negatives could also result when *Campylobacter* is not evenly distributed throughout the water sample or when the amount of *Campylobacter* is simply below the detection limit of the method (Ferrari et al. 2019). Another difficulty with dead-end ultrafiltration is that its filter can be clogged by highly turbid water (Ferrari et al. 2019). Since the clogging of the filter prevents accurate detection, dead-end ultrafiltration is best used with water samples with low or moderate turbidity.

An alternate detection method, whole-genome sequencing has been used to identify *Campylobacter* based on the presence of virulence genes and antimicrobial resistance markers and through phylogenetic analysis. This method was effective in covering some of the shortcomings of enrichment and culture analysis and showed high specificity for cluster detection. However, it too had its shortcomings with regard to the detection of *Campylobacter*. For example, when analyzing for the presence of virulence genes through whole-genome sequencing, different species cannot be identified (Redondo et al. 2019). Since all pathogenic species of *Campylobacter* carry these genes, it is difficult to separate the species but nevertheless is useful in the detection of *Campylobacter* at the genus level. Additionally, when testing for antimicrobial resistance markers, simply the presence of these markers does not guarantee antimicrobial resistance, phenotypic tests for resistance would be more reliable (Redondo et al. 2019). The degree of specificity and representativeness of whole-genome analysis is limited by the database of isolates that it is based on (Redondo et al. 2019). Therefore, this type of genomic analysis should be approved by regulatory agencies and standardized to be more accurate and reliable.

Another method that has been developed is the utilization of a sensitive DNA microarray to detect *Campylobacter*. This method has been compared to the standard use of culture and PCR analysis in the detection of *Campylobacter* and revealed some defects in these techniques. In a study conducted by Keramas et al. (2004), chicken feces were analyzed for their contents, specifically for the presence of

Campylobacter. When conventional cell culture was used, only 60% of the samples were positive for any strain of *Campylobacter*. However, when PCR analysis was utilized, 95% of samples were positive and when the DNA microarray was used, all the samples returned positive (Keramas et al. 2004). It has been explained that this was because the PCR and DNA microarray techniques were not only able to detect alive and viable *Campylobacter* cells, but they were also able to detect dead and uncultivatable *Campylobacter* cells (Keramas et al. 2004).

Lack of Expertise and Training Opportunities

The reliability of the detection of *Campylobacter* is also related to the methods by which the samples are collected and processed. Without proper collection and processing procedures, contamination and false positives/negatives result could be present, skewing the data. The lack of expertise in sample collection and processing is present in most methods of *Campylobacter* detection. The way that the samples are processed is unique to the method of *Campylobacter* detection, as different methods require different inputs.

In one study comparing the standard enrichment/culturing and the real-time PCR techniques, there were a handful of samples that returned positive after enrichment and culturing but returned negative through real-time PCR (De Boer et al. 2015). This was unexpected since the overall conclusion of the experiment was that real-time PCR was not only faster but more accurate than culturing and enrichment. These false negatives could be attributed to the lack of expertise in sample processing before subjecting the samples to the PCR.

As with real-time PCR, conventional PCR techniques also require specific processing of samples before detection and analysis. In a study comparing the enrichment/culturing and conventional PCR detection methods by Lawson et al. (1999), there were a significant number of samples that were culture-negative but PCR-positive and another set of samples that were culture-positive but PCR-negative. The culture-positive/PCR-negative abnormalities could be attributed to the handling of the samples in the period of 10 days between culture and DNA extraction. Given this long amount of time, it is possible that the DNAs were degraded and unable to be amplified by the PCR procedure, and therefore resulted in a negative PCR reading. The culture-negative/PCR-positive abnormalities may be resultant of the presence of VBNC or dead *Campylobacter* that would still be detectable via PCR previously discussed. These presumptions were supported by a study by Kulkarni et al. (2002) since the reduction of the period between culture and DNA extraction from 10 days to 24 hours resulted in the elimination of culture-positive/PCR-negative sample readings. Therefore, the lack of expertise in the handling and time duration of the processing of the samples can skew the data and generate inaccurate results.

Time Constraints and Limitations of Commonly Allocated Resources

As previously stated, there are many defects associated with the various methods for *Campylobacter* detection. Two significant ones are time constraints and the limitation of resources. The classic enrichment and culturing method has been widely accepted in scientific communities around the world. However, when looking at this method in terms of time constraints, an obvious problem arises. Conventional enrichment and culturing of samples usually require at least 72 hours to produce results (De Boer et al. 2015). This is a great deal of time and renders the method incredibly inefficient. With regards to resources, the enrichment and culturing method requires media for the enrichment and growth specific to *Campylobacter*. These materials come with significant costs and must be continually bought and replaced according to use. Therefore, the classic enrichment and culture method is lengthy and relatively resource-consuming.

To solve the time constraints associated with the culturing method, real-time PCR utilizes genetic information to quickly analyze and identify the presence of *Campylobacter* and related species. Many studies have shown that when the whole procedure was conducted, results were available in only 4 hours (De Boer et al. 2015). Compared to the enrichment and culturing method, real-time PCR was around 20 times more efficient at detecting *Campylobacter* in terms of time. In terms of resources, real-time PCR utilizes molecular techniques to amplify the DNA of interest. There are certain materials necessary for this procedure, including primers, enzymes, and other necessary biomolecules. However, in the large picture, real-time PCR is relatively cheap and can be conducted with little handling. As with real-time PCR, conventional PCR has its advantages when compared to the enrichment and culturing method. The method involves nucleic acid extraction, DNA amplification, and other analytical techniques (Giesendorf et al. 1992). The time required for the whole procedure is more than real-time PCR but is becoming more streamlined over time. Comparatively, the resources needed for conventional PCR and real-time PCR are similar, although more handling efficiency is needed for conventional PCR methods.

Another valid detection method, dead-end ultrafiltration for water samples also has its respective limitations with regard to time and resources. The timeline for this method includes the processing of the contaminated water samples. The amount of time needed to process the samples is proportional to the volume of contaminated water. In one study, Ferrari et al. (2019) found that the time to process 60 liters of water was around 25 minutes. This is extremely quick; however, it can only be used with water samples and cannot be used with solids. With regards to resources, the method necessitates specific pumps and filtration systems that may not be easily acquirable. Overall, dead-end ultrafiltration can be an efficient, time-saving method for the detection of *Campylobacter* in only water samples.

Whole-genome sequencing is another technique used to detect *Campylobacter* and is useful for specific cluster detection. The procedure is largely dominated by computational and genetic analysis. Given this, the method requires little handling past the initial sampling and processing period (Redondo et al. 2019). After this period, whole-genome sequencing is conducted, and the sequences of the samples are determined. Much of the data analysis after whole-genome sequencing was done with specific computer programs and databases that are publicly available. Therefore, apart from sample processing, the procedure for this technique is quick. However, specific technology for whole-genome sequencing is necessary. This method is a time-efficient way to detect *Campylobacter* given that the technology is present.

In addition to these methods, a DNA microarray can be utilized to detect *Campylobacter*. The technique consists of several steps including the preparation of the microarray, the hybridization process, and the optical readout. Combining all the steps, the method is very quick and a rapid, accurate detection technique. In one study, Keramas et al. (2004) compared the use of a specialized DNA microarray to culture method and PCR analysis for the detection of two species of *Campylobacter*. From the results of this experiment, Keramas et al. concluded that the use of DNA microarrays is more time-efficient and more resource efficient than both classic culturing and PCR analysis. Specifically, when DNA microarrays were used, the presence of *Campylobacter* was detected in samples within 3 hours (Keramas et al. 2004). Not only was the procedure timeline hastened, but the number of reagents and materials required for detection was also lessened when using the microarray. Though DNA microarrays serve as both a timely and resource-efficient method for the detection of *Campylobacter*, further studies are needed to validate the procedure.

Future Recommendations for Better Practice

Current methods of detection have adequately succeeded in the identification of *Campylobacter*. However, these methods have limitations, such as the inability to differentiate between closely related species due to high sequence similarity (Magana et al. 2017). Additional limitations include inhibitors that prevent assays from obtaining precise quantification and a high number of false positives. Improved molecular methods of containment and detection can help to bridge the gap left by the current methods.

Immune-based methods of foodborne pathogen detection, such as flow cytometry, enzyme-linked immunosorbent assays (ELISA), and quantitative immunofluorescence have been well-established. Compared to conventional culturing techniques, the enzyme immunoassay (EIA) kits demonstrated greater sensitivity and specificity (Ricke et al. 2019). However, the EIA kits may result in false positives due to cross-reactivity (Ricke et al. 2019). Attempts to address these issues have led to the innovation of technology that has produced self-contained immune-based biosensors and nano-based assays. The biosensors would allow for a more precise evaluation of the pathogen due to the ability to convert the binding activities of antibodies into electrical signals (Ricke et al. 2019). The Surface Plasma Resonance (SPR) sensor was developed to detect *C. jejuni* by covalently attaching rabbit polyclonal antibodies to gold chips, resulting in an improved limit of detection (LOD) as compared to a direct assay (Masdor et al. 2017). A cotton-swab immunoassay was developed by immersing swabs into cocktails of various colored nano-beaded conjugated assays and a high LOD score was achieved with no observable cross-reactivity (Alamer et al. 2018). Immunoassays could also be improved with the use of proteomics to aid in antigen target refinement. Proteomics are used to demonstrate that different proteins are expressed under different conditions, such as between the microaerophilic and aerobic culture conditions (Rodrigues et al. 2016). Similarly, a difference in protein expression was detected using proteomics when *C. jejuni* changed from the stationary to the exponential growth phases (Turonova et al. 2017). By determining differences in protein expressions under various environments and growth phases, novel antibodies may be discovered to use in immunoassays.

The methods for *Campylobacter* detection in meat have undergone many alterations. For example, initially the reference methods called for enriching meat in a broth under microaerobic conditions (Zhou et al. 2011). While these methods could detect *Campylobacter*, they were not optimized to obtain the true number of positive samples in meat. A change in the reference methods occurred, which now enrich meat in a broth under aerobic conditions because it is a simplified and cost-effective procedure, but still is not optimized to determine the true number of positive samples (Zhou et al. 2011). An alternative method has recently been developed where meat samples have been rinsed in buffered peptone water with selective antimicrobials and incubated under aerobiosis (Oyarzabal et al. 2013). Compared to the reference methods, this alternate method of rinsing meat was found to be less time-consuming, required less sample preparation, and was more sensitive in the isolation of naturally occurring *Campylobacter* (Oyarzabal et al. 2013).

Quantifying bacteria can be challenging because bacteria can lose the ability to grow in culture due to cold or oxygen stress during storage (Pacholewicz et al. 2019). This highlights the need to develop culture-independent quantification methods. One method that has been used as an alternative is a real-time PCR which discriminates between live and dead cells to detect and quantify various bacteria. The problem with real-time PCR is that the discrimination between live and dead cells is not finite enough; an insufficient reduction of signal from dead cells leads to false positives and skews the quantification of the bacteria (Pacholewicz et al. 2019). To improve real-time PCR, an internal sample process control (ISPC) has been developed. In a recent study, a number of peroxide-killed *C. sputorum* cells were added to the samples and a species-specific gene fragment (similar to the *C. jejuni*, *C. coli* or *C. lari* target) was used as the target for the real-time PCR (Pacholewicz et al. 2019). This ISPC was used to monitor the level of reduction of the signal from dead cells and DNA losses during sample processing (Pacholewicz et al. 2019). Real-time PCR using the ISPC was found to be lacking in false positives, thus presenting an increased sensitivity and efficiency to the previous methods (Pacholewicz et al. 2019). Again, real-time PCR uses intercalating fluorescent dyes to quantify pathogens by introducing fluorescence through non-specific dyes that bind to the DNA or by using probes with different fluorescent labels (Ricke et al. 2019). Fluorescent dyes accumulate with an intensity directly proportional to the amount of target template DNA. Compared to conventional PCR, real-time PCR has increased speed and specificity. However, real-time PCR is limited in its ability to detect *Campylobacter* from a diverse sample due to the presence of inhibitors or other biological challenges (Ricke et al. 2019). One way to overcome this limitation is to use the same paramagnetic beads to separate the DNA and determine cell concentration. This integrated approach allowed for a fully automated and rapid sample preparation and DNA extraction method (Rudi et al. 2004).

Conventional PCR has been used for decades to detect foodborne pathogens by exploiting the genomic variation of a particular organism to differentiate it from related and unrelated organisms (Ricke et al. 2019). However, as new *Campylobacter* species are discovered, conventional PCR presents challenges in determining species specificity and avoiding false positives. Innovations in PCR-based techniques could address many limitations.

Multiplex PCR can be used for the rapid identification of multiple species from one food production sample. Compared to conventional PCR, which can only amplify one target sequence, multiplex PCR is capable of amplifying more than one target sequence using up to six pairs of primers (Elnifro et al. 2000). Multiplex assays are able to specifically determine if *Campylobacter* is present in a complex sample. A study used multiplex PCR without pre-enrichment of the bovine fecal samples to determine the identity of the pathogens in the sample (Inglis and Kalischuk 2003). Genus-level identification of *Campylobacter* was found to be more sensitive than other microbiological isolation techniques, showing that multiplex assays have more sensitivity and specificity (Inglis and Kalischuk 2003). In the same study, the QIAamp DNA stool mini kit was found to be an effectual method. This mini kit effectively removed inhibitors to extract added control DNA from bovine feces, which has been a challenge for many PCR-based techniques (Inglis and Kalischuk 2003). Multiplex PCR has been shown to be faster than conventional PCR assays because it does not rely on the enrichment step. It is also a more cost-effective method due to the decreased number of tests that must be run to identify multiple pathogens at once.

Digital PCR (dPCR) is able to determine the absolute copy number of a gene target by utilizing the fraction of negative replicates and a Poisson statistical algorithm (Baker 2012). First, a sample is separated into many small reaction chambers. After the amplification is complete, the ratio of positive and negative reactions indicates the actual number of gene copies present in the test sample (Baker 2012). Compared to real-time PCR, which uses fluorescence intensity during amplification, dPCR is far more precise. Another advantage of dPCR is that it is less vulnerable to enzyme amplification inhibitors that are often present in the samples (Baker 2012). A variation of dPCR is droplet dPCR (ddPCR). In ddPCR, the samples are mixed with the necessary reagents and then dispersed into droplets (Baker 2012). Then, a droplet reading machine, which functions similarly to a flow cytometer, analyzes each sample droplet to determine whether a reaction has taken place (Baker 2012). A study compared real-time PCR with ddPCR using the same primers and probes and found that ddPCR detected more pathogens at each sampling point, indicating that ddPCR has higher sensitivity than real-time PCR (Rothrock et al. 2013). These digital techniques do not require the calibration and internal controls necessary for qPCR. However, the disadvantages to these PCR-based techniques are that they are more expensive and have a narrower dynamic range than real-time PCR.

Many of the methods that have been discussed in this section have improved on existing methods. Novel techniques have been developed with the goal of increasing specificity and precision, as well as decreasing the time required to obtain the results. Further research can determine the feasibility of these techniques, but they have all been shown to improve the detection, identification, and quantification of the *Campylobacter* genus.

Conclusion

In recent years, research has rapidly developed to explain much of the information about *Campylobacter* (Taboada et al. 2013). However, even with the technology that has been used recently, there are still many disagreements surrounding the methods used as they are not completely precise. Therefore, it is yet a challenge to identify and characterize *Campylobacter* with time constrain and given resources. So much more research is necessary in order to identify and characterize *Campylobacter* in a stable, time and cost-efficient way.

References

Abeyta, C., P.A. Trost, D.H. Bark, J.M. Hunt, C.A. Kaysnet and M.M. Wekell. 1997. The use of bacterial membrane fractions for the detection of *Campylobacter* species in shellfish. J. Rapid Methods Autom. Microbiol. 5: 223–247.

Alamer, S., S. Eissa, R. Chinnappan and M. Zourob. 2018. A rapid colorimetric immunoassay for the detection of pathogenic bacteria on poultry processing plants using cotton swabs and nanobeads. Microchimica Acta 185(3).

Baker, M. 2012. Digital PCR hits its stride. Nat. Methods 9(6): 541–544.

Baylis, C.L., S.A. MacPhee, K.W. Martin, T.J. Humphrey and R.P. Betts. 2000. Comparison of three enrichment media for the isolation of *Campylobacter* spp. from foods. J. Appl. Microbiol. 89: 884–891.

Beery, J.T., M.B. Hugdahl and M.P. Doyle 1988. Colonization of gastrointestinal tracts of chicks by *Campylobacter jejuni*. Appl. Environ. Microbiol. 54: 2365–2370.

Berrang, M.E., R.J. Buhr, J.A. Cason and J.A. Dickens. 2001. Broiler Carcass Contamination with *Campylobacter* from Feces during Defeathering. J. Food Protect. 64(12): 2063–066.

Bradbury, W.C., M.A. Marko, J.N. Hennessy and J.L. Penner. 1983. Occurrence of plasmid DNA in serologically defined strains of *Campylobacter jejuni* and *Campylobacter coli*. Infect. Immun. 40: 460–463.

Butzler, J.-P. and J. Oosterom. 1991. *Campylobacter*: Pathogenicity and significance in foods. I. J. Food Microbiol. 12(1): 1–8.

Centers for Disease Control and Prevention. August 30, 2017. *Campylobacter* (Campylobacteriosis). https://www.cdc.gov/campylobacter/index.html (Accessed July 17, 2019).

Centers for Disease Control and Prevention. June 21, 2019. Multistate Outbreak of Gastrointestinal Illnesses Linked to Oysters Imported from Mexico. https://www.cdc.gov/vibrio/investigations/rawoysters-05-19/index.html (Accessed July 21, 2019).

Corry, J.E.L., D.E. Post, P. Colin and M.J. Laisney. 1995. Culture media for the isolation of *Campylobacter*. Int. J. Food Microbiol. 26: 43–76.

Davis, K.R., A.C. Dunn, C. Burnett et al. 2014. *Campylobacter jejuni* infections associated with raw milk consumption—Utah. Morbidity and Mortality Weekly Report (MMWR) 65(12): 301–05. doi:10.15585/mmwr.mm6512a1 (Accessed July 17, 2019).

De Boer, P., H. Rahaoui, R.J. Leer, R.C. Montijn and J.M. Van der Vossen. 2015. Real-time PCR detection of *Campylobacter* spp: a comparison to classic culturing and enrichment. Food Microbiol. 51: 96–100.

Del Collo, L.P., J.S. Karns, D. Biswas et al. 2017. Prevalence, antimicrobial resistance, and molecular characterization of *Campylobacter* spp. in bulk tank milk and milk filters from US dairies. J. Dairy Sci. 100(5): 3470–79.

Elnifro, E.M., A.M. Ashshi, R.J. Cooper and P.E. Klapper. 2000. Multiplex PCR: optimization and application in diagnostic virology. Clin. Microbiol. Review 13(4): 559–570.

Endtz, H.P., G.J. Ruijs, A.H. Zwinderman, R.T. van der, M. Biever and R.P. Mouton. 1991. Comparison of six media, including a semisolid agar, for the isolation of various *Campylobacter* species from stool specimens. J. Clin. Microbiol. 29: 1007–1010.

Epps, S.V., R.B. Harvey, M.E. Hume, T.D. Phillips, R.C. Anderson and D.J. Nisbet. 2013. Foodborne *Campylobacter*: infections, metabolism, pathogenesis and reservoirs. Int. J. Environ. Res. Public Health 10(12): 6292–6304.

Federighi, M., C. Magras, M.F. Pilet, D. Woodward, W. Johnson, F. Jugiau and J.L. Jouve. 1999. Incidence of thermotolerant *Campylobacter* in foods assessed by NF ISO 10272 standard: results of a two-year study. Food Microbiol. 16: 195–204.

Ferrari, S., S. Frosth, L. Svensson, L.L. Fernström, H. Skarin and I. Hansson. 2019. Detection of *Campylobacter* spp. in water by dead-end ultrafiltration and application at farm level. J. Appl. Microbiol. 127: 1270–1279.

Food Safety. April 12, 2019. Bacteria and Viruses. https://www.foodsafety.gov/food-poisoning/bacteria-and-viruses#campylobacter (Accessed July 17, 2019).

Galanis, E. 2007. *Campylobacter* and bacterial gastroenteritis. Can. Med. Assoc. J. 177(6): 570–71.

Garénaux, A., F. Jugiau, F. Rama et al. 2008. Survival of *Campylobacter jejuni* strains from different origins under oxidative stress conditions: effect of temperature. Curr. Microbiol. 56: 293–297.

Giesendorf, B.A.J., W.G.V. Quint, M.H.C. Henkens, H. Stegeman, F.A. Huf and H.G.M. Niesters. 1992. Rapid and sensitive detection of *Campylobacter* spp. in chicken products by using the polymerase chain reaction. Appl. Environ. Microbiol. 58: 3804–3808.

Guerry, P. 2007. *Campylobacter* Flagella: Not just for motility. Trend Microbiol. 15(10): 456–61.

Hendrixson, D.R. and V.J. Dirita. 2004. Identification of *Campylobacter jejuni* genes involved in commensal colonization of the chick gastrointestinal tract. Mol. Microbiol. 52(2): 471–84.

Hofreuter, D. 2014. Defining the metabolic requirements for the growth and colonization capacity of *Campylobacter jejuni*. Front. Cell Infect Microbiol. 4: 137–37.

Holler, C., D. Witthuhn and B. Janzen-Blunck. 1998. Effect of low temperatures on growth, structure, and metabolism of *Campylobacter coli* SP10. Appl. Environ. Microbiol. 64: 581–587.

Humphrey, T.J. and P. Beckett. 1987. *Campylobacter jejuni* in dairy cows and raw milk. Epidem. Infect. 98(3): 263–69.

Humphrey, T., S. Obrien and M. Madsen. 2007. Campylobacters as zoonotic pathogens: a food production perspective. Inter. J. Food Microbil. 117(3): 237–57.

Inglis, G.D. and L.D. Kalischuk. 2003. Use of PCR for direct detection of *Campylobacter* species in bovine feces. Appl. Environ. Microbiol. 69(6): 3435–3447.

Kaakoush, N.O., W.G. Miller, H. De Reuse and G.L. Mendz. 2007. Oxygen requirement and tolerance of *Campylobacter jejuni*. Res. Microbiol. 158(8-9): 644–50.

Kakoyiannis, C.K., P.J. Winter and R.B. Marshall. 1984. Identification of *Campylobacter coli* isolates from animals and humans by bacterial restriction endonuclease DNA analysis. Appl. Environ. Microbiol. 48: 545–49.

Keramas, G., D. Bang, M. Lund et al. 2004. Use of culture, PCR analysis, and DNA microarrays for detection of *Campylobacter jejuni* and *Campylobacter coli* from chicken feces. J. Clin. Microbiol. 42: 3985–3991.

Korhonen, L.K. and P.J. Martikalnon. 1991. Survival of *Escherichia coli* and *Campylobacter jejuni* in untreated and filtered lake water. J. Appl. Bacteriol. 71(4): 379–82.

Kulkarni, S.P., S. Lever, J.M.J. Logan et al. 2002. Detection of campylobacter species: a comparison of culture and polymerase chain reaction based methods. J. Clinic Pathol. 55: 749–753.

Langenberg, W., E.A.J. Rauws, A. Widjojokusumo and G.N.J. Tytgat. 1986. Identification of *Campylobacter pyloridis* isolates by restriction endonuclease DNA analysis. J. Clinic Microbiol. 3(24): 414–17.

Lawson, A.J., J.M. Logan, G.L. O'neill, M. Desai and J. Stanley. 1999. Large-scale survey of *Campylobacter* species in human gastroenteritis by PCR and PCR-enzyme-linked immunosorbent assay. J. Clinic Microbiol. 37(12): 3860–3864.

Levin, R. 2007. *Campylobacter jejuni*: A review of its characteristics, pathogenicity, ecology, distribution, subspecies characterization and molecular methods of detection. Food Biotechnol. 21(4): 271–347.

Lindblom, G.-B., E. Sjögren and B. Kaijser. 1986. Natural *Campylobacter* colonization in chickens raised under different environmental conditions. J. Hyg. 96: 385–391.

Lior, H., D.L. Woodward, J.A. Edgar, L.J. Laroche and P. Gill. 1982. Serotyping of *Campvlobacter jejuni* by slide agglutination based on heat-labile antigenic factors. J. Clin. Microbiol. 15: 761–768.

Lopes, G.V., M. Landgraf and M.T. Destro. 2018. Occurrence of *Campylobacter* in raw chicken and beef from retail outlets in São Paulo, Brazil. J. Food Safety 38(3): 12442.

Lund, M., S. Nordentoft, K. Pedersen and M. Madsen. 2004. Detection of *Campylobacter* spp. in chicken fecal samples by real-time PCR. J. Clin. Microbiol. 42(11): 5125–5132.

Magana, M., S. Chatzipanagiotou, A.R. Burriel and A. Ioannidis. 2017. Inquiring into the Gaps of *Campylobacter* surveillance methods. Veter Sci. 4(4): 36.

Masdor, N., Z. Altintas and I. Tothill. 2017. Surface plasmon resonance immunosensor for the detection of *Campylobacter jejuni*. Chemosensors 5(2): 16.

Mezher, Z., S. Saccares, R. Marcianò et al. 2016. Occurrence of *Campylobacter* spp. in poultry meat at retail and processing plants' levels in Central Italy. Ital. J. Food Safety 5(1): 5495.

Mohammadpour, H., E. Berizi, S. Hosseinzadeh, M. Majlesi and M. Zare. 2018. The prevalence of *Campylobacter* spp. in vegetables, fruits, and fresh produce: a systematic review and meta-analysis. Gut Pathogens 10(1)2.

Munroe, D.L., J.F. Prescott and J.L. Penner. 1983. *Campylobacter jejuni* and *Campylobacter coli* serotypes isolated from chickens, cattle, and pigs. J. Clin. Microbiol. 18: 877–881.

Nachamkin, I., X.H. Yang and N.J. Stern. 1993. Role of *Campylobacter jejuni* flagella as colonization factors for three-day-old chicks: analysis with flagellar mutants. Appl. Environ. Microbiol. 59: 1269–1273.

Oyarzabal, O.A., A. Williams, P. Zhou and M. Samadpour. 2013. Improved protocol for isolation of *Campylobacter* spp. from retail broiler meat and use of pulsed field gel electrophoresis for the typing of isolates. J. Microbiol. Method 95(1): 76–83.

Pacholewicz, E., C. Buhler, I.F. Wulsten et al. 2019. Internal sample process control improves cultivation-independent quantification of thermotolerant *Campylobacter*. Food Microbiol. 78: 53–61.

Park, S.F. 2002. The physiology of *Campylobacter* species and its relevance to their role as foodborne pathogens. Inter. J. Food Microbiol. 74(3): 177–88.

Penner, J.L. and J.N. Hennessy. 1980. Passive hemagglutination technique for serotyping *Campylobacter fetus* subsp. *jejuni* on the basis of soluble heat-stable antigens. J. Clin. Microbiol. 12: 732–737.

Persson, S. 2005. Multiplex PCR for identification of *Campylobacter coli* and *Campylobacter jejuni* from pure cultures and directly on stool samples. J. Med. Microbiol. 54(11): 1043–047.

Portner, D.C., R.G.K. Leuschner and B.S. Murray. 2007. Optimizing the viability during storage of freeze-dried cell preparations of *Campylobacter jejuni*. Cryobiology 54: 265–270.

Rautelin, H., J. Jusufovic and M.L. Hanninen. 1999. Identification of hippurate-negative thermophilic *Campylobacters*. Diagn. Microbiol. Infect. Dis. 35: 9–12.

Reddy, S. and O.T. Zishiri. 2017. Detection and prevalence of antimicrobial resistance genes in *Campylobacter* spp. isolated from chickens and humans. Onderstepoort J. Vet. Res. 84(1).

Redondo, N., A. Carroll and E. McNamara. 2019. Molecular characterization of *Campylobacter* causing human clinical infection using whole-genome sequencing: Virulence, antimicrobial resistance, and phylogeny in Ireland. Plos One 14(7): e0219088.

Ricke, S.C., K.M. Feye, W.E. Chaney, Z. Shi, H. Pavlidis and Y. Yang. 2019. Developments in rapid detection methods for the detection of foodborne *Campylobacter* in the United States. Front Microbiol. 9: 3280.

Rodrigues, R.C., N. Haddad, D. Chevret, J.M. Cappelier and O. Tresse. 2016. Comparison of proteomics profiles of *Campylobacter jejuni* strain Bf under microaerobic and aerobic conditions. Front Microbiol. 7: 1596.

Rosef, O. and G. Kapperud. 1982. Isolation of *Campylobacter fetus* subsp. *jejuni* from feces of Norwegian poultry. Acta Vet Scand 23: 128–134.

Rothrock, M.J., K.L. Hiett, B.H. Kiepper, K. Ingram and A. Hinton. 2013. Quantification of zoonotic bacterial pathogens within commercial poultry processing water samples using droplet digital PCR. Adv. Microbiol. 03(05): 403–411.

Rudi, K., H.K. Høidal, T. Katla, B.K. Johansen, J. Nordal and K.S. Jakobsen. 2004. Direct real-time PCR quantification of *Campylobacter jejuni* in chicken fecal and cecal samples by integrated cell concentration and DNA purification. Appl. Environ. Microbiol. 70(2): 90–797.

Sahin, O., I.I. Kassem, Z. Shen, J. Lin, G. Rajashekara and Q. Zhang. 2015. In poultry: ecology and potential interventions. Avi Diseases 59(2): 185–200.

Schallenberg, M., P.J. Bremer, S. Henkel, A. Launhardt and C.W. Burns. 2005. Survival of *Campylobacter jejuni* in water: effect of grazing by the freshwater Crustacean *Daphnia Carinata* (Cladocera). Appl. Environ. Microbiol. 71(9): 5085–088.

Scherer, K., E. Bartelt, C. Sommerfeld and G. Hildebrandt. 2006. Comparison of different sampling techniques and enumeration methods for the isolation and quantification of *Campylobacter* spp. in raw retail chicken legs. Inter. J. Food Microbiol. 108(1): 115–19.

Sellars, M.J., S.J. Hall and D.J. Kelly. 2002. Growth of *Campylobacter jejuni* supported by respiration of fumarate, nitrate, nitrite, trimethylamine-n-oxide, or dimethyl sulfoxide requires oxygen. J. Bacteriol. 184(15): 4187–96.

Silva, J., D. Leite, M. Fernandes, C. Mena, P.A. Gibbs and P. Teixeira. 2011. *Campylobacter* spp. as a foodborne pathogen: A review. Front Microbiol. 2: 200.

Skirrow, M.B. and J. Benjamin. 1980. *Campylobacters*: cultural characteristics of intestinal *Campylobacter* from man and animals. J. Hyg. 85: 427–442.

Stahl, M., J. Butcher and A. Stintzi. 2012. Nutrient acquisition and metabolism by *Campylobacter jejuni*. Front Cell Infect. Microbiol. 2: 5.

Taboada, E.N., C.G. Clark, E.L. Sproston and C.D. Carrillo. 2013. Current methods for molecular typing of *Campylobacter* species. J. Microbiol. Methd. 95(1): 24–31.

Teunis, P., A. Havelaar, J. Vliegenthart and G. Roessink. 1997. Risk assessment of *Campylobacter* species in shellfish: identifying the unknown. Wat Sci. Technol. 35(11-12): 29–34.

Turonova, H., N. Haddad, M. Hernould, D. Chevret, J. Pazlarova and O. Tresse. 2017. Profiling of *Campylobacter jejuni* proteome in exponential and stationary phase of growth. Front Microbiol. 8: 913.

Wassenaar, T.M., B.A.M. van der Zeijst, R. Ayling and D.G. Newell. 1993. Colonization of chicks by motility mutants of *Campylobacter jejuni* demonstrates the importance of flagellin A expression. J. Gen. Microbiol. 139: 1171–1175.

Weingarten, R.A., M.E. Taveirne and J.W. Olson. 2009. The dual-functioning fumarate reductase is the sole succinate: quinone reductase in *Campylobacter jejuni* and is required for full host colonization. J. Bacteriol. 191(16): 5293–5300.

World Health Organization. January 23, 2018. Campylobacter. https://www.who.int/news-room/fact-sheets/detail/campylobacter (Accessed July 17, 2019).

Zhou, P., S.K. Hussain, M.R. Liles, C.R. Arias, S. Backert, J. Kieninger and O.A. Oyarzabal. 2011. A simplified and cost-effective enrichment protocol for the isolation of *Campylobacter* spp. from retail broiler meat without microaerobic incubation. BMC Microbiol. 11: 175.

Part II

Diagnosis of Human-Pathogenic Gram-Positive Bacteria

Chapter 3

Detection of
Listeria monocytogenes in
Food and Environment

Atin. R. Datta[1],* and *Laurel S. Burall*[2]

Introduction

Listeria monocytogenes (*Lm*), a Gram-positive, facultative aerobic, and non-spore-forming bacterium, is the causative organism of foodborne human and animal listeriosis. The human disease is characterized by both invasive illness, which includes septicemia, meningitis, abortion, and death (Vazquez-Boland et al. 2001b), and also a relatively rare but well-documented gastroenteritis characterized by symptoms, including diarrhea, nausea, and fever (Norton and Braden 2007). The invasive form of listeriosis is associated with more than 95% hospitalization and 20–30% mortality. According to Scallan et al. (2011), about 1,600 cases of human listeriosis and 250 deaths have been reported in the USA each year. The global burden of human listeriosis, as estimated for 2010, is about 23,150 illnesses, resulting in 5,463 deaths and 172,823 DALYs (disability-adjusted life year) (de Noordhout et al. 2014). The susceptible population includes neonates, the elderly, pregnant women, people with underlying health conditions, and people with weakened immunity. However, a few recent reports indicate that a small percentage of invasive illnesses were not associated with these risk factors (Angelo et al. 2017; Charlier et al. 2017). The febrile gastroenteritis listeriosis cases, on the other hand, have been reported to affect all individuals, irrespective of their health status, with high attack rates and no mortality (Norton and Braden 2007). The economic burden of human listeriosis,

[1] Office of Food Safety, Center for Food Safety and Applied Nutrition, U.S. Food and Drug Administration, 5001 Campus Drive, College Park, MD 20740, USA.

[2] Office of Applied Research and Safety Assessment, Center for Food Safety and Applied Nutrition, U.S. Food and Drug Administration, 8301 Muirkirk Road, Laurel, MD 20708, USA.

* Corresponding author: atin.datta@fda.hhs.gov

including the burden on the food industry, has been estimated to be billions of dollars (Hoffmann et al. 2012).

Lm, the only human pathogen of the genus *Listeria*, is ubiquitous in the environment and as a result, can contaminate foods and food processing environments. The organism has been serologically classified into 13 serotypes based on both somatic (O) and flagellar (H) antigens (Seeliger and Hohne 1979). Recently, a new serotype of *Lm*, termed 4h, has been isolated from ovine outbreaks in China (Yin et al. 2019). The major contributors to human infections, including major outbreaks (Table 1), and food contamination events belong to serotypes 1/2a, 1/2b, and 4b. A genetic variation of serotype 4b, termed 4bV (Burall et al. 2016) or IVb-v1 (Leclercq et al. 2011; Lee et al. 2012), has also been recognized to be associated with several outbreaks (Burall et al. 2017b). Serotype 4b appears to be the dominant outbreak-causing serotype; although, a few large listeriosis outbreaks have been reported as caused by serotypes 1/2a and 1/2b (Table 1). Based on genomic sequences, *Lm* strains can be grouped into four distinct lineages, some consisting of multiple overlapping serotypes (Orsi et al. 2011). Lineage I strains are predominantly

Table 1. Major invasive listeriosis outbreaks.*

Year	Location	Cases (% Mortality)	Food	Serotype
1980–81	Canada	41 (34)	Coleslaw	4b
1983	USA	49 (29)	Pasteurized Milk	4b
1984	Switzerland	57 (32)	Soft Cheese	4b
1985	USA	142 (34)	Jalisco Cheese	4b
1987–89	UK	823 (?)	Pate	4b
1992	France	279 (30)	Rillettes (Pork)	4b
1998–99	USA	101 (21)	Hot Dogs, Deli Meats	4b
1998–99	Finland	25 (24)	Butter	3a
2000	USA	29 (7)	Deli Turkey Meat	1/2a
2008	Canada	57 (40)	Deli Meat	1/2a
2010	USA	10 (50)	Celery	1/2a
2011	USA	147 (33)	Cantaloupe	1/2a, 1/2b
2012	USA	22 (25)	Ricotta Salata Cheese	1/2a
2014	USA	35 (20)	Caramel Apple	4b, 4bV
2010–15	USA	10 (30)	Ice Cream	1/2b
2015	USA	30 (10)	Soft Cheese	Unknown
2016	USA	19 (5)	Packaged Greens	4bV
2015–18	EU	47 (19)	Frozen Corn	4b
2017–18	South Africa	1,060 (20)	Polony	4b
2015–18	EU	47 (19)	Frozen Corn/Vegetable	4b
2018–19	Germany	112 (6)	Blood Sausage	4b
2019–20	USA	36 (11)	Enoki Mushroom	1/2a, 1/2b

* Outbreaks with a number of reported illnesses ten and above.

responsible for sporadic cases and outbreaks, while lineage II strains are frequently associated with foods and natural environments (Orsi et al. 2011). Based on gene sequences, *Lm* strains have been grouped into 22 clonal complexes and several epidemic clones (genetically similar strains involved in separate outbreaks) (Ragon et al. 2008). Distinct genetic, phenotypic, and virulence characteristics have been found to be associated with these lineages and clonal complexes (Hingston et al. 2017; Pirone-Davies et al. 2018). In addition, *Lm* strains involved in gastrointestinal disease outbreaks (Table 2) were found to be genetically distinct from the invasive outbreak strains (Laksanalamai et al. 2012) and also differed in their virulence potential in insect larvae *Galleria mellonella* model (Rakic Martinez et al. 2017).

Since 1985, following a large foodborne outbreak in California involving Jalisco cheese (Linnan et al. 1988), research activities in many countries have generated a vast resource of information on *Lm* growth and survival, physiology, virulence mechanisms, and genomics (Glaser et al. 2001; Vazquez-Boland et al. 2001a; Radoshevich and Cossart 2018; Marik et al. 2019). These activities resulted in a greater understanding of the genomic architecture of the organisms and key genetic attributes of *Lm* that are responsible for infection and disease processes (Hilliard et al. 2018), survival under stress conditions whether inside the human host or outside the host in foods, food processing environment (Kathariou et al. 2002), and the natural environment (Vivant et al. 2013). It is important to recognize that many of these genes have overlapping roles in these important phenotypes, and many of the purported house-keeping genes have also been shown to have important roles in pathogenesis, stress tolerance, and biofilm formation (Bucur et al. 2018). *Lm* can form a biofilm on a variety of surfaces, either by itself or in combination with other microorganisms. As biofilms are often resistant to cleaning and sanitizing efforts (Pan et al. 2019; Hua et al. 2019), *Lm* in these biofilms may lead to persistence in the food processing environment and recurring food contamination (Ferreira et al. 2014). The knowledge gained from these studies has led to the development of better detection strategies, e.g., chromogenic agars based on the detection of the *plc*A gene product

Table 2. Major non-invasive (gastrointestinal illness) listeriosis outbreaks.*

Year	Country	Cases	Food	Serotype
1993	Italy	18	Rice Salad	1/2b
1994	USA	45	Chocolate Milk	1/2b
1997	Italy	1,566	Corn, Tuna Salad	4b
2000	New Zealand	32	Corned Beef, Ham	1/2
2001	USA	16	Sliced Turkey Meat	1/2a
2001	Sweden	48	Raw Milk Cheese	1/2a
2001	Japan	34	Soft Cheese	1/2b
2002	Canada	86+46**	Cheese	4b
2008	Austria	12	Jellied Pork	4b

* Outbreaks with more than ten reported cases.
** Two separate outbreaks including five cases of invasive symptoms.

phosphatidylinositol-specific phospholipase C (PI-PLC), and the identification of high-risk foods in terms of *Listeria* survival and growth (USDA 2003). *Lm* is a psychrophilic organism that grows reasonably well at refrigerated temperatures (Chan and Wiedmann 2008) and often outcompetes associated mesophilic microbiota at refrigerated temperatures, further emphasizing the need for strict adherence to time-temperature guidelines for foods. With the rapid improvements in DNA sequencing technologies and computational tools, it was possible to develop whole genome sequencing (WGS) based methods for subtyping and outbreak investigation (Jackson et al. 2016).

During the initial series of listeriosis outbreaks, several laboratories were involved in the development of cultural/biochemical methods for the detection of *Lm* in food samples. It was clear from the onset that a single method would not work for different foods, e.g., dairy and meat products, as these products are chemically different and may contain different microbiota. Globally, regulatory agencies and food industries initiated analyzing foods both during outbreaks, as well as for surveillance, to reduce the burden of listeriosis. Risk analyses on foodborne listeriosis (USFDA 2003; FAO 2004) indicated that although most of the population is exposed to *Lm* through food, only a few people develop invasive infections. Dose-response studies based on the data from animal models showed that the successful infection by *Lm* requires a relatively high dose of the microorganism, roughly > 8 log cfu for healthy individuals and 2–3 log cfu for susceptible individuals (Takeuchi et al. 2006; Williams et al. 2007). As most naturally contaminated foods contained low numbers of organisms, it was reasoned that some kind of time temperature abuse was probably needed to attain a level suitable for an infectious dose, even for susceptible populations, although recent listeriosis outbreaks involving frozen foods challenge that notion (Chen et al. 2016; Pouillot et al. 2016; Weissfeld 2017; ECDC 2018). An emphasis on risk was given to ready-to-eat (RTE) foods as these foods do not go through a kill step before consumption. Currently, two different standards for *Lm* in RTE foods are followed. In the USA, detection of *Lm* in an analytical unit of any food, including RTE, is violative (zero tolerance) and therefore subject to regulatory actions (USFDA 2008). However, in the EU, Canada, and other countries, some RTE foods have an enforced zero tolerance while *Lm* < 100 cfu/g is permitted in some other RTE foods at the end of shelf life. For example, in Canada (Health Canada 2011), RTE foods are divided into two categories: one, in which *Lm* can grow and the shelf life of these products is more than five days, has a zero tolerance. Second, for all other RTE foods, the tolerance limit is 100 cfu/g. In the EU, all RTE foods, irrespective of their ability to support *Lm* growth, other than those intended for infants which are held to the zero-tolerance standard, < 100 cfu/g during marketing are allowed (Luber 2011). In Chile, zero tolerance is enforced in foods that can support *Lm* growth and foods intended for children, while in other foods < 100 cfu/g is the enforced limit (Moreno Switt 2020). These regulatory policies are often the driving force behind the development of detection and enumeration methods of *Lm* in foods. In the US in 2011, the adoption of more preventive approaches for the reduction of foodborne listeriosis following the Food Safety Modernization Act (US Government 2011) has led to environmental sampling requirements for *Listeria* spp.

and *Lm* in food processing plants. These sampling requirements have necessitated the development of detection methods specific to environmental samples so that corrective actions can be taken to mitigate the possibility of food contamination in affected plants. The following paragraphs deal with the development of detection methods in foods and environmental samples, as well as the enumeration of *Lm* in these samples. Focus is given to more recent developments as older methods are slowly being phased out and replaced by newer methods. The main objective of this review is not to discuss all the methods currently used but to focus on the basic principles and challenges associated with these methods.

Detection of *Lm* in Food

Detection of *Lm* in foods poses a unique challenge as the level of contamination appears to be low for direct detection by microbiological and/or molecular methods, especially relative to other microbes naturally present in many foods. This is different from the detection of *Lm* from sterile sites, e.g., blood and cerebro-spinal fluid (CSF) from clinical samples, which are normally free of background microorganisms and have higher numbers of target organisms. The strategy chosen was the development of selective enrichment followed by the use of selective screening of the enrichment cultures. A review of Table 1 and Table 2 reveals that the majority of outbreaks are caused by contaminated dairy or meat products; although, recently several major outbreaks have been caused by vegetables and fruits. The earlier efforts on detection methods, therefore, were geared toward the detection of *Lm* in these food categories. Initial attempts at the development of selective enrichment broths were carried out by the US Food and Drug Administration (FDA) for dairy products and the United States Department of Agriculture (USDA) for meat products, based on their regulatory authority in the US. The research leading to the development of these selective enrichment broths identified the various concentrations of several antibacterial and anti-fungal ingredients critical to enrichment for *Lm*. The idea was that most of the *Listeria* strains would have a higher tolerance to these antimicrobials at the concentrations present in these selective broths than competing microorganisms present in the foods. Particularly challenging were foods containing many diverse background microorganisms, e.g., different cheeses, fruits and vegetables, sprouts, etc. To overcome the challenge between the expansion of very low numbers of *Listeria* and suppression of background microflora, it was recognized early in the development of these protocols that the use of selective enrichment was needed. The selective enrichment broths ideally would suppress the outgrowth of background microorganisms while supporting *Listeria* growth within a reasonable amount of time. These considerations have led to the development of three major selective enrichment media and several different protocols (varying media, time, and temperature for incubation). Table 3 summarizes some major differences in the three major enrichment broths. These enrichment broths utilize nutrient-rich ingredients as basal media, e.g., extracts and enzymatic digests of animal tissues and yeast extract, and different combinations of phosphates for buffering. Fraser broth formulation is unique; it allows a visual assessment of *Listeria* growth because the esculin

Table 3. Major differences in *Listeria* selective enrichment broths commonly used.

Ingredient/1,000 ml	BLEB	UVM	Fraser*
Acriflavine	10 mg	12 mg	25 mg
Nalidixic acid	40 mg	20 mg	20 mg
Cycloheximide**	50 mg	0	0
Lithium chloride	0	0	3 g
Esculin	0	1 g	1 g
Ferric ammonium citrate	0	0	0.5 g
Sodium pyruvate	1.11 g	0	0
Sodium chloride	0	20 g	20 g

* Half-Fraser broth contains half the amount of acriflavine and nalidixic acid compared to Fraser broth.
** Natamycin (Pimaricin) may be an alternative to more toxic cycloheximide.

hydrolysis product, 6, 7-dihydroxy-coumarin, produces a black precipitate when reacted with ferric (Fe^{3+}) ion, turning culture media black. However, *Enterococci*, which hydrolyzes esculin, also produced a black color, when present in food, causing false-positive assessments. To minimize this problem, LiCl was added to specifically inhibit *Enterococci* (Fraser and Sperber 1988). However, *Lm* strains carrying a mutation in *lmo* 1930 have been shown to possess increased sensitivity to lithium chloride and also an inability to hydrolyze esculin and reduced growth rate in BHI (Brain Heart Infusion) and TSAYE (Tryptic Soy Agar-Yeast Extract). Such strains, therefore, most probably would be missed in standard enrichment protocols using Fraser broth or other media that utilize a similar mechanism for detection (Parsons et al. 2019). The selection of enrichment broth, as well as the time and temperature of incubation for food analysis, varies among laboratories, but the majority follow one of three protocols: the FDA Bacteriological Analytical Manual (BAM) (Hitchins et al. 2016), the USDA method (USDA 2016), or the International Organization for Standardization (ISO) method (ISO 2012). The steps of these three methods are summarized in Table 4. Many commercial detection kits follow one of these three protocols for enrichments for a seamless workflow with their detection kits.

Table 4. Major *Listeria* enrichment protocols.

Enrichment	FDA	USDA	ISO
Primary	BLEB @30°C for 4h followed by the addition of antimicrobials and further incubation for the remainder of 24 or 48 hours	m-UVM @30°C for 20–26 h	Half-Fraser broth@30°C for 26 ± 2 h
Secondary	None	1:100 dilution of primary enriched medium in MOPS-*mBLEB @ 35°C for 18–24 h	1:100 dilution of primary enriched medium to Fraser broth @37°C for 24 ± 2 h

* mBLEB contains 15 mg/1,000 ml of acriflavine.

Effects of Antimicrobials on Detection of Sub-Lethally Injured *Lm*

Lm in food, whether naturally grown or processed, undergoes both physical and chemical stresses of various proportions. These stresses often trigger a plethora of biochemical and physiological changes, leading to the development of injured cells. These injuries may result from food additives/preservatives (e.g., $NaNO_2$, NaCl, and lactic acids), processing (e.g., heating, freezing, drying, smoking), anti-microbial treatments (e.g., sanitizers in processing plants), or the natural environment (e.g., harsh condition, chemicals, microflora in soil, lack of nutrients, etc.).

Some stresses cause irreversible damage, leading to the death of the microorganism, while others cause "sub-lethally" injured microorganisms, leading to the formation of a viable but non-culturable (VBNC) state. For example, exposure to sanitizers like chlorine or benzalkonium chloride can shift *Lm* to VBNC without affecting pathogenicity (Afari and Hung 2018; Highmore et al. 2018; Noll et al. 2020). The presence of a VBNC state poses a significant public health challenge as these cells cannot be detected by standard cultural methods. Whether an organism survives a stress factor depends on the successful expression of many genes controlled by *sig*B and other regulatory genes. The ability to successfully revive these "stressed" microorganisms during an enrichment process poses an enormous challenge as these enrichment broths contain various antibacterial and anti-fungal chemicals in differing amounts. Allowing microorganisms to outgrow in a non-selective environment for a brief period, followed by the addition of selective agents, is often considered to be a pragmatic approach. The FDA BAM, in addition to allowing 4 hours of incubation in a non-selective environment, also includes pyruvate in its formulation to help in the recovery process via a nutritive enhancement. Some experimental protocols completely avoid any antimicrobials in their enrichment broths and depend on immunomagnetic separation (IMS) for selective enrichment after culture in a non-selective broth (Gorski et al. 2014). Both UVM and Fraser broth formulations exclude cycloheximide, and the half-Fraser broth formulation contains half the concentrations of nalidixic acid and acriflavine as used in Fraser broth, which is supposed to help the recovery of stressed microorganisms by reducing the stress presented initially. Lithium chloride in Fraser and half-Fraser broths also interfere with injured *Lm* growth. It is also important to realize that different stressors cause different types of injury, thereby a single protocol may not work for all kinds of sub-lethal injuries an organism may experience. Several authors have compared the efficacies of commonly used enrichment broths for the detection of artificially injured *Lm* cells, either by freeze or heat treatment, and concluded that these enrichment broths are comparable in their ability to recover stressed cells (Busch and Donnelly 1992; Ryser et al. 1996; Donnelly 2002).

Comparison of Standard Enrichment Broths With Selective Foods: Effect of Food Microbiota

Food-specific enrichment protocols, using different formulations of enrichment broths, were the main driving force behind the development of BLEB (FDA), UVM (University of Vermont), and Fraser (USDA) broths. From the beginning, it was appreciated that different food categories, in this case, dairy and meat products, may provide different challenges for *Lm* detection and hence required optimization of inhibitory compounds, time, and temperature for incubation, as well as downstream selection using selective agars. The food-specific interaction emanated from food composition, food additives, microbial load including other *Listeria* species present in the foods, and inhibitors that might be produced by one or more members of the food's microbial community. For example, whole cantaloupe, fresh sprouts, cheese, and other processed foods often contain large numbers of background microflora compared to pasteurized milk. Even after selective enrichment, some background microflora might persist, which could complicate the detection of typical *Lm* colonies when plated on selective agar plates, e.g., Rapid L'mono or ALOA. Under these circumstances, a molecular-based approach may serve better in detecting *Lm*, presuming the target is readily detected from the background. It has been shown that many naturally occurring microbes, including many lactic acid bacteria, may produce small molecular weight peptides, collectively known as bacteriocins, some of which may inhibit *Lm* outgrowth. Therefore, the presence of such organisms may interfere with *Lm* growth during enrichment, thereby resulting in false-negative detection (Nilsson et al. 2004).

Comparison of Standard Enrichment Broths for Supporting Different Serotype and Lineages of *Lm* Strains

Isolation of major disease-causing serotypes belonging to various lineages is important from both public health and epidemiological standpoints. This is particularly relevant as *Lm* isolation from clinical samples, e.g., blood, CSF, etc., do not undergo similar protocols as are often used for food and environmental samples. Food and environmental samples, in general, contain very small numbers of *Lm* and in some samples a large amount of non-*Lm* microflora, especially when compared to the relevant clinical samples. The challenge presented is how to enrich all *Lm* strains, while simultaneously suppressing the growth of the background microflora. From an epidemiological perspective, foods and environmental samples may contain multiple genotypes of *Lm*; thus, it is critical that our detection methods isolate all strains equally well. Outbreaks caused by multiple genotypes and serotypes (Walsh et al. 2014; Angelo et al. 2017) have been reported, highlighting that thorough efforts to identify contaminating strains are critical when determining which foods are linked to specific cases. That raises questions about whether current enrichment protocols can detect all genotypes of *Lm* equally efficiently and how such differences affect the epidemiological investigation. For example, although 4b is associated with the majority of disease cases, the majority of food and environmental isolates are often found to be 1/2a or 1/2b (Loncarevic et al. 1996). Additionally, it has been observed that *Lm* strains isolated from foods using a direct plating method are more genetically diverse than when isolated by a selective enrichment method indicating bias in enrichment protocol. Similarly, Ryser et al. (1996) demonstrated that different ribotypes of *Lm*, indicative of different genotypes, showed different recovery rates in UVM and *Listeria* Recovery broth (LRB) from samples of ground meat. Bruhn et al. (2005) showed that lineage II strains of *Lm* outcompete lineage I strains in both UVM I and UVM II broths. When inoculated at identical cell densities, *L. innocua* outcompeted lineage I strains but not lineage II strains. Interestingly, when grown in BHI, a standard non-selective broth, no difference in growth was observed between lineage I, lineage II, and *L. innocua* strains. Work with watershed samples (Gorski et al. 2014) demonstrated that a selective enrichment protocol recovers more serotype 1/2a strains compared to non-selective enrichment. Although it is not clear why certain serotypes/lineages do well in certain selective media, the serotype-specific genetic footprints have been well documented (Zhang et al. 2003; Laksanalamai et al. 2012). Picking multiple colonies from selective agar plates and identifying their serotypes may alleviate part of this problem, presuming all strains have been able to reach levels suitable for plate-based detection. In addition, an outgrowth advantage of *L. innocua* over *Lm* in selective enrichment broths (Cornu et al. 2002) poses serious problems with false-negative results leading to increased public health concerns.

Detection of *Lm* from Environmental Samples

Detection of *Lm* from the environment encompasses a broad spectrum of sample types and environments. These samples can refer to swabs taken in the environment of a food processing facility, as well as samples taken from the natural environment. Swabs in a processing facility may encompass a variety of surfaces, including food contact, floor drains, ladders, and various types of tools and equipment, as well as a wide variety of facilities that may be enclosed or partially open, such as packing sheds. The natural environment includes a broad spectrum of samples including fecal material, water, soil, vegetative material, and air. Each of these samples, because of their varied physicochemical properties, can affect the recovery and detection of *Lm*. Careful consideration of these factors is necessary before deciding on the optimal method for the specific investigative need.

Lm in the Food Production Environment

Samples taken from food processing facilities have been subjected to varying nutrient conditions, sanitizers, and cleaning procedures that can include harsh chemicals and mechanical disruption. These conditions can trigger different stress responses or serve as drivers for the formation of biofilms, both of which can impact recovery. Different environments are also likely to have different and/or more diverse background flora. Floor drains, a common contamination site for *Listeria* spp. (Estrada et al. 2020), collect drainage from throughout the facility and maintain a moist environment that is ideal for *Lm* and other members of the *Listeria* genus. Sampling these sites for pathogen presence in the facility can lead to the pooling of a wide variety of organisms from throughout the facility. Conversely, food contact surfaces would be subjected to regular cleaning routines that would reduce the microbiota load in those samples. The routine use of sanitizers, along with other control mechanisms, throughout the food production environment is likely to result in the presence of *Lm* that is sub-lethally injured. This can complicate recovery as these cells may not be readily cultured, especially as the selective conditions in enrichments can lead to further loss of these cells. These factors can hinder the recovery of the pathogen via enrichments.

Lm in the Natural Environment

While most efforts associated with *Lm* detection have focused on clinical and food-related detection, it is important to consider that *Lm* is naturally present in the environment. Its presence in this environment has critical implications for *Lm* food contamination, especially regarding fresh produce. Reservoirs in the natural environment may result in the introduction of the pathogen to the production environment. For example, domestic and wild animals may introduce *Lm* via their waste (Skovgaard and Morgen 1988; Bouttefroy et al. 1997; Parsons et al. 2020). Separately, the presence of *Lm* in the soil or water may result in transfer directly via soil contact, irrigation, or watershed transfer (Oliveira et al. 2011). When evaluating the natural environment, studies have shown that recovery is highest in moist environments (Chapin et al. 2014), suggesting that the potential for watershed transfer is significant.

Due to annual weather patterns, temperature and moisture can vary widely within the same location. Temperature and moisture changes seen in the natural environment are unlikely to be present in most food processing facilities due to the presence of climate control systems but represent a unique stress for *Lm* in the natural environment that may affect recovery. Soil contamination is variable, depending on the geographic region, usage, soil type, and study methods (Dowe et al. 1997; Fox et al. 2009; Locatelli et al. 2013; Vivant et al. 2013; Harrand et al. 2020). This detection variability based on soil characteristics highlights the potential for interplay between environmental stress associated with a source and the method used for detection. Therefore, evaluation of these diverse samples can provide a critical understanding of *Lm* survival in the environment to aid disease control but also present a wide array of unique challenges.

Impact of Stress and Sub-lethal Injury on the Recovery of *Lm*

While it has been stated that many methods allow 24-hour recovery for stresses organisms, data suggest that recovery of stressed bacteria could take longer as the lag phase is increased in most recovery media (Silk et al. 2002). Stresses can encompass a variety of conditions including sanitizer exposure, temperature, nutrient deprivation, and osmotic pressure, which can lead to sub-lethal injury or growth alterations of the cells (Giotis et al. 2007; Vail et al. 2012; Noll et al. 2020). Culture-based methods rely on the use of selective pressures to recover *Lm* but these pressures can result in the failure to recover injured cells that are unable to initiate replication events for a variety of reasons. The failure to consider the possibility of sub-lethal injury when developing detection methods has the consequence of under-detecting *Listeria* contamination events (Donnelly 2002; Giotis et al. 2007). This has led to the development of various recovery broths to aid the recovery of *Listeria* from environmental samples. One evaluation comparing various selective media, including Fraser, UVM, and FDA *Listeria* enrichment broth (LEB), suggested that none of these media provided a significant advantage over the other, though the absence of selection in Trypticase soy broth did result in a significant shorter lag phase than observed for Fraser (Silk et al. 2002). However, other work led to the development of *Listeria* Repair Broth (LRB), which is used as a brief recovery enrichment prior to selective enrichment media (Busch and Donnelly 1992). A review of studies analyzing this broth against FDA and USDA enrichment protocols suggested that the incorporation of LRB into detection protocols could result in improved detection of *Listeria* spp. (Donnelly 2002; D'Amico and Donnelly 2008). While other differences were observed suggesting possible benefits, the differences noted in these works raise concern as the study was done in the absence of competing microorganisms. A significantly longer lag time could result in out-competition by background flora, reducing the likelihood of detecting microorganisms with sub-lethal cell injury. The effect of this lag time was observed when comparing three enrichment schemes for the recovery of desiccation-injured cells. This study found that enrichment schemes using 24-hour incubations had reduced detection rates compared to ones using the same media with a 48-hour incubation (Sheth 2018).

Sampling Considerations

Understanding where, when, and how to sample is just as important as choosing the correct detection method when detecting *Listeria* spp. in the environment. These considerations are especially critical in food processing facilities, as robust sampling for the presence of *Listeria* spp. and *Lm* is key for the prevention of contamination of the final food product. Within facilities, it is best to sample equipment that is active or immediately after a production run is completed, as opposed to immediately after cleaning. Vibration and movement associated with equipment activity can introduce pathogens to food contact areas from hard-to-sample areas. Therefore, evaluation after cleaning, prior to equipment activity, can result in a false sense of security as readily reached surfaces are likely to be free of contamination. Furthermore, it is possible that bacteria may have sub-lethal injuries that can reduce their recovery but not reduce the potential for downstream contamination events. Swab types and neutralizing broth can affect the ability of the pathogen to be collected from the equipment and neutralize sanitizers that could complicate pathogen recovery. A survey of processing plants in the EU found that most of the responding facilities preferred friction-based sampling (e.g., swabs, sponges and pads) versus contact-based methods (e.g., contact plates and Petrifilm™) (Brauge et al. 2020). The latter detection method, using contact-based sampling, is exemplified by Petrifilms (3M), which have been found to yield similar results to other detection methods, though independent studies comparing Petrifilm™ to the BAM method found that Petrifilm™ had much lower recovery yields (Cruz et al. 2012; Benesh et al. 2013; Hitchins et al. 2016). This result was verified when a comparison, evaluating three friction-based methods and contact plates, found no significant difference in the recovery of *Lm* between the methods, based on colony-forming units (Brauge et al. 2020). A variety of transport media are available but the incorporation of a neutralizing agent to prevent further loss of microorganisms due to sanitizer presence is critical (Dey and Engley 1983; Zhu et al. 2012). Downstream detection methods, however, should be evaluated to verify that they are not inhibited by components of the transport medium selected.

When creating a sampling plan for a facility, the focus should be designed to detect contamination at points in the processing line after a lethal treatment, such as heating, and before packaging as this is the highest risk period for contamination. Previous knowledge suggested that frozen foods were unlikely to represent a risk, as no outgrowth was likely, and sampling did not need to be as extensive (Tompkin 2002); however, recent outbreaks linked to frozen foods suggest that this is not true (CDC 2015; ECDC 2018). As the presence of moisture generally correlates with the presence of *Listeria* spp. (Hsu et al. 2005; D'Amico and Donnelly 2008; Chapin et al. 2014; Estrada et al. 2020), frozen food production with its potential for condensation may represent a particularly challenging environment for the control of *Listeria* spp. Food production environments are also subject to routine cleaning protocols to prevent contamination of the facility and the food products. The presence of these sanitizers as residues on the surfaces may be transferred to the sample, further abrogating the ability of a method to detect the pathogen if the transport medium lacks compounds to neutralize the effect of these sanitizers as noted earlier.

Considerations for Methods When Evaluating Environmental Samples

In broad terms, there are three main categories of detection methods, culture-based, molecular, and biosensors. Validation of these methods, where available, is a critical consideration when choosing a method for a specific purpose. However, it should be noted that most validations are performed with spiked samples and it is possible that naturally contaminated samples may lead to different detection rates than observed during method validation testing (Warburton et al. 1991). Each of these detection methods has advantages related to detection limits, hands-on time, potential automation, overall processing times, and cost. Additionally, dependent on facility-specific requirements, tests may need to be performed on-site rather than off-site so that results can be rapidly relayed to address affected areas. This also reduces turnaround time on analysis by eliminating transport of the sample to the off-site facility, which can be a critical determinant when considering the release of a production lot to the market.

Another consideration when choosing a method for the detection of *Listeria* spp. is whether the method results in the recovery of a viable organism or simply indicates the presence of the pathogen. Many rapid methods rely on the amplification of an input template derived via a variety of approaches. This amplification may be performed using nucleic acid amplification or it may be performed via antibody or phage detection recognizing multiple protein epitopes. However, these methods often lead to the destruction of the target organism to release the analyte. As a result, rapid methods, like biosensors, which eschew prior culture enrichment to reduce the time to results, may not yield culturable microorganisms. These approaches can be beneficial when rapid results for food contact surfaces are needed to verify food safety prior to product market release. However, the lack of an isolate can hinder outbreak and traceback investigations and the determination of the persistence of a strain in an environment. WGS of isolates detected in food processing environments has revealed the presence of continuing contamination by an original source strain, indicating the need for improved mitigation strategies and/or repeated reintroduction from an external source (Orsi et al. 2008; Stasiewicz et al. 2015). Additionally, the lack of an enrichment step may complicate the detection of stressed cells and/or low numbers of cells in a high-background sample.

A final consideration when choosing a method for environmental sampling is what target organism will be detected. Many of the rapid tests evaluated here focus on the detection of *Listeria* spp. This is likely due to advantages when performing environmental sampling for detecting *Listeria* spp., as opposed to specifically the pathogen *Lm*. When testing food samples, it is critical to identify *Lm* both to facilitate the removal of adulterated foods from markets and to rapidly identify the correct mitigation strategy. However, when evaluating environmental samples, a different set of considerations alters the range of options. For food contact surfaces, the identification of *Lm* is critical for the same reasons as detection in the foods themselves as contamination of these surfaces is highly likely to indicate contamination of foods. However, surfaces in a facility in zones 2, 3, and 4, while critical for controlling contamination, allow the consideration of simply identifying *Listeria* spp. which would indicate conditions that have the potential to support the presence of *Lm*, allowing producers to take corrective action. This approach has key regulatory considerations, depending on the region and food product, that must be evaluated carefully before choosing a method.

As noted, many of the rapid tests only detect *Listeria* spp. Several of these tests allow producers to further evaluate a *Listeria* spp.-positive sample to determine if *Lm* is present, in order to rule out potential risk to the food product, which is a critical step. This is possible because these tests utilize an enrichment, prior to the rapid detection method, that can subsequently be used for culture on selective agar. Alternatively, species-specific tests are available, although most of these have been evaluated only for food. This mixed approach, used as part of a seek-and-destroy strategy for eliminating *Lm* in the food production environment, can provide faster results due to faster processing times (Table 5). The concern with this detection approach is that detection of *Listeria* spp. may result in the failure to detect *Lm*, even if it is present due to the growth biases that may occur in methods optimized for *Listeria* spp. versus *Lm*. Growth biases preventing the detection of *Lm* in the presence of *L. welshimeri* and *L. innocua* have already been observed, suggesting that this potential is worthy of careful consideration (Zitz et al. 2011; Dailey et al. 2015). This could mean that a facility could fail to detect *Lm* from a reserved enrichment sample even after secondary analysis due to competition from other members of the genus. Therefore, it is possible that a *Listeria* spp. positive result may indicate an even higher risk for *Lm* than just as a potential indicator.

Table 5. Select rapid methods for the detection of *Listeria* in environmental samples.

Assay Name	Company	Target Organism	Technology	Enrichment*
ANSR® *Listeria monocytogenes*	Neogen	*Lm*	Isothermal nucleic acid amplification	16–24 hr
Listeria RightNow™	Neogen	*Listeria*	Isothermal nucleic acid amplification	No
RapidChek®	Romer Labs	*Lm*	LFI	44–48 hr
ANSR *Listeria*	Neogen	*Listeria*	Isothermal nucleic acid amplification	16–24 hr
Veriflow® *Listeria* Species	Invisible Sentinel	*Listeria*	PCR & lateral flow device	24 hr
MicroSEQ®	Applied Biosystems	*Listeria*	Real-time PCR	28–32 hr
Solus One *Listeria*	Solus Scientific	*Listeria*	ELISA	22–30 hr
iQ-Check *Listeria* spp.	Bio-Rad	*Listeria*	Real-time PCR	18 hr
Sample6 DETECT/L™	Weber	*Listeria*	Bioillumination	22 hr
Reveal® *Listeria* 2.0	Neogen	*Listeria*	Immunodiagnostic	27–30 hr
3M Molecular Detection Assay	3M	*Listeria*	Isothermal nucleic acid amplification	
3M™ Petrifilm™	3M	*Listeria*	Culture detection	28 hr
PDX-LIB	Paradigm Diagnostics	*Listeria*	Culture detection	48 hr
Path-Chek	Microgen Bio	*Listeria*	Culture detection	24–48 hr
InSite *Listeria*	Hygiena	*Listeria*	Culture detection	24–48 hr

* Indicates if the methods require enrichment and, if so, the length of the incubation recommended.

Culture-Based *Listeria* Detection in Environmental Samples

Culture-based methods are slower than molecular approaches due to the generation time required to enrich for the presence of *Lm* or *Listeria* spp. from within the background flora. However, these methods are optimal for certain sample types. For example, bacterial levels in soil appear to be low and often require the use of culture-based methods to reliably detect them. A study by Locatelli et al. found that while culture-based methods were able to detect *Lm* in 17% of the soil samples evaluated, a parallel analysis of those samples using molecular detection was only able to detect *Lm* in only 2% of the samples (Locatelli et al. 2013). The authors attributed this to levels of *Lm* in the soil below 10^4 cfu/g. It is also possible that the detection of *Lm* and other pathogens from the soil via molecular techniques is hindered by the presence of various compounds in the soil that may interfere with preparations of the sample and downstream detection (Watson and Blackwell 2000; Locatelli et al. 2013). These methods use a combination of nutrients and antimicrobials to reduce or eliminate the ability of background flora to grow. The utility of these media, therefore, will be impacted by which flora are likely to be present in the sample. Given that background flora and chemical compositions will vary depending on sample type, no one medium will be ideal for all potential samples. Additionally, it is worth noting that while mixed contaminations in foods and clinical cases are not generally considered the norm, in environmental samples this potential is increased. Therefore, it is wise to consider that each medium has the potential to introduce bias in the recovered organisms from environmental samples. As noted earlier, lineage II strains have been found to have an outgrowth advantage over lineage I strains when co-cultured in UVM, as has *L. innocua* over *Lm* (Bruhn et al. 2005). This latter growth advantage for various other members of the *Listeria* genus over *Lm*, as well as genetic subtypes of *Lm* over others, has been seen in other media (Ryser et al. 1996; Donnelly 2002; Keys et al. 2013; Dailey et al. 2015). In fact, UVM was shown to result in the failure to detect isolates linked previously to outbreaks (Donnelly 2002). Awareness of these potential biases is critical as the presence of these other members can be indicators of the conditions for contamination, but their presence could also hinder the detection of *Lm* in the event of co-contamination (D'Amico and Donnelly 2008). Early comparison of culture methods utilizing modifications of early USDA and US FDA methods, found no notable difference between the two methods, though the authors noted slightly higher recoveries when the FDA enrichment broth was coupled with a secondary Fraser enrichment (Warburton et al. 1991). This observation may indicate that the ideal enrichment scheme could involve coupling different enrichment media and potential multiple enrichment protocols to optimize the reduction of the background flora.

Most methods rely on a 24–48 hour culture time, with an increasing trend to shorter incubation times, particularly for environmental samples, to reduce turnaround time to results. However, studies have shown that the duration of enrichment prior to sampling from an enrichment can affect whether a positive result is observed. While logic might suggest longer incubation times are better and this has been shown particularly for Fraser broth (Donnelly 2002), converse data were observed by Fortes et al. showed that, while most samples gave similar results at 22 hours and 48 hours, some yielded positive results at 22 hours and others at 48 hours (Fortes et al. 2013). A separate study found a similar pattern with 22% of the positive samples being identified at 24 hours or 48 hours but not both (Sullivan and Wiedmann 2020). These studies indicate that neither incubation time is likely to detect all positive samples and that a mixed approach, evaluating for the presence of *Listeria* spp. and/or *Lm* at 24 hours and 48 hours, would lead to the detection of more positive events and result in a more robust detection effort (Ryser et al. 1996). This approach would work readily for a solid or liquid medium that uses color changes to identify the presence of the organism, as it would simply require a visual assessment at a later time point. However, this approach could be problematic for methods that lack an integrated detection system or for phenotypes that are variable, depending on the cost and time. Many selective media have incorporated visual alterations that rely on esculin hydrolysis to darken the medium or phospholipase C activity to hydrolyze compounds, resulting in either a haze such as is seen with

ALOA or blue color as seen with Rapid L'mono agar (Bio-Rad). However, these phenotypes may not be uniformly expressed in all isolates, which could result in false-negative results for both phenotypes (Donnelly 2002; Leclercq 2004; Burall et al. 2014; Maury et al. 2017). Reliance on medium darkening, such as when using Fraser broth, is likely to result in false negatives and all such enrichments should be streaked on a secondary medium to confirm the absence of *Listeria* spp. (Donnelly 2002). These studies suggest that reliance on such observable phenotypes may result in the failure to detect *Lm*. The pairing of these methods with downstream molecular identification methods would help identify atypical strains that may still be capable of causing disease (Maury et al. 2017). One such mechanism is the BAX® System Real-Time PCR for *Listeria* (Hygiena). An analysis comparing this system to selective Oxoid medium culture detection found a slightly higher detection rate using a culture-based method for the identification of *Listeria* spp. (Norton et al. 2000). A key limit of the BAX® analysis was the identification of false-negative results (9 of 89 culture-positive samples) (Norton et al. 2000). This was presumed to be due to poor enrichment and the authors suggested that the use of a two-step enrichment, as recommended by the manufacturer, could correct this problem (Norton et al. 2000). Conversely, the detection of *Listeria* spp. from samples with culture-negative results (5 of 125 culture-negative samples) suggested the possibility of the detection of non-viable cells, a problem associated with molecular techniques (Norton et al. 2000). Separate work pairing a PCR-based screening with culture enrichment found that detection was generally equivalent with the culture-based approach but yielded faster results (D'Amico and Donnelly 2008). The use of these screening methods could also eliminate the concerns associated with atypical phenotypes.

Separate from the enrichment media considerations, culture-based methods are paired with a final selective agar medium to isolate colonies from the enrichment. These media can also impact the ability to recover *Lm*. A separate study, evaluating different ALOA-type agars ability to detect *Lm*, found that, while 24 hours was enough for most instances, several isolates needed an additional 24 hours to fully develop the typical *Lm* colony morphology (Stessl et al. 2009). Further work from Leclercq suggests that up to 96 hours may be needed for atypical isolates to yield colonies suitable for further analysis, independent of stress or background flora (Leclercq 2004). While this method relied on isothermal nucleic acid amplification for final detection, a preliminary enrichment was necessary for the detection of low levels of *Listeria* spp. These authors postulated that the failure to detect *Listeria* at 48 hours may be due to the overgrowth of competing organisms, inhibiting detection (Fortes et al. 2013). This overgrowth of background flora hindering the detection of *Lm* has been seen elsewhere, especially when using half-Fraser brother, further highlighting the need to consider a sample's native microbial population when selecting an enrichment scheme (Stessl et al. 2009). Further, different *Listeria* species and, even lineages, have differing growth rates in selective media, depending on the media and compared *Listeria* members (Bruhn et al. 2005).

Previous sections have discussed the traditional methods associated with enrichment from food, which are often used for environmental detection as well. However, newer media have also been developed to reduce incubation time and have focused on environmental surfaces, which, due to regular cleaning and sanitation, may have reduced organic and microbial load compared to foods. These media are often formulated for the detection of *Listeria* spp. One such medium is ACTERO™ *Listeria* Enrichment Media (FoodChek Systems, Inc.) which was shown to be better at recovering *Listeria* from plastic surfaces with a 24-hour enrichment than a UVM-BLEB 48-hour enrichment protocol (Claveau et al. 2014). *Listeria* Indicator Broth (PDX-LIB, Paradigm Diagnostics, Inc.) is another such enrichment medium designed and recently modified for rapid culture-based enrichment and detection of *Listeria* spp. from the food production environment and was compared with the original formulation (Olstein and Feirtag 2019) and found to have higher sensitivity and reduced false positives. However, a thorough analysis comparing this medium to other standard media is still needed and the manufacturer notes that it is only intended to detect four species of *Listeria* (www.pdx-inc.com). Another such medium is RapidChek *Listeria* enrichment media (Romer Labs), developed for use in conjunction with a rapid screening lateral flow detection device (Muldoon et al. 2012; Juck et al. 2018). Unfortunately, assessment of the enrichment capabilities of the proprietary media, which are commonly associated with various rapid screening methods, is complicated because these methods also use a proprietary endpoint detection system. However, other proprietary enrichment systems have been developed that incorporate detection within the culture via color change, like seen with Fraser broth, which can allow a more direct assessment of their ability to enrich *Listeria* spp. A comparison of two of these, designed for the detection of *Listeria* spp., found a high rate of false positives which would be problematic in an environment where a positive result requires significant interventions (Schirmer et al. 2012). These methods likely rely on the same principles used for other chromogenic media and, therefore would have the same concerns raised earlier, regarding atypical phenotypes or background flora that have confounding phenotypes.

A trend in the method development has been the development of enrichment and detection methods to detect multiple pathogens, frequently *Lm*, *Salmonella*, and *E. coli* O157:H7 (Kawasaki et al. 2005; Germini 2009). These methods pair PCR with a multi-target enrichment medium. However, a key gap in this principal is that the growth of *Lm* is generally much reduced compared to the two Gram-negative pathogens and may also be reduced compared to other background flora. Indeed, the medium may compromise the ability to detect low-level contamination events because it is not optimized for *Lm*. This was seen in a study of SEL, a selective enrichment medium developed to enrich the three main pathogens, *Salmonella*, *E. coli* O157:H7, and *Lm*. However, there was an observation of reduced growth of *Lm* in independent culture, compared to *Salmonella* and *E. coli* O157:H7. Additionally, the only error when evaluating the ability to detect pathogens occurred when detecting low levels of *Lm* from a meat sample (Kim and Bhunia 2008). For these reasons, while the convenience of a single enrichment protocol to detect these three pathogens may be appealing, there is an elevated risk of missing *Lm* when utilizing these approaches. Additionally, a second study, evaluating the ability of SEL to recover the three pathogens after sub-lethal injury, found that SEL had a reduced ability to recover *Lm* compared to the recovery observed for *Salmonella* and *E. coli* (Suo and Wang 2013).

Molecular Detection Methods

In an effort to develop more rapid detection methods, research shifted toward molecular techniques to detect *Lm*, some protocols are paired with partial enrichments. Molecular detection of *Lm* or *Listeria* spp. has key benefits for the industry. The use of a "test and hold" approach can provide significant savings for companies as they help avoid costly recalls as well as any negative effects on the company's reputation associated with a public recall. However, this approach is only practical if testing can be completed quickly, relative to the shelf life of the product. Molecular approaches, which amplify signals specifically via PCR or antibody recognition, can reduce or eliminate the need for culture-based signal enrichment. These molecular assays, such as PCR and immune-based assays, can therefore be very sensitive and rapid, facilitating rapid determination of the safety of a food product prior to market release. However, these techniques may not be able to discriminate between viable cells and cells that have been properly neutralized, which is a possibility on surfaces that have been treated to eliminate pathogens. The development of methods that can rapidly identify the presence of viable *Lm* in the food product or food processing environment can help companies move quickly to determine whether a lot is safe for market release. These methods use a variety of approaches and may use a preliminary enrichment to partially separate out *Lm* or *Listeria* spp. from the background flora to provide a more detectable signal. Critically, many of these methods focus on the detection of *Listeria* spp., not *Lm*, which may affect when they can be used and/or require additional typing methods after detection to determine if *Lm* was the detected organism. Varying targets may be selected including rRNA (Wendorf et al. 2013; Caballero et al. 2016b), DNA (Petrauskene et al. 2012; Klass et al. 2019), or proteins (Juck et al. 2018; Tonner et al. 2019).

These methods can be divided into two categories, as noted earlier. This is dependent on whether there is a pre-enrichment step or not, which additionally also affects whether a method will have the potential to yield a culturable organism for downstream subtyping analyses. There are a limited number of methods that do not rely on at least a partial enrichment, as enrichment is a useful tool when attempting to amplify a specific signal to detectable levels above a background signal. One of these methods is *Listeria* Right Now™ (LRN)(Neogen Corp.), which relies on the numerous copies of rRNA present in a bacterial cell as its target (Le et al. 2019). The signal is then amplified using an isothermal nucleic acid amplification approach with a fluorescent marker for final detection, leading to results within 1 hour (Le et al. 2019). Validation studies of this assay with the FDA BAM method (Hitchins et al. 2016) found no significant difference when evaluating samples from environmental surfaces typical of food production facilities, suggesting that this method could be a fast alternative to culture methods (Le et al. 2019; Roman et al. 2019). An alternate approach is seen with Sample6 DETECT/L™ (Weber Scientific), which does not rely on enrichment, though does have extended processing time compared to LRN with 6–10 hours before results are achieved (Cappillino et al. 2015; Banerjee et al. 2018). This method couples the bacteriophage with a luciferase assay and is designed to only detect viable cells, though assays evaluating non-replicative or injured cells have not been performed (Banerjee et al. 2018). A modified version of this method was compared in a validation study to an enrichment scheme the USDA method uses (Table 4) and found equivalent performance between the two methods, suggesting that both are equally capable of detecting *Listeria* spp. from stainless steel surfaces (Banerjee et al. 2018). While both methods provide rapid results that are comparable to culture-based methods, they detect *Listeria* as a genus and do not differentiate *Lm* from other members of the genus. Furthermore, sample processing for the analysis makes it challenging to further assess the contamination event if that becomes necessary.

Most commercial molecular detection methods partner with an enrichment step that is often proprietary. The enrichment-detection schemes are inherently linked in their validation studies, which, as mentioned earlier, requires them to be considered as a joint process. These methods pair an enrichment step that is usually no longer than 24 hours with a variety of downstream rapid molecular detection approaches. These include lateral flow immunoassay devices (LFI), PCR, and isothermal nucleic acid amplification. ANSR® uses this latter approach with rRNA as the target, like LRN, that can provide results in 40 minutes, not counting the 16–24 hr enrichment (Wendorf et al. 2013). While the assay was found to be comparable in the initial validation to the reference culture method, a reduction in the probability of detection for positive samples was observed with the 16 hours incubation timepoint, suggesting that a 24-hour enrichment would provide a more robust evaluation (Wendorf et al. 2013; Caballero et al. 2016a). Several systems have paired real-time or conventional PCR assays with an enrichment protocol. Two examples of this are BAX, discussed earlier, and MicroSEQ (Life Technologies). MicroSEQ has the advantage of being developed using the non-proprietary BAM enrichment protocols and was found to be reliable when detecting *Listeria* spp. from various environmental surfaces but required a longer processing time (Petrauskene et al. 2012). Bio-Rad's iQ-Check *Listeria* spp. the system also uses real-time PCR as its final detection after enriching in proprietary *Listeria* special broth (LSB) for 18–26 hours (Klass et al. 2019). Comparison with the USDA method (Table 4) found comparable results using the iQ-Check protocol for several environmental surfaces (Klass et al. 2019).

Lateral flow devices as a detection method offer ease of use to facilities, as well as rapid detection after an initial enrichment. These assays may rely on the antibody recognition of a target or PCR coupled with a visual flow assay. The latter method is the principle behind the Veriflow® *Listeria* spp. assay that, when compared to the USDA method (Table 4), showed no significant difference in detection ability (Joelsson et al. 2017). This assay has the drawback, compared to other lateral flow devices, that PCR must be performed. Rather than detection via a real-time machine or gel electrophoresis, the reaction is loaded into a cassette that, when the amplicon is present, produces a chromogenic band as the positive indicator (Joelsson et al. 2017). This approach provides amplification of the signal via enrichment and PCR, allowing rapid, specific identification of the target organism. Conversely,

Reveal® 2.0 (Neogen) and RapidChek® (Romer Labs) both rely on antibodies and an immunochromatographic signal to indicate the presence of *Listeria* spp. (Alles et al. 2012; Juck et al. 2018). These methods rely on proprietary enrichments, but RapidCheck® requires a 48-hour incubation versus the 27–30 hour incubation for Reveal® 2.0 (Alles et al. 2012; Juck et al. 2018). Intriguingly, validation studies indicated that the Reveal® 2.0 system had an equivalent, if not slightly improved, detection compared to the USDA method (Table 4) (Alles et al. 2012). A similar result when evaluating samples from environmental surfaces were observed for the RapidCheck® method when compared to the USDA method (Table 4) but only with a 40-hour enrichment (Juck et al. 2018). These results suggest that LFI are likely to provide an important advantage in environments needing faster results as they couple ease of use, potentially shorter enrichment times, and high accuracy. Another approach used that also uses immune-based detection is Solus One, which relies on enzyme-linked immunoassay (ELISA) that takes just under 3 hours but requires a proprietary enrichment of 22–30 hours (Tonner et al. 2019). This method was shown to be able to detect *Listeria* spp. from spiked plastic and stainless steel surfaces with an equivalent probability of detection to the FDA BAM method via manual and automated methods (Tonner et al. 2019). Another variant of this is employed by the VIDAS LMO2, which is designed to specifically detect *Lm* (Johnson and Mills 2013).

Unfortunately, there are a limited number of studies that compare these rapid methods and, as these methods rely on proprietary technology, it can be challenging to draw conclusions based on these studies. Furthermore, there are limitations to drawing conclusions about a method's utility based on validation studies that lack incorporation of stress, surface wear, and diverse background flora. A comparison between the VIDAS LMO2 method, RT-PCR, and direct plating, after a brief pre-enrichment, suggested that the VIDAS may have reduced sensitivity to detect dry stressed biofilm cells compared to RT-PCR (Rios-Castillo et al. 2020). This would suggest that the use of this assay and other assays based on ELISA may be impaired when detecting cells subjected to desiccation stress. A study comparing the FDA BAM method (Hitchins et al. 2016) to various rapid tests found that none of the rapid tests evaluated, which included Reveal®, had as high a recovery rate when evaluating via a most probable number (MPN) approach, suggesting that these methods may not perform well with samples that have lower concentrations of cells, injured cells, or inhibitory backgrounds (Cruz et al. 2012). Critically, this study assessed an environmental surface that had been used in industry and might therefore more accurately represent some of the challenges that might occur when sampling from a used surface. A separate study compared the detection rates of three methods, Petrifilm™, Reveal®, and BAX as representatives of three classes of detection methods, using ISO 11290-1 as the basis of comparison, found that the Petrifilm™ had greatly reduced sensitivity compared to the other two methods, suggesting that this method is insufficient for the needs of a food production facility (Kovacevic et al. 2009). The Reveal® system had reduced sensitivity compared to the BAX and reference methods, suggesting that while LFI may offer some advantages, they come at a cost when evaluating naturally contaminated samples (Kovacevic et al. 2009). Taken together, these efforts highlight the need to further evaluate available methods in more direct comparisons with samples better representing natural contamination events.

Enumeration

Enumeration of *Lm* in food and/or environmental samples is important to understand the extent of contamination, verification, and assessment of any disinfection and food processing technique to establish regulatory thresholds where required and for estimation of exposure and infectious dose assessment. Enumeration provides the amount of *Lm* found in the food or environmental samples at the time of collection, provided there is no loss or gain of the target organism between collection and analysis. The biggest challenge is the enumeration of low numbers of organisms generally present in food and environmental samples. Two general enumeration methods are currently employed: direct plating of sample preparation on selective agars or MPN using selective broths. Direct plating and MPN are used by various regulatory agencies (Pagotto 2011; Hitchins et al. 2016; ISO 2017). Both methods have their advantages and disadvantages and the selection of a specific method should be decided based on the nature of the sample (solid vs. liquid food, environmental swabs, etc.). Studies with artificially contaminated (Gnanou-Besse et al. 2004) and naturally contaminated foods using a membrane filtration system showed that such methods could be a useful alternative to direct plating on selective agars. The ability to filter solid food, however, poses an additional challenge as the sample must be homogeneously suspended in a liquid, thereby further compromising the ability to enumerate a very low number of *Lm*. Several recent studies with naturally contaminated ice cream samples (Chen et al. 2016; Burall et al. 2017a) showed that MPN methods are more sensitive and reliable (less variability) when the contamination level is low (1–100 cfu/g or ml) while the direct plating methods appear to be a more reliable method where contamination levels are high (> 1,000 cfu/g or ml). Concerns are raised regarding the effect of stressed *Lm* cells in these samples when subjected to immediate challenge to the variety of antimicrobials present in selective agars and/or selective broths. In addition to failure to recover *Lm* due to injury, the potential for induction of filamentous cells by various stresses raises the possibility of underestimating numbers present in a food (Giotis et al. 2007). Lavieri et al. (2014) have shown that a thin layer of a non-selective agar, e.g., TSA on top of MOX agar, can help the recovery of heat-stressed *Lm* 2–3 fold better than direct plating onto MOX agar but had no significant effect on the recovery of pressure stressed *Lm*. Additional research is thus needed to achieve the maximum resuscitation of injured cells without changing their numbers before enumeration.

Future

During an ongoing foodborne outbreak investigation, rapid analysis of suspected foods and/or environmental samples with the aid of epidemiological investigation is crucial to identify the contaminated food and remove it from circulation. Rapid detection of contamination events is key to minimizing consumer exposure and reducing the incidence of illness. Standard microbiological analysis of food and environmental samples is time-consuming, as cultural techniques require several days before one obtains suspected bacterial colonies and identifies them by a battery of biochemical, serological, and molecular biological tools. The entire process is hard to reduce as bacteria require time to multiply and produce single colonies for further analysis. Molecular tools can provide some amount of rapidity if we target specific markers, either protein, RNA, or DNA. This can be achieved by limited initial growth of the samples in a suitable broth, followed by utilization of a so-called rapid diagnostic kit. A detailed discussion of these methods is described under "molecular detection methods." Recently, the Centers for Disease Control and Prevention (CDC) in the USA has been utilizing so-called "culture-independent diagnostics tests" (CIDT) in parallel with the standard cultural method to assess the relevancy of such CIDT tests in foodborne outbreaks (CDC 2019). With the rapid improvement of DNA sequencing technologies, it is now possible to investigate the use of a metagenomics approach, as it provides several advantages over the standard culture-based method or molecular methods. Detection of pathogens based on a metagenomic approach is pathogen agnostic in the sense that initial sequencing information can be achieved by standard procedures applicable to the 16S rRNA gene or of entire genomic content (shotgun). The latter has the advantage of being more versatile as one can obtain additional information beyond genus and/or species, including information about the presence or absence of virulenc-related genes, e.g., *hly*A, *plc*A, *inl*A, of *Lm* and other genetic information. Besides being used as an identification tool, sequence-based approaches can also be utilized for the establishment of genetic relatedness aiding epidemiological investigation during outbreaks. The metagenomics approach, however, has plenty of challenges: the improvement and optimization of enrichment protocols, standardization of the lysis protocol for a diverse set of organisms, improvement and standardization of DNA amplification, and the need for improvement and standardization of bioinformatics tools for data extraction, analysis, and interpretation (Grutzke et al. 2019). Research and future investments in these areas, however, holds great promise to make the entire endeavor of finding "a needle in the haystack" much more sensitive and rapid than the current detection methods.

Acknowledgments

We thank our respective Office management for their continued support. The findings and conclusions in this work are those of the authors and do not necessarily represent the official position of the USFDA. Mention of products and/or methods is not an endorsement by USFDA.

References

Afari, G.K. and Y.C. Hung. 2018. Detection and Verification of the Viable but Nonculturable (VBNC) state of *Escherichia coli* O157:H7 and *Listeria monocytogenes* using flow cytometry and standard plating. J. Food Sci. 83(7): 1913–1920.

Alles, S., S. Curry, D. Almy, B. Jagadeesan, J. Rice and M. Mozola. 2012. Reveal *Listeria* 2.0 test for detection of *Listeria* spp. in foods and environmental samples. J. AOAC Intl. 95(2): 424–434.

Angelo, K.M., A.R. Conrad, A. Saupe, H. Dragoo, N. West, A. Sorenson et al. 2017. Multistate outbreak of *Listeria monocytogenes* infections linked to whole apples used in commercially produced, prepackaged caramel apples: United States, 2014–2015. Epidemiol. Infect. 145(5): 848–856.

Banerjee, K., B. Pierson, E. Carrier, L. Malsick, C. Hu, S. Daudenarde et al. 2018. Validation of workflow changes, phage concentration and reformatted detection threshold for the Sample6 DETECT/L Test: Level 3 modification. J. AOAC Intl. 101(6): 1895–1904.

Benesh, D.L., E.S. Crowley and P.M. Bird. 2013. 3M Petrifilm environmental *Listeria* plate. J. AOAC Intl. 96(2): 225–228.

Bouttefroy, A., J.P. Lemaitre and A. Rousset. 1997. Prevalence of *Listeria* sp. in droppings from urban rooks (*Corvus frugilegus*). J. Appl. Microbiol. 82(5): 641–647.

Brauge, T., L. Barre, G. Leleu, S. Andre, C. Denis, A. Hanin et al. 2020. European survey and evaluation of sampling methods recommended by the standard EN ISO 18593 for the detection of *Listeria monocytogenes* and *Pseudomonas fluorescens* on industrial surfaces. FEMS Microbiol. Lett. 367(7): fnaa057.

Bruhn, J.B., B.F. Vogel and L. Gram. 2005. Bias in the *Listeria monocytogenes* enrichment procedure: lineage 2 strains outcompete lineage 1 strains in University of Vermont selective enrichments. Appl. Env. Micro. 71(2): 961–967.

Bucur, F.I., L. Grigore-Gurgu, P. Crauwels, C.U. Riedel and A.I. Nicolau. 2018. Resistance of *Listeria monocytogenes* to stress conditions encountered in food and food processing environments. Front. Micro. 9: 2700–2700.

Burall, L.S., C. Grim, G. Gopinath, P. Laksanalamai and A.R. Datta. 2014. Whole-genome sequencing identifies an atypical *Listeria monocytogenes* strain isolated from pet foods. Genome Announc 2(6).

Burall, L.S., C.J. Grim, M.K. Mammel and A.R. Datta. 2016. Whole genome sequence analysis using JSpecies tool establishes clonal relationships between *Listeria monocytogenes* strains from epidemiologically unrelated listeriosis outbreaks. PLoS One 11(3): e0150797.

Burall, L.S., Y. Chen, D. Macarisin, R. Pouillot, E. Strain, A.J. de Jesus et al. 2017a. Enumeration and characterization of *Listeria monocytogenes* in novelty ice cream samples manufactured on a specific production line linked to a listeriosis outbreak. Food Cont. 82: 1–7.

Burall, L.S., C.J. Grim and A.R. Datta. 2017b. A clade of *Listeria monocytogenes* serotype 4b variant strains linked to recent listeriosis outbreaks associated with produce from a defined geographic region in the US. PLoS One 12(5): e0176912.

Busch, S.V. and C.W. Donnelly. 1992. Development of a repair-enrichment broth for resuscitation of heat-injured *Listeria monocytogenes* and *Listeria innocua*. Appl. Env. Micro. 58(1): 14–20.

Caballero, O., S. Alles, Q.N. Le, R.L. Gray, E. Hosking, L. Pinkava et al. 2016a. Validation of modifications to the ANSR® *Listeria* method for improved ease of use and performance. J. AOAC Intl. 99(1): 98–111.

Caballero, O., S. Alles, Q.N. Le, R.L. Gray, E. Hosking, L. Pinkava et al. 2016b. Validation of the ANSR® *Listeria monocytogenes* method for detection of *Listeria monocytogenes* in selected food and environmental samples. J. AOAC Intl. 99(1): 112–123.

Cappillino, M., R.P. Shivers, D.R. Brownell, B. Jacobson, J. King, P. Kocjan et al. 2015. Sample6 DETECT/L: an in-plant, in-shift, enrichment-free *Listeria* environmental assay. J. AOAC Intl. 98(2): 436–444.

CDC. 2015. Multistate Outbreak of Listeriosis Linked to Blue Bell Creameries Products (Final Update). http://www. cdc. gov/listeria/outbreaks/ice-cream-03-15/index.html.

CDC. 2019. Culture-independent Diagnostic Tests. from www.cdc.gov/foodsafety/challenges/cidt.html.

Chan, Y.C. and M. Wiedmann. 2008. Physiology and genetics of *Listeria monocytogenes* survival and growth at cold temperatures. Crit. Rev. Food Sci. Nut. 49(3): 237–253.

Chapin, T.K., K.K. Nightingale, R.W. Worobo, M. Wiedmann and L.K. Strawn. 2014. Geographical and meteorological factors associated with isolation of *Listeria* species in New York State produce production and natural environments. J. Food Prot. 77(11): 1919–1928.

Charlier, C., E. Perrodeau, A. Leclercq, B. Cazenave, B. Pilmis, B. Henry et al. 2017. Clinical features and prognostic factors of listeriosis: the MONALISA national prospective cohort study. Lancet Infect. Dis. 17(5): 510–519.

Chen, Y., L.S. Burall, D. Macarisin, R. Pouillot, E. Strain, A.J. De Jesus, A. Laasri et al. 2016. Prevalence and level of *Listeria monocytogenes* in ice cream linked to a listeriosis outbreak in the United States. J. Food Prot. 79(11): 1828–1832.

Claveau, D., S. Olishevskyy, M. Giuffre and G. Martinez. 2014. Detection of *Listeria* spp. using ACTERO *Listeria* enrichment media. J. AOAC Intl. 97(4): 1127–1136.

Cornu, M., M. Kalmokoff and J.P. Flandrois. 2002. Modelling the competitive growth of *Listeria monocytogenes* and *Listeria innocua* in enrichment broths. Intl. J. Food Microbiol. 73(2-3): 261–274.

Cruz, C.D., J.K. Win, J. Chantarachoti, A.N. Mutukumira and G.C. Fletcher. 2012. Comparing rapid methods for detecting *Listeria* in seafood and environmental samples using the most probable number (MPN) technique. Intl. J. Food Microbiol. 153(3): 483–487.

D'Amico, D.J. and C.W. Donnelly. 2008. Enhanced detection of *Listeria* spp. in farmstead cheese processing environments through dual primary enrichment, PCR, and molecular subtyping. J. Food Prot. 71(11): 2239–2248.

Dailey, R.C., L.J. Welch, A.D. Hitchins and R.D. Smiley. 2015. Effect of *Listeria seeligeri* or *Listeria welshimeri* on *Listeria monocytogenes* detection in and recovery from buffered Listeria enrichment broth. Food Microbiol. 46: 528–534.

de Noordhout, C.M., B. Devleesschauwer, F.J. Angulo, G. Verbeke, J. Haagsma, M. Kirk et al. 2014. The global burden of listeriosis: a systematic review and meta-analysis. Lancet Infect. Dis. 14(11): 1073–1082.

Dey, B.P. and F.B. Engley, Jr. 1983. Methodology for recovery of chemically treated *Staphylococcus aureus* with neutralizing medium. Appl. Env. Micro. 45(5): 1533–1537.

Donnelly, C.W. 2002. Detection and isolation of *Listeria monocytogenes* from food samples: implications of sub-lethal injury. J. AOAC Intl. 85(2): 495–500.

Dowe, M.J., E.D. Jackson, J.G. Mori and C.R. Bell. 1997. *Listeria monocytogenes* survival in soil and incidence in agricultural soils. J. Food Prot. 60(10): 1201–1207.

ECDC. 2018. Multi-country outbreak of *Listeria monocytogenes* serogroup IVb, multi-locus sequence type 6, infections linked to frozen corn and possibly to other frozen vegetables – first update. EFSA Supporting Publications 15(7): 1448E.

Estrada, E.M., A.M. Hamilton, G.B. Sullivan, M. Wiedmann, F.J. Critzer and L.K. Strawn. 2020. Prevalence, persistence, and diversity of *Listeria monocytogenes* and *Listeria* species in produce packinghouses in Three U.S. States. J. Food Prot: 277–286.

FAO. 2004. Risk assessment of *Listeria monocytogenes* in ready-to-eat foods.

Ferreira, V., M. Wiedmann, P. Teixeira and M.J. Stasiewicz. 2014. *Listeria monocytogenes* persistence in food-associated environments: epidemiology, strain characteristics, and implications for public health. J. Food Prot. 77(1): 150–170.

Fortes, E.D., J. David, B. Koeritzer and M. Wiedmann. 2013. Validation of the 3M molecular detection system for the detection of *Listeria* in meat, seafood, dairy, and retail environments. J. Food Prot. 76(5): 874–878.

Fox, E., T. O'Mahony, M. Clancy, R. Dempsey, M. O'Brien and K. Jordan. 2009. *Listeria monocytogenes* in the Irish dairy farm environment. J. Food Prot. 72(7): 1450–1456.

Fraser, J.A. and W.H. Sperber. 1988. Rapid detection of *Listeria* spp. in food and environmental samples by esculin hydrolysis. J. Food Prot. 51(10): 762–765.

Germini, A., A. Masola, P. Carnevali and R. Marchelli. 2009. Simultaneous detection of *Escherichia coli* O175:H7, *Salmonella* spp., and *Listeria monocytogenes* by multiplex PCR. Food Cont. 20(8): 733–738.

Giotis, E.S., I.S. Blair and D.A. McDowell. 2007. Morphological changes in *Listeria monocytogenes* subjected to sublethal alkaline stress. Intl. J. Food Microbiol. 120(3): 250–258.

Glaser, P., L. Frangeul, C. Buchrieser, A. Amend, F. Baquero et al. 2001. Comparative genomics of *Listeria* species. Science 294(5543): 849–52.

Gorski, L., S. Walker, A.S. Liang, K.M. Nguyen, J. Govoni, D. Carychao et al. 2014. Comparison of subtypes of *Listeria monocytogenes* isolates from naturally contaminated watershed samples with and without a selective secondary enrichment. PLOS ONE 9(3): e92467.

Grutzke, J., B. Malorny, J.A. Hammerl, A. Busch, S.H. Tausch, H. Tomaso et al. 2019. Fishing in the soup - pathogen detection in food safety using metabarcoding and metagenomic sequencing. Front. Microbiol. 10: 1805.

Harrand, A.S., L. Strawn, P.M. Illas-Ortiz, M. Wiedmann and D. Weller. 2020. *Listeria monocytogenes* prevalence varies more within fields than between fields or over time on conventionally farmed New York produce fields. J. Food Prot. 83(11): 1958–1966.

Health Canada. 2011. Policy on *Listeria monocytogenes* in Ready-to-Eat Foods.

Highmore, C.J., J.C. Warner, S.D. Rothwell, S.A. Wilks and C.W. Keevil. 2018. Viable-but-Nonculturable *Listeria monocytogenes* and *Salmonella enterica* Serovar thompson induced by chlorine stress remain infectious. mBio 9(2): e00540–00518.

Hilliard, A., D. Leong, A. O'Callaghan, E.P. Culligan, C.A. Morgan, N. DeLappe et al. 2018. Genomic characterization of *Listeria monocytogenes* isolates associated with clinical listeriosis and the food production environment in Ireland. Genes (Basel) 9(3): 171.

Hingston, P., J. Chen, B.K. Dhillon, C. Laing, C. Bertelli, V. Gannon et al. 2017. Genotypes associated with *Listeria monocytogenes* isolates displaying impaired or enhanced tolerances to cold, salt, acid, or desiccation stress. Front. Micro. 8: 369.

Hitchins, A.D., K. Jinneman and Y. Chen. 2016. BAM: Detection and enumeration of *Listeria monocytogenes*. Jinneman, K.W.B., M. Davidson, P. Feng, B. Ge, G. Gharst, T. Hammack, S. Himathongkham, J. Kase and P. Regan. Bacteriological Analytical Manual. U.S. FDA.

Hoffmann, S., M.B. Batz and J.G. Morris, Jr. 2012. Annual cost of illness and quality-adjusted life year losses in the United States due to 14 foodborne pathogens. J. Food Prot. 75(7): 1292–1302.

Hsu, J.L., H.M. Opitz, R.C. Bayer, L.J. Kling, W.A. Halteman, R.E. Martin et al. 2005. *Listeria monocytogenes* in an Atlantic salmon (*Salmo salar*) processing environment. J. Food Prot. 68(8): 1635–1640.

Hua, Z., A.M. Korany, S.H. El-Shinawy and M.-J. Zhu. 2019. Comparative evaluation of different sanitizers against *Listeria monocytogenes* biofilms on major food-contact surfaces. Front. Micro. 10: 2462.

ISO. 2012. Microbiology of food and animal food—Horizontal method for the detection and enumeration of *Listeria monocytogenes* and of *Listeria* species—Part 1: Detection method. ISO 11290-1.

ISO. 2017. Microbiology of food and animal feed—Horizontal method for the detection and enumeration of *Listeria monocytogenes* and of *Listeria* species Part 2 Enumeration method. ISO 11290-2:2017.

Jackson, B.R., C. Tarr, E. Strain, K.A. Jackson, A. Conrad, H. Carleton et al. 2016. Implementation of nationwide real-time whole-genome sequencing to enhance listeriosis outbreak detection and investigation. Clin. Infect. Dis. 63(3): 380–386.

Joelsson, A.C., S.P. Terkhorn, A.S. Brown, A. Puri, B.J. Pascal, Z.E. Gaudioso et al. 2017. Comparative evaluation of Veriflow® *Listeria* species to USDA culture-based method for the detection of *Listeria* spp. in food and environmental samples. J. AOAC Intl. 100(5): 1434–1444.

Johnson, R. and J. Mills. 2013. VIDAS *Listeria monocytogenes* II (LMO2). J. AOAC Intl. 96(2): 246–250.

Juck, G., V. Gonzalez, A.O. Allen, M. Sutzko, K. Seward and M.T. Muldoon. 2018. Romer Labs RapidChek® *Listeria monocytogenes* test system for the detection of *L. monocytogenes* on selected foods and environmental surfaces. J. AOAC Intl. 101(5): 1490–1507.

Kathariou, S. 2002. *Listeria monocytogenes* virulence and pathogenicity, a food safety perspective. J. Food Prot. 65(11): 1811–1829.

Kawasaki, S., N. Horikoshi, Y. Okada, K. Takeshita, T. Sameshima and S. Kawamoto. 2005. Multiplex PCR for simultaneous detection of *Salmonella* spp., *Listeria monocytogenes*, and *Escherichia coli* O157:H7 in meat samples. J. Food Prot. 68(3): 551–556.

Keys, A.L., R.C. Dailey, A.D. Hitchins and R.D. Smiley. 2013. Postenrichment population differentials using buffered Listeria enrichment broth: implications of the presence of *Listeria innocua* on *Listeria monocytogenes* in food test samples. J. Food Prot. 76(11): 1854–1862.

Kim, H. and A.K. Bhunia. 2008. SEL, a selective enrichment broth for simultaneous growth of *Salmonella enterica*, *Escherichia coli* O157:H7, and *Listeria monocytogenes*. Appl. Env. Microbiol. 74(15): 4853–4866.

Klass, N., B. Bastin, E. Crowley, J. Agin, M. Clark, J.P. Tourniaire et al. 2020. Modification of the Bio-Rad iQ-Check *Listeria* spp. kit for the detection of *Listeria* species in environmental surfaces. J. AOAC Intl. 103(1): 216–222.

Kovacevic, J., V.M. Bohaychuk, P.R. Barrios, G.E. Gensler, D.L. Rolheiser and L.M. McMullen. 2009. Evaluation of environmental sampling methods and rapid detection assays for recovery and identification of *Listeria* spp. from meat processing facilities. J. Food Prot. 72(4): 696–701.

Laksanalamai, P., S.A. Jackson, M.K. Mammel and A.R. Datta. 2012. High density microarray analysis reveals new insights into genetic footprints of *Listeria monocytogenes* strains involved in listeriosis outbreaks. PLoS One 7(3): e32896.

Lavieri, N.A., J.G. Sebranek, J.C. Cordray, J.S. Dickson, S. Jung, D.K. Manu et al. 2014. Evaluation of the thin agar layer method for the recovery of pressure-injured and heat-injured *Listeria monocytogenes*. J. Food Prot. 77(5): 828–831.

Le, Q.N., S. Alles, B. Roman, E. Tovar, E. Hosking, L. Zhang et al. 2019. Validation of the *Listeria* Right Now(TM) test for detection of *Listeria* spp. from selected environmental surfaces without enrichment. J. AOAC Intl. 102(3): 926–935.

Leclercq, A. 2004. Atypical colonial morphology and low recoveries of *Listeria monocytogenes* strains on Oxford, PALCAM, Rapid'L.mono and ALOA solid media. J. Micro. Meth. 57(2): 251–258.

Leclercq, A., V. Chenal-Francisque, H. Dieye, T. Cantinelli, R. Drali, S. Brisse et al. 2011. Characterization of the novel *Listeria monocytogenes* PCR serogrouping profile IVb-v1. Intl. J. Food Micro. 147(1): 74–77.

Lee, S., T.J. Ward, L.M. Graves, L.A. Wolf, K. Sperry, R.M. Siletzky et al. 2012. Atypical *Listeria monocytogenes* serotype 4b strains harboring a lineage II-specific gene cassette. Appl. Env. Microbiol. 78(3): 660–667.

Linnan, M.J., L. Mascola, X.D. Lou, V. Goulet, S. May, C. Salminen et al. 1988. Epidemic listeriosis associated with Mexican-style cheese. N. Eng. J. Med. 319(13): 823–828.

Locatelli, A., G. Depret, C. Jolivet, S. Henry, S. Dequiedt, P. Piveteau et al. 2013. Nation-wide study of the occurrence of *Listeria monocytogenes* in French soils using culture-based and molecular detection methods. J. Micro. Meth. 93(3): 242–250.

Loncarevic, S., W. Tham and M.L. Danielsson-Tham. 1996. Prevalence of *Listeria monocytogenes* and other *Listeria* spp. in smoked and 'gravad' fish. Acta Vet. Scan. 37(1): 13–18.

Luber, P., C. Scott, C. Dufour, J. Farber, A. Datta and E.C.D. Todd. 2011. Controlling *Listeria monocytogenes* in ready-to-eat foods: Working towards global scientific consensus and harmonization—Recommendations for improved prevention and control. Food Cont. 22(9): 1535–1549.

Marik, C.M., J. Zuchel, D.W. Schaffner and L.K. Strawn. 2019. Growth and Survival of *Listeria monocytogenes* on intact fruit and vegetable surfaces during postharvest handling: a systematic literature review. J. Food Prot. 83(1): 108–128.

Maury, M.M., V. Chenal-Francisque, H. Bracq-Dieye, L. Han, A. Leclercq, G. Vales et al. 2017. Spontaneous loss of virulence in natural populations of *Listeria monocytogenes*. Infect. Immun. 85(11): e00541-17.

Moreno Switt, A. 2020. Personal Communication.

Muldoon, M.T., A.C. Allen, V. Gonzalez, M. Sutzko and L. Klaus. 2012. SDIX RapidChek *Listeria* F.A.S.T. environmental test system for the detection of *Listeria* species on environmental surfaces. J. AOAC Intl. 95(3): 850–859.

NicAogáin, K. and C.P. O'Byrne. 2016. The role of stress and stress adaptations in determining the fate of the bacterial pathogen *Listeria monocytogenes* in the food chain. Front. Micro. 7: 1865.

Nilsson, L., Y.Y. Ng, J.N. Christiansen, B.L. Jorgensen, D. Grotinum and L. Gram. 2004. The contribution of bacteriocin to inhibition of *Listeria monocytogenes* by *Carnobacterium piscicola* strains in cold-smoked salmon systems. J. Appl. Micro. 96(1): 133–143.

Noll, M., K. Trunzer, A. Vondran, S. Vincze, R. Dieckmann, S. Al Dahouk et al. 2020. Benzalkonium Chloride induces a VBNC state in *Listeria monocytogenes*. Microorganisms 8(2): 184.

Norton, D.M., M. McCamey, K.J. Boor and M. Wiedmann. 2000. Application of the BAX for screening/genus *Listeria* polymerase chain reaction system for monitoring *Listeria* species in cold-smoked fish and in the smoked fish processing environment. J. Food Prot. 63(3): 343–346.

Norton, D.M. and C.R. Braden. 2007. Foodborne listeriosis. pp 305–356. *In:* Ryser, E.T. and E.H. Marth (ed.). *Listeria*, Listeriosis and Food Safety. CRC Press, Boca Raton, FL.

Oliveira, M., J. Usall, I. Vinas, C. Solsona and M. Abadias. 2011. Transfer of *Listeria innocua* from contaminated compost and irrigation water to lettuce leaves. Food Micro. 28(3): 590–596.

Olstein, A.D. and J.M. Feirtag. 2019. Improved positive predictive performance of *Listeria* indicator broth: a sensitive environmental screening test to identify presumptively positive swab samples. Microorganisms 7(5): 151.

Orsi, R.H., M.L. Borowsky, P. Lauer, S.K. Young, C. Nusbaum, J.E. Galagan, B.W. Birren et al. 2008. Short-term genome evolution of *Listeria monocytogenes* in a non-controlled environment. BMC Genomics 9: 539.

Orsi, R.H., H.C. den Bakker and M. Wiedmann. 2011. *Listeria monocytogenes* lineages: Genomics, evolution, ecology, and phenotypic characteristics. Intl. J. Med. Micro. 301(2): 79–96.

Pagotto, F., Y.-L. Trottier and I. Iugovaz. 2011. Enumeration of *Listeria monocytogenes* in foods. Government of Canada. MFLP-74.

Pan, Y., F. Breidt, Jr. and S. Kathariou. 2006. Resistance of *Listeria monocytogenes* biofilms to sanitizing agents in a simulated food processing environment. Appl. Env. Micro. 72(12): 7711–7717.

Parsons, C., M. Jahanafroozi and S. Kathariou. 2019. Requirement of lmo1930, a Gene in the Menaquinone Biosynthesis Operon, for Esculin hydrolysis and lithium chloride tolerance in *Listeria monocytogenes*. Microorganisms 7(11): 539.

Parsons, C., J. Niedermeyer, N. Gould, P. Brown, J. Strules, A.W. Parsons et al. 2020. *Listeria monocytogenes* at the human-wildlife interface: black bears (*Ursus americanus*) as potential vehicles for *Listeria*. Micro. Biotechnol. 13(3): 706–721.

Petrauskene, O.V., Y. Cao, P. Zoder, L.Y. Wong, P. Balachandran, M.R. Furtado et al. 2012. Evaluation of applied biosystems MicroSEQ real-time PCR system for detection of *Listeria* spp. in food and environmental samples. J. AOAC. Intl. 95(4): 1074–1083.

Pirone-Davies, C., Y. Chen, A. Pightling, G. Ryan, Y. Wang, K. Yao et al. 2018. Genes significantly associated with lineage II food isolates of *Listeria monocytogenes*. BMC Genomics 19(1): 708.

Pouillot, R., K.C. Klontz, Y. Chen, L.S. Burall, D. Macarisin, M. Doyle et al. 2016. Infectious dose of *Listeria monocytogenes* in outbreak linked to ice cream, United States, 2015. Emerg. Infect. Dis. 22(12): 2113–2119.

Radoshevich, L. and P. Cossart. 2018. *Listeria monocytogenes*: towards a complete picture of its physiology and pathogenesis. Nat. Rev. Microbiol. 16(1): 32–46.

Ragon, M., T. Wirth, F. Hollandt, R. Lavenir, M. Lecuit, A. Le Monnier et al. 2008. A new perspective on *Listeria monocytogenes* Evolution. PLOS Pathogens 4(9): e1000146.

Rakic Martinez, M., M. Wiedmann, M. Ferguson and A.R. Datta. 2017. Assessment of *Listeria monocytogenes* virulence in the *Galleria mellonella* insect larvae model. PLoS One 12(9): e0184557.

Rios-Castillo, A.G., C. Ripolles-Avila and J.J. Rodriguez-Jerez. 2020. Detection of *Salmonella typhimurium* and *Listeria monocytogenes* biofilm cells exposed to different drying and pre-enrichment times using conventional and rapid methods. Intl. J. Food Micro. 324: 108611.

Roman, B., M. Mozola, R. Donofrio, B. Bastin, N. Klass, P.M. Bird et al. 2019. Matrix extension study: Listeria right now test for detection of *Listeria* spp. from selected environmental surfaces without enrichment. J. AOAC Intl. 102(5): 1589–1594.

Ryser, E.T., S.M. Arimi, M.M. Bunduki and C.W. Donnelly. 1996. Recovery of different *Listeria* ribotypes from naturally contaminated, raw refrigerated meat and poultry products with two primary enrichment media. Appl. Env. Micro. 62(5): 1781–1787.

Scallan, E., R.M. Hoekstra, F.J. Angulo, R.V. Tauxe, M.A. Widdowson, S.L. Roy et al. 2011. Foodborne illness acquired in the United States-major pathogens. Emerg. Infect. Dis. 17(1): 7–15.

Schirmer, B.C., S. Langsrud, T. Moretro, T. Hagtvedt and E. Heir. 2012. Performance of two commercial rapid methods for sampling and detection of *Listeria* in small-scale cheese producing and salmon-processing environments. J. Micro. Meth. 91(2): 295–300.

Seeliger, H.P. and K. Hohne. 1979. Serotyping of *Listeria monocytogenes* and related species. Method. Microbiol. 13: 13–49.

Sheth, I., F. Li, M. Hur, A. Laasri, A.J. De Jesus, H.J. Kwon et al. 2018. Comparison of three enrichment schemes for the detection of low levels of desiccation-stressed *Listeria* spp. from select environmental surfaces. Food Cont. 84: 493–498.

Silk, T.M., T.M. Roth and C.W. Donnelly. 2002. Comparison of growth kinetics for healthy and heat-injured *Listeria monocytogenes* in eight enrichment broths. J. Food Prot. 65(8): 1333–1337.

Skovgaard, N. and C.A. Morgen. 1988. Detection of *Listeria* spp. in faeces from animals, in feeds, and in raw foods of animal origin. Intl. J. Food Micro. 6(3): 229–242.

Stasiewicz, M.J., H.F. Oliver, M. Wiedmann and H.C. den Bakker. 2015. Whole-genome sequencing allows for improved identification of persistent *Listeria monocytogenes* in food-associated environments. Appl. Env. Micro. 81(17): 6024–6037.

Stessl, B., W. Luf, M. Wagner and D. Schoder. 2009. Performance testing of six chromogenic ALOA-type media for the detection of *Listeria monocytogenes*. J. Appl. Micro. 106(2): 651–659.

Sullivan, G. and M. Wiedmann. 2020. Detection and prevalence of *Listeria* in US produce packinghouses and fresh-cut facilities. J. Food Prot. 83(10): 1656–1666.

Suo, B. and Y. Wang. 2013. Evaluation of a multiplex selective enrichment broth SEL for simultaneous detection of injured *Salmonella*, *Escherichia coli* O157:H7 and *Listeria monocytogenes*. Braz. J. Micro. 44(3): 737–742.

Takeuchi, K., N. Mytle, S. Lambert, M. Coleman, M.P. Doyle et al. 2006. Comparison of *Listeria monocytogenes* virulence in a mouse model. J. Food Prot. 69(4): 842–846.

Tompkin, R.B. 2002. Control of *Listeria monocytogenes* in the food-processing environment. J. Food Prot. 65(4): 709–725.

Tonner, E., S. Kelly, S. Illingworth, N. Perera, B. Bastin, P. Bird et al. 2019. Evaluation of the Solus One *Listeria* method for the detection of *Listeria* species on environmental surfaces. J. AOAC Intl. 102(2): 570–579.

USDA. 2016. Isolation and Identification of *Listeria monocytogenes* from Red Meat, Poultry, Ready-to-Eat Siluriformes (Fish) and Egg Products, and Environmental Samples.

USDA and USFDA. 2003. Quantitative Assessment of the Relative Risk to Public Health from Foodborne *Listeria monocytogenes* Among Selected Categories of Ready-to-Eat Foods.

USFDA. 2003. Quantitative assessment of relative risk to public health from foodborne *Listeria monocytogenes* among selected categories of ready-to-eat foods. http://www.fda.gov/Food/FoodScienceResearch/RiskSafetyAssessment/ucm183966.htm.

USFDA. 2008. Compliance Policy Guide. Sec 555.320 *Listeria monocytogenes*.

U.S. Government. 2011. Food Safety Modernization Act. https://www.fda.gov/food/guidance-regulation-food-and-dietary-supplements/food-safety-modernization-act-fsma.

Vail, K.M., L.M. McMullen and T.H. Jones. 2012. Growth and filamentation of cold-adapted, log-phase *Listeria monocytogenes* exposed to salt, acid, or alkali stress at 3 degrees C. J. Food Prot. 75(12): 2142–2150.

Vazquez-Boland, J.A., G. Dominguez-Bernal, B. Gonzalez-Zorn, J. Kreft and W. Goebel. 2001a. Pathogenicity islands and virulence evolution in Listeria. Micro. Infect. 3(7): 571–584.

Vazquez-Boland, J.A., M. Kuhn, P. Berche, T. Chakraborty, G. Dominguez-Bernal, W. Goebel et al. 2001b. *Listeria* pathogenesis and molecular virulence determinants. Clin. Micro. Rev. 14(3): 584–640.

Vivant, A.-L., D. Garmyn and P. Piveteau. 2013. *Listeria monocytogenes*, a down-to-earth pathogen. Front. Cell. Infect. Micro. 3: 87.

Walsh, K.A., S.D. Bennett, M. Mahovic and L.H. Gould. 2014. Outbreaks associated with cantaloupe, watermelon, and honeydew in the United States, 1973–2011. Food. Path. Dis. 11(12): 945–952.

Warburton, D.W., J.M. Farber, A. Armstrong, R. Caldeira, N.P. Tiwari, T. Babiuk et al. 1991. A Canadian comparative study of modified versions of the "FDA" and "USDA" methods for the detection of *Listeria monocytogenes*. J. Food Prot. 54(9): 669–676.

Watson, R.J. and B. Blackwell. 2000. Purification and characterization of a common soil component which inhibits the polymerase chain reaction. Can. J. Micro. 46(7): 633–642.

Weissfeld, A.S., N. Landes, H. Livesay and E. Trevino. 2017. *Listeria monocytogenes* contamination of ice cream: a rare event that occurred twice in the last two years. Clin. Micro. News. 39(3): 19–22.

Wendorf, M., E. Feldpausch, L. Pinkava, K. Luplow, E. Hosking, P. Norton et al. 2013. Validation of the ANSR *Listeria* method for detection of *Listeria* spp. in environmental samples. J. AOAC Intl. 96(6): 1414–1424.

Williams, D., E.A. Irvin, R.A. Chmielewski, J.F. Frank and M.A. Smith. 2007. Dose-response of *Listeria monocytogenes* after oral exposure in pregnant guinea pigs. J. Food Prot. 70(5): 1122–1128.

Yin, Y., H. Yao, S. Doijad, S. Kong, Y. Shen, X. Cai et al. 2019. A hybrid sub-lineage of *Listeria monocytogenes* comprising hypervirulent isolates. Nature Comm. 10(1): 4283.

Zhang, C., M. Zhang, J. Ju, J. Nietfeldt, J. Wise, P.M. Terry et al. 2003. Genome diversification in phylogenetic lineages I and II of *Listeria monocytogenes*: identification of segments unique to lineage II populations. J. Bact. 185(18): 5573–5584.

Zhu, L., D. Stewart, K. Reineke, S. Ravishankar, S. Palumbo, M. Cirigliano et al. 2012. Comparison of swab transport media for recovery of *Listeria monocytogenes* from environmental samples. J. Food Prot. 75(3): 580–584.

Zitz, U., M. Zunabovic, K.J. Domig, P.T. Wilrich and W. Kneifel. 2011. Reduced detectability of *Listeria monocytogenes* in the presence of *Listeria innocua*. J. Food Prot. 74(8): 1282–1287.

Chapter 4

Diagnosis of Human-Pathogenic *Staphylococcus* in Surveillance and Outbreak Detection

Nabanita Mukherjee,[1] *Goutam Banerjee,*[2] *Saumya Agarwal*[2] and *Pratik Banerjee*[2,*]

Introduction

The name *Staphylococcus* originated from the Greek word *staphyle* (meaning a bunch of grapes) and *kokkos* (meaning berry). *Staphylococcus* was first described by Sir Alexander Ogston in 1880. The Scottish surgeon isolated the microorganism from the pus of a patient, who was suffering from a surgical abscess in the knee joint area. In 1884, Friedrich Julius Rosenbach, a German physician, described the two species of *Staphylococcus* based on the colors of the colony. The bacterial colony that appeared yellow was named *Staphylococcus aureus* (originated from the Latin word "aurum" meaning gold). The bacterial colonies with white pigments were named *Staphylococcus albus* (originated from the Latin word that means white). Later, *Staphylococcus albus* was renamed *S. epidermidis* because it is predominantly found on the surface of human skin. Many *Staphylococcus* species reside on the skin surface and in the nasopharynx of humans and usually do not cause any health problems. However, a few strains of *Staphylococcus* can cause mild to severe infections in humans under certain conditions. *S. aureus*, *S. epidermidis*, and *S. saprophyticus* are the most clinically important species known to be responsible for human infections.

---

[1] Department of Infectious Diseases, Center of Excellence for Influenza Research and Surveillance (CEIRS), St. Jude Children's Research Hospital, Memphis, TN 38105, USA.

[2] Department of Food Science and Human Nutrition, University of Illinois at Urbana-Champaign, Urbana, IL 61801, USA.

* Corresponding author: pratik@illinois.edu

Staphylococcus has been recognized as the most frequently reported agent of nosocomial infection. *S. aureus* is known to be associated with skin infections, including skin abscess, boil, and carbuncle (Mukherjee et al. 2014). In addition, *S. aureus* is known to be responsible for food poisoning with severe vomiting and diarrhea. In 1914, Barber described the association of food poisoning and *Staphylococcus* in a case where the food poisoning was due to the ingestion of unrefrigerated milk from a cow suffering from mastitis caused by *Staphylococcus*. Later, it was discovered that enterotoxins produced by *Staphylococcus* are responsible for *Staphylococcus*-associated food poisoning. In addition to this, *S. aureus* can cause bacteremia, toxic shock syndrome, meningitis, deep abscesses, pneumonia, endocarditis, and surgical wound infection in hospitalized patients. *S. aureus* is responsible for causing many diseases ranging in a milder degree of severity to life-threatening infections. It is also infamous for causing disease both in nosocomial and community settings. *S. aureus* infection was fatal until the introduction of penicillin as an antibiotic. The mortality rate of *S. aureus* infection was approximately 80% before the introduction of penicillin. The first cases of penicillin resistance *S. aureus* were reported in a hospital setting (Barber and Rozwadowska-dowzenko 1948; Kirby 1944) and later it was spread into the community. During this period, the other β-lactam antibiotics, which are resistant to penicillinases such as methicillin, oxacillin, cloxacillin, dicloxacillin, and flucloxacillin were widely used to treat *S. aureus* infections.

Both *S. epidermidis* and *S. saprophyticus* are part of the normal flora of the human body and are known as opportunistic pathogens. *S. epidermidis* is known to colonize on human skin surfaces, predominantly in the axillae, head, and nares. *S. epidermidis* is known to be associated with infections related to indwelling medical devices, including peripheral or central intravenous catheters, central nervous system shunts, cardiovascular devices, and artificial heart valves. In addition to this, *S. epidermidis* has been reported to be associated with bloodstream infections in patients in the intensive care unit. *S. epidermidis* is responsible for a high rate of mortality due to prosthetic valve endocarditis infections with severe intracardiac abscesses. *Staphylococcus saprophyticus*, which colonizes the perineum, rectum, urethra, cervix, and gastrointestinal tract, is the second leading cause of bacterial urinary tract infections after *Escherichia coli*. *S. saprophyticus* has been associated with acute cystitis in young women (Raz et al. 2005). Many species of *Staphylococcus* are commensals of many animals and can be transmitted to the human body through the ingestion of contaminated food or occupational exposure.

Genus *Staphylococcus*: Physiological and Morphological Characteristics

Classification

Staphylococcus has been described as a member of the family Micrococcaceae in Bergy's Manual of Determinative Bacteriology. From a taxonomical point of view, *Staphylococcus* genus has been observed to form a coherent group upon DNA-ribosomal rRNA hybridization and comparative oligonucleotide analysis of

16S rRNA. *Staphylococcus* has low genomic guanine and cytosine (G+C) content of DNA (approximately 33 mol%) that helps to differentiate it from its closely related family members. There are more than 30 *Staphylococcus* species classified based on several biochemical analyses and by DNA-DNA hybridization methods. *S. aureus*, *S. epidermidis,* and *S. saprophyticus* are the human normal flora and can cause infections under certain conditions. Other clinically relevant staphylococcal species include *S. intermedius, S. chromogens, S. cohnii, S. caprae, S. delphini, S. felis, S. gallinarum, S. hemolyticus, S. hyicus, S. lugdunensis, S. lentus, S. sciuri, S. simulans, S. warneri,* and *S. xylosus.*

Habitat

Most of the *Staphylococcus* species are part of the normal flora of humans and animals. In humans, it normally resides on the skin surface and in the nasopharynx area. For example, *S. aureus* resides in the nasal passage and axillae. *S epidermidis* colonizes in axillae, head, nares, and human skin surface. *S. saprophyticus* resides in the perineum, rectum, urethra, cervix, and gastrointestinal tract. Many *Staphylococcus* species are commensals of other animals.

Morphology

Staphylococci are Gram-positive and spherical-shaped (approximately 0.5–1.0 μm in diameter) bacteria that appear as grape-like clusters under a light microscope. This bacterium appears spherical with a smooth surface under the electron microscope. *Staphylococcus* appears as bunches as they can divide into two planes and this characteristic helps to distinguish micrococci and staphylococci from streptococci. *Staphylococcus* are non-spore-forming and non-motile bacteria.

Biochemical Characteristics

Staphylococcus are classified into two groups, coagulase-positive (*S. aureus* and *S. intermedius*) and coagulase-negative (*S. epidermidis*), based on their ability to react to the blood clotting (the coagulase reaction). Most of the coagulase-positive *Staphylococcus* species are known as pathogenic to humans (*S. aureus*), and animals (*S. intermedius* and *S. hyicus*). There are more than 30 species known to belong to coagulase-negative staphylococci and among them, many are associated with human infections, including *S. epidermidis, S. saprophyticus,* and *S. lugdunesis.*

Staphylococci are catalase-positive and oxidase-negative. The catalase test helps to differentiate *Staphylococcus* (catalase-positive) from its closely related genus *Streptococcus* (catalase-negative). Bacteria belonging to the *Staphylococcus* genus exhibit tolerance characteristics against high salt concentration and heat. *Staphylococcus* are known as facultative anaerobes since they can grow by aerobic respiration or by fermentation at temperature ranges between 18°C and 40°C. Most of the *Staphylococcus* species require nitrogen, vitamin B, thiamine, nicotinamide, and certain essential amino acids, including arginine, valine, etc., for their growth and survival. *S. aureus* produces golden-yellow colonies and hemolysis in Tryptic Soy Agar and Blood Agar plates, respectively. The golden-yellow pigmented colonies represent the presence of Staphyloxanthin, a virulence factor that helps *Staphylococcus* to evade the host immune system.

Sources of Pathogenic *Staphylococcus*

Foodborne Source

Staphylococcus aureus (*S. aureus*) is carried by approximately 30–50% of the human population, and it is the causative pathogen for staphylococcal food poisoning (SFP). It is often associated with symptoms like nausea, vomiting, cramps, and diarrhea within 6 hours of the consumption of food contaminated by the strains of *S. aureus*. The quantity of Staphylococcal enterotoxin (SE) can be as small as 20–100 ng to cause SFP in humans. The severity and onset of the symptoms vary depending on the individual and the enterotoxin amount (Hennekinne et al. 2012). The production of SEs is promoted by the growth of *S. aureus* strains in foods with high protein content like meat and dairy products at a temperature of around 10–46°C with other suitable environmental conditions like pH (5–9.6) and water activity (0.8–0.9) (Schelin et al. 2011). SEs sustain after food processing operations like drying and freezing because they are heat stable and tolerant to protein denaturation. Also, the pathogenic activity tends to get retained in the digestive treat after consuming the contaminated food (Fisher et al. 2018).

In the last few years, there has been an increase in incidences of Methicillin-resistant *S. aureus* (HA-MRSA, CA-MRSA, and LA-MRSA) contaminating human foods causing SFP. The enterotoxin activity of both Methicillin susceptible *S. aureus* (MSSA), and MRSA is the same and under the conditions discussed above, they are foodborne pathogens. Outbreaks due to MRSA are assumed to be less severe because of their low presence in food (Sergelidis and Angelidis 2017). An SFP outbreak on April 13, 2013, was due to enterotoxins present in the ice cream at a christening party in Germany. The ice cream was freshly produced at the hotel venue. Out of 31 guests, 13 developed SFP symptoms within 3–4 hours post-lunch. Seven people were provided with treatment, and they recovered within a day. Twelve food samples were collected from the leftovers which included five ice cream samples of different flavors and seven other dishes. The strain isolated from ice cream samples and human cases produced SEA and also had the same *spa*-type (t127). After consideration of the phenotypic and genotypic typing results, it was clear that ice cream was the SFP vehicle. The primary source can either be the ice cream production equipment or any contaminated ingredient since the hotel employees did not carry the strain responsible for the outbreak (Fetsch et al. 2014). Egg products are an excellent source of protein both nutritionally and economically. They are retailed as shell eggs, and other kinds of liquid, frozen, or dried products (Munoz et al. 2015). Eggs have various other functional qualities like emulsification, gelling, and foaming due to which they find a lot of usage in food manufacturing industries. However, there is a high possibility of contamination of shell eggs and other egg products as they have suitable attributes to harbor a host of pathogenic bacteria and thus they are often associated with several food poisoning outbreaks (Moyle et al. 2016). Therefore, Sánchez et al. (2019) simulated the contamination of egg and structured egg products by the genus *Staphylococcus*. In the current study, *S. warneri* is regarded as the representative of the *Staphylococcus* genus, and its survival and development are analyzed on model egg foods (egg white, yolk, whole egg, and potato omelet). Multiple parameters like cooking and storage temperature and availability of oxygen were examined to find the growth rate (μ) of *S. warneri* in different conditions. Besides, rheological studies were done on solid egg food to identify any changes in the food structure due to bacterial growth. The study concluded that to lower the chance of SFP, the egg products should be cooked for a couple of minutes and high temperature should be provided (> 100°C) followed by storage at refrigerated temperature.

Chicken meat and giblets require extensive processing operations and thus more time in temperatures above refrigeration (greater than 4°C) can promote SE production. Also, broiler chicken products are moist and rich in proteins and glycogen and are known to favor enterotoxin production. A study was conducted by Abolghait et al. (2020) which confirmed the prevalence of MRSA (*mecA* positive and *mecC* negative) and three-fourth of MRSA isolates initiated the production of SEB within 24 hours when kept at a temperature higher than 8°C.

The SFP outbreak occurred at a sports event in Luxembourg in which 31 people were hospitalized. The suggested vehicle of infection was pasta salad with pesto, but this food had no leftovers and thus cannot be sampled. However, the ingredients of this dish (basil leaves, cheese, and pine nuts) were all negative for any *S. aureus* contamination. Also, strains unrelated to the outbreak strain were isolated from other food samplers and staff. Thus, disruption of the cold chain after food preparation can be the cause of this outbreak.

A study done by Pu et al. (2009) investigated the presence of *S. aureus* and MRSA in 120 samples of retail meat collected from 30 grocery stores across Louisiana, USA. Out of 90 pork samples, 45.6% confirmed the presence of *S. aureus* strains, and 5.6% were contaminated with the strains of MRSA. Whereas in the case of 30 beef samples, the prevalence was found to be 20% and 3.3% for *S. aureus* and MRSA, respectively. Thus, six meat samples harbored the MRSA strains, which were mainly of two types, USA100 and USA300.

Castro et al. (2020) recently investigated the presence of *S. aureus* in cheese production farms situated in rural areas. The food samples consisted of raw milk, starter cultures, and Minas artisanal cheese. The samples showed a high prevalence of *strains* producing SEA and tsst-1. Toxigenic isolates collected from samples of raw milk and workers' hand indicated that strict guidelines ought to be followed on dairy farms. Another SFP outbreak happened at a boarding school in Switzerland after the consumption of raw milk cheese. All 14 persons, including 10 children and four staff members, demonstrated SFP symptoms within 7 hours of consumption. The sampled cheese had high levels of SED (> 200 ng/g of cheese) and low levels of SEA (> 6 ng/g of cheese). Similarly, another extensive outbreak happened in Japan affecting around 13,000 consumers of low-fat milk, which was manufactured from the incriminated powdered skim milk. The thermal operation destroyed the staphylococcal strains but SEs sufficient enough to cause the SFP retained in the milk (Asao et al. 2003).

RTE foods of all categories were sampled with a focus on investigating the prevalence of MRSA in 1,128 samples. They were categorized into animal-based products and fresh-cut fruits and vegetables. Samples also consisted of ice desserts, rice products, and low-water activity (LWA) foods. The overall staphylococcal contamination was 62.5% and the greatest in fresh-cut fruits/vegetables whereas the lowest (10.8%) was in the case of LWA foods (Wang et al. 2019).

Animal Origin

Staphylococcus aureus is a contagious and common udder pathogen that is often associated with the clinical mastitis of dairy cows. A study conducted by Capurro et al. (2010), indicated that the *S. aureus* PFGE pulsotype found in raw milk samples were mostly the same ones that got isolated from the extra-mammary sites of lactating cows spread during milking. Thus, the study aimed to find the sources of infection in dairy herds. It highlighted that hock skin is a major reservoir of *S. aureus* besides the other extra-mammary sites. The common reservoirs of *Staphylococcus* in dairy environments are teat skin, milking liners, milkers, and other body sites of lactating cows or other fomites (Haveri et al. 2008; Piccinini et al. 2009).

There has been a wide usage of antimicrobial drugs in animal husbandry resulting in increased incidences of isolation of drug-resistant strains from food animals (Kwon et al. 2006; Kwon et al. 2005). Samples were collected from the noses and cloacae of chickens in 2006 and were compared to the lyophilized samples of diseased chickens of 1970 which indicated that the overall antimicrobial resistance of the recent isolates increased. This was the first study in which LA-MRSA was isolated from healthy poultry (Nemati et al. 2008). Another investigation in Korea, comprising 1,913 samples from major food animals like cattle, pigs, and chickens resulted in the isolation of *S. aureus* from around 421 (22%) samples. Out of these, *mecA*-positive MRSA was isolated from 15 samples (12 dairy cows and three poultry) (Lee 2003).

In 2008, Khanna et al. studied MRSA colonization in pig farms and found that out of 20 farms, 9 were contaminated with MRSA (Khanna et al. 2008). The results of the study reflected that MRSA is quite common in pigs and poses a high risk of transmission between pigs and pig farmers. The prevalence of MRSA in pigs and pig farmers was 24.9% and 20%, respectively. Studies have shown the same kind of high MRSA prevalence in Dutch slaughterhouses and other than the pig farmers they can be frequently found in persons who are in close contact with pigs like pig transporters, personnel working in slaughterhouses, and veterinarians (De Neeling et al. 2007). Few other studies have reported incidences of MRSA colonization in which they were responsible for causing infection in humans (Van Rijen et al. 2007; Vandenbroucke-Grauls and Beaujean 2006; Voss et al. 2005).

Environment

The outdoor, as well as indoor environmental surfaces, play a significant role in *S. aureus* and MRSA transmission from infected to healthy individuals. The staphylococcal species can survive on environmental surfaces (Beard-Pegler et al. 1988) and therefore there is an increased possibility of their persistence in household settings. Eells et al. (2014) surveyed the household fomites collected from the home environment of infected people. They found that even after three months of skin infection the *S. aureus* isolates persisted in the abiotic environment and continued to be an important reservoir that had the potential to cause the infection in the future. The factors affecting their survival are the composition of dust, temperature, humidity, the surface of the material, and the staphylococcal strain (Dietze et al. 2001; Wagenvoort et al. 2000). Besides direct contact, *S. aureus* transmission can also take place through an indirect route via aerosols, dust, and fomites. Some previous studies have isolated MRSA from the following indoor sites, such as doorknobs and computer desks (Masterton et al. 1995), sweep plates including bedding and carpets (Allen et al. 1997), kitchen sponges, toys (Scott et al. 2008), toilet areas (Medrano-Félix et al. 2011), sofa (Uhlemann et al. 2011), top of the shelf (Davis et al. 2012), etc. MRSA and other drug-resistant strains have also been isolated from samples collected from outdoor sites, like sand and water samples from recreational beaches and university and community environmental surfaces (Roberts et al. 2013).

Dairy cows suffering from intra-mammary infection (IMI) due to *S. aureus* often shed the infecting bacteria in the milk and can therefore contaminate the bulk tank milk (BTM). The biofilm formation on the various milking equipment can also be a potential source of BTM contamination (Latorre et al. 2020). A study was done by collecting samples from BTM and adherences from the surface of milking equipment to assess the virulence profiles. *S. aureus* isolates were prevalent in the adherences from rubber liners, milk collectors, collector valves, short milk tubes, long milk hoses, bulk tank outlets, agitator blades, milk cans, jars, jar covers, milk hoses, and milk line sections (Pacha et al. 2020).

Waterborne transmission is also often linked with the spread of the pathogen in the human species. MRSA and other resistant genes have been isolated from various wastewater sites and wastewater treatment plants (Goldstein et al. 2012; Wan and Chou 2014). Studies have demonstrated that MRSA can survive up to 14 days after inoculation in water and surface water is a major reservoir for MRSA (Tolba et al. 2008). In a similar study, MRSA isolates mostly LA-MRSA was isolated from water in the Puzih River basin in Taiwan, indicating the tributary's proximity to the livestock farms. Waste and wastewater discharge have suggested suggest the presence of MRSA isolates in water bodies which pose a huge threat to human health and thus drainage from industries and farms should be consistently kept under monitoring (Tsai et al. 2020).

Food Handlers

SFP outbreaks, which are a result of food contamination through food handlers like the hotel staff and caterers, are caused when infected handlers have their skin uncovered in case of skin infection or they cough or sneeze over food that will not be exposed to high temperature or cooked further. The risk of growth of *S. aureus* and enterotoxin production increases when the food is left in certain temperature conditions that are conducive to the proliferation of the bacteria.

In 2012, a buffet served at a sporting event caused SFP poisoning to 22 individuals, out of which six were hospitalized. The limitation to study this outbreak was that no food samples were available for analysis. The findings indicated that chicken stir-fry and/or fried rice could be the contaminated food vehicle. Although the catering company reported no evidence of time/temperature abuse and no infected staff members, there were many limitations associated with the investigation. The known epidemiology of this SFP outbreak suggested that food was contaminated by the handler and the food might have undergone further temperature abuse while catering (Pillsbury et al. 2013).

In 2016, Castro et al. assessed the prevalence of *S. aureus* in the anterior nares and hands of 160 food handlers working in the food industry. Out of 162 subjects, 40 were found to be *S. aureus* carriers (Castro et al. 2016). Furthermore, 18% of the handlers carried the bacteria in their hands, 32% in the nose, and 10% in both hands and nose. Out of all 50 isolates, none were MRSA but 82% of them were resistant to at minimum one antibiotic. The isolation of MRSA from food handlers is generally rare but prevalent (De Boer et al. 2009; Udo et al. 2009).

Companion Animals

Methicillin-resistant *Staphylococcus aureus*, usually known as a human pathogen is also getting known as a pathogen for companion animals. It is responsible for some serious health conditions in animals like tissue infection, wound infection, joint problems, or even death in certain cases (Van Duijkeren et al. 2003). It has been reported that MRSA infection in companion animals can pose a health risk to humans around them and vice versa (Van Duijkeren et al. 2004). Rankin et al. (2005) have reported the detection of PVL-positive, MRSA isolates not only from dogs and rabbits but also cats and African gray parrots. This toxin is responsible for severe illnesses requiring clinical treatment.

Samples were collected from the environment of horse-riding centers to identify the dominant species and the drug-resistant profile. Air, manure, and nostrils of horses showed the presence of 408 strains of Staphylococcus. Most of the collected strains were MDR type and it should be noted that the horses tested had never been hospitalized or treated with any kind of antibiotic. Thus, it can be inferred that horses are a natural reservoir of drug-resistant staphylococci. *S. vitulinus* was the dominant species in this study (Wolny-Koładka 2018).

Mink farms are another important reservoir of LA-MRSA. It was isolated from air, gloves, cages, paws, and pharynx of minks, carcasses, etc., and four out of five farms showed persistence of LA-MRSA. The study hypothesis suggests the route of colonization of mink by ingestion of by-products from contaminated pigs. They also pose a substantial health hazard to farmers handling them as they are prone to bites and scratches from colonized minks (Fertner et al. 2019).

Virulence Factors, Drug Resistance, and Biofilm Formation

Virulence Factors

S. aureus is responsible for causing different diseases by producing a wide range of virulence factors, surface components (capsule, teichoic acid, protein A, etc.), enzymes (coagulase, esterases, Hyaluronidase, lipases, deoxyribonuclease, β-lactamase, staphylokinase, lipase, etc.), and several toxins (cytolytic toxins such as α, β, γ, δ-hemolysins; leukocidins such as Panton-Valentine leukocidin or PVL; enterotoxins such as staphylococcal enterotoxins or SEs; exfoliative toxins such as ETA and ETB; and the toxin associated with toxic shock syndrome-1 or TSST-1). These virulence factors help *S. aureus* to adhere to the host cells, evade the host immune system and colonize the host cells.

Hemolysins

Staphylococcus secretes several cytolytic exotoxins that are responsible to damage the integrity of the cell by destroying the plasma membrane. Hemolysins play an important role in destroying the host immune cell, which is very helpful for the bacteria to spread out within the host (Kaneko and Kamio 2004). Table 1 shows the different types of hemolysins produced by several *Staphylococcus* spp. and the disease caused by them.

Table 1. Staphylococcal Hemolysins.

Staphylococcus Species	Types of Hemolysins	Disease	Reference
S. aureus	α, β, γ, and δ	Sepsis	(Watkins et al. 2012)
S. auricularis	α	Sepsis	(Becker et al. 2014)
S. carnosus	Lack of hemolysins	-	(Löfblom et al. 2017)
S. epidermidis	α and δ	Catheter-associated blood stream infections	(Pinheiro et al. 2015)
S. haemolyticus	α and δ	Responsible for blood infections in newborns	(Becker et al. 2014)
S. hyicus	δ	Exudative epidermitis in pigs	(Fudaba et al. 2005)
S. lugdunensis	δ	Commonly infect Skin and soft tissue	(Böcher et al. 2009)
S. saprophyticus	α	Acute cystitis in young women	(Raz et al. 2005)
S. sciuri	β, δ	Responsible for endocarditis, septic shock, peritonitis, pelvic inflammatory disease, and wound infections	(Stepanović et al. 2001)
S. simulans	δ	Urinary tract and wound infections	(Hébert and Hancock 1985)
S. warneri	δ	Septic arthritis	(Becker et al. 2014)

Leukotoxins

Leukotoxins are pore-forming toxins capable of channeling through the plasma membrane of host cells. These bi-component toxins can cause cell lysis by osmotic dysregulation. *Staphylococcus aureus* strains associated with infection in humans are known to produce four leukotoxins each having a specialized function, but they have sequence similarity. Each of these two subunits is termed slow (S) or fast (F) for the corresponding leukotoxin. It is based on the component which binds first (F) or which binds second (S). The four bi-component leukotoxins are Panton-Valentine leukocidin (subunit S: LukS-PV, F: LukF-PV), LukAB/LukGH (subunit S: LukA/LukH, F: LukB/LukG), LukED (subunit S: LukE, F: LukD), and γ-Hemolysin (subunit S: HlgA, HlgC, F: HlgB). After localization on the host cell surface, subunits combine to form an octameric pore (β-barrel) with four F and four S components arranged in an alternative manner (Aman et al. 2010). The leukotoxins can attack and lyse the leukocytic cells, but there are differences observed in the susceptibility of different host cells. All four leukotoxins can lyse the human neutrophils but not red blood cells. γ-Hemolysin and LukED can only affect the red blood cells that too on a lesser extent. The almost similar functions explain the near structural similarity of the four leukotoxins. The susceptibility to leukotoxins also varies among different animal species. For instance, neutrophils in mice are resistant to PVL and LukAB/GH, unlike human neutrophils. The susceptibility also varies from cell to cell in the same animal; for example, LukAB/GH can lyse the monkey neutrophils but not PVL leukotoxin.

γ-Hemolysin is the only leukotoxin capable of lysing red blood cells explaining the upregulation of *hlg* genes in human blood. The *ex vivo* findings in mouse models indicate that it aids in the survival of *S. aureus* in blood. However, it might be having different modes of infection because almost 99% of *S. aureus* strains have *hlg* gene, even in cases other than bloodstream infection. LukED has demonstrated its toxicity against mouse phagocytes and rabbit red blood cells besides human neutrophils. It promoted lethality in mouse models having *S. aureus* infection by lysing the phagocytes. This leukotoxin is found to be prevalent in 87% of the clinical strains including the resistant strains. LukAB/GH works synergistically with PVL, thereby further increasing its toxicity. It can kill most human leukocytes like neutrophils, monocytes, macrophages, and dendritic cells. It is not just secreted like other toxins but is also one of the only leukotoxins that are shown to get attached to the surface of *S. aureus*. It is primarily toxic to neutrophils, and it might be possible that it helps in *S. aureus* survival not just by lysis but also by helping the bacteria to escape from the phagosome. Unlike γ-Hemolysin it shows no hemolytic activity. The virulence of this toxin was confirmed using a systemic model, which resulted in a higher bacterial load in the infected kidney in presence of LukAB/GH. Since it is identified recently compared to other leukotoxins, not much analysis is done about its prevalence in the *S. aureus* clinical strains but the limited number of strains isolated by the researchers were all shown to be positive with this leukotoxin (DuMont et al. 2011).

PVL is associated more with community-acquired MRSA strains than the *S. aureus* clinical strains. Although it is linked to virulent strains which cause serious infections like pneumonia and skin/tissue, its virulent factors are still not completely proven due to divergent conclusions on animal models. Even after decades of research done on PVL production and its effect on the pathogenicity of *S. aureus* strains, it remains uncertain (Sharma-Kuinkel et al. 2012). PVL is known to activate innate immunity and turn up the immune defenses of the host cell. In some studies, it outweighed its lytic action and instead helped the host cells to fight the infection by activation in immunity (Perret et al. 2012). LukAB/GH also showed an inflammatory response in the rabbit study model. In addition, the expression and production of leukotoxin vary with different tissue sites and growth conditions, increasing the complexity of leukotoxin studies. Overall, the contribution of leukotoxin during *S. aureus* infection is dependent on multiple factors including toxin concentration (Yoong and Torres 2013).

Exfoliative Toxin

The exfoliative toxin also called epidermolytic toxins are serine proteases that recognize and then cleave desmosome proteins in the outer layers of skin causing staphylococcal scalded skin syndrome (SSSS). This disease causes the infants to lose the superficial skin layer and dehydration along with other forms of secondary infections. Early SSSS symptoms are fever, lethargy, etc., which gets aggravated by large, big rashes and fragile blisters filled with fluid substances. When the blisters burst due to some mechanical action, it leaves the skin without an epidermis. A disease with a similar etiology but more localized to a site of infection is called "bullous impetigo." Until 2019, four isoforms of exotoxins were identified as ETA, ETB, ETC, and ETD with exfoliative toxin C not associated with human infection. It was first isolated from horses and known to cause exfoliation in mice and chickens. However, in 2019, a new exfoliative toxin ETE was identified by Imanishi et al. (2019). It was initially named as ETD-like protein because it showed 59% identity with ETD toxin and differed from ETA and ETB. But further studies confirmed its toxin activity like ETA and ETB, which made the researchers propose this newly identified toxin as ETE. It is currently associated with ewe mastitis and subsequent research will clarify more about the animal, which can get infected with this strain. Although ETs were getting studied from 1970 until the 21st century, no conclusive theory about their proteolytic activity and association with skin exfoliation was established. Different studies favored this hypothesis but there were also many contradictory results. ETA and ETB have a molecular mass of approximately 27 kDa and contain 242 and 246 amino acids respectively.

In SSSS, the blistering takes place only in the superficial layer of the skin. The ETs are highly selective and only hydrolyze Desmoglein 1 (Dsg-1). It is a desmosomal cadherin that is responsible for maintaining cell-to-cell adhesivity. Dsg-1 is present in all skin layer strata along with Dsg-3 except stratum granulosum. Therefore, the deeper layer of skin gets protected due to the additional presence of Dsg-3, but the stratum granulosum gets severely affected. This also explains why ETs cause blisters in epidermal layers in SSSS conditions, although they are even produced in upper respiratory organs and cause toxemia. It is rare to occur in adults and occurs mainly in neonates and children who have weak immune responses. In addition, renal clearance of the toxins and in general a higher microbial load of microorganisms makes them more susceptible to SSSS than adults. The mortality rate is usually low if diagnosed and treated immediately. Other than *S. aureus* there are other strains like *Staphylococcus chromogenes* and *Staphylococcus hyicus* causing exfoliative skin disorders in other animals like pigs, mice, etc. The toxin production and resistance mechanism differs among different animal models but the interaction between ET and Dsg-1 is mainly considered the reason behind SSSS (Bukowski et al. 2010).

Staphylococcus aureus enterotoxins are an important food safety concern and are often associated with SFP. Besides SFP they also cause toxic shock syndrome

(TSS) which is a much more severe condition. To date, SEs are the superfamily of 26 superantigens and are further categorized into SEs and staphylococcal enterotoxin-like proteins (SEls). They belong to the family of superantigens (SAgs) because they release a high quantity of inflammatory cytokines due to the sudden activation of T-cells. Out of these 26 SAgs, one is toxic shock syndrome toxin (TSST-1), which was earlier referred to as SEF, 11 are staphylococcal enterotoxins (SEA–SEE, SEG–SEI, SER–SET) and 14 are SE-like proteins (SElJ–SElQ, SElU–SElZ) (Abdurrahman et al. 2020). It is expected that in coming years the exotoxin family will keep on growing with technological advancement in the detection and characterization of these toxins. The categorization is done based on the toxin's activity of invoking emesis. SEs typically attack the intestine cells causing gastroenteritis, diarrhea along with emesis. However, the more recently identified SEls are the staphylococcal superantigens that have not been tested or do not invoke emesis. The growth and SE production of the *S. aureus* strains is dependent mainly on temperature, pH, water activity, NaCl concentration, redox potential and atmosphere, and also preferred aerobic.

SEs and SEls are single-chain globular proteins having a low molecular weight of approximately 20 to 30 kDa. Their amino acid sequence varies greatly between 21–83% but the crystallographic analysis indicates its structural similarity (Rödström et al. 2015). They are known to be carried by different MGEs like plasmids, prophages, transposons, and *S. aureus* pathogenicity islands (SaPIs). However, there is an exception in the case of *selx* and *selw*, which are located on the core chromosome instead (Benkerroum 2018).

There is not enough conclusive evidence about the spread of TSS via food, but this transmission route cannot be eliminated. Numerous studies were done to study the relationship between TSS and SFP. They both are caused by staphylococcal enterotoxins, but TSS can only be caused when SEs manage to damage the epithelial cells of the organs either through the oral route or otherwise. The consensus after decades of research is that the dissemination of SEs leading to TSS after intestinal exposure depends on how much toxin transcytosis took place, which again primarily depends on the type of SEs and the amount of toxin ingestion. SFP, on the other hand, is confined to gastrointestinal infection, but the high dosage can trigger a systemic TSS-like effect. It is mainly linked to food products like meat and milk because of the high protein content, which provides a suitable condition for the growth of *S. aureus*. Classical SEs (SEA–SEE) and mostly SEAs (80%) are implicated in SFP outbreaks. Other SEs like SEG, SEH and SEI are also reported to cause SFP incidents (Cao et al. 2012).

SEs/SEls impact on human health is difficult to assess due to a highly diverse amino acid sequence, their action mechanism, immunological reaction, and food matrix interactions. The environmental factors, which are discussed above also need to be considered to understand and apply the control measures to curb *S. aureus*-associated outbreaks and infections. There has been a paucity of reports, especially on newer SEs and SEls.

TSS Toxin 1

As discussed above, the TSS toxin is a member of the staphylococcal superantigen family. TSST-1 and a few SEs are the superantigens associated with Staph toxic shock syndrome. Superantigens are consequential because they do not follow the conventional way of invoking an immune response. The conventional process of T-cell activation is that first the antigen molecules interact with antigen-presenting cells (APC) and get hydrolyzed inside them. The antigen peptides formed are bound to the APC along with the MHCII complex and attract T-cell receptors which have variable α and β chains. The Vβ chain specifically recognizes this complex of peptide, TCR, and MHCII stimulating 1 out of 10,000 or 1,000,000 (10^6) cells. However, superantigens like TSST-1 and SEs can skip this entire specific interaction and directly bind to activate up to 20% of the total T-cells. This results in almost 5,000 times increased T-cell activation. This massive outburst of cytokine, also called "cytokine storm" puts the body in a shock which depending on the severity can lead to multiorgan failure or sometimes even death (Que and Moreillon 2015).

In earlier pieces of literature, TSS was reported as staphylococcal scarlet fever (Stevens 1927). This condition gained attention when in the 1980s many young women started suffering from TSS on using tampons while on their menses. This condition is categorized into menstrual and nonmenstrual TSS. The *agr* gene that regulates the production of TSST-1 requires certain protein, oxygen, and carbon dioxide conditions to be fulfilled to produce the toxin. All these requirements are met with the use of high-absorbency tampons, which does the final job of introducing a high concentration of oxygen in the vaginal environment. This stimulates the production of TSST-1 and the associated toxic shock syndrome. The nonmenstrual TSS is comparatively less popular but it can happen with anyone. Besides TSST-1, SEB and SEC enterotoxins which are regulated by the same *agr* gene are responsible for nonmenstrual TSS. Colonization can happen at different body sites, like surgical wounds, lungs, skin, catheters, etc. The important characteristic of this syndrome that distinguishes it from other infections is that the affected tissues will not undergo inflammation because TSST-1 prevents macrophage influxion. There is a higher risk of TSS in patients who lack the specific antibody against SAgs. The most common symptoms are fever, hypotension, and macular rash. Different organ systems are involved like a lever, blood, renal, CNS, mucous membrane, muscular, and intestines because it is a systemic syndrome (Stevens and Bryant 2011).

Drug Resistance

Staphylococcus species are known for their ability to express several drug-resistant genes and virulence factors, which in turn help the bacteria to evade the host defense system and survive. *Staphylococcus* species often acquire resistance against various antibiotics via horizontal gene transfer as well as chromosomal mutation. Infections caused by antibiotic-resistant *Staphylococcus aureus* have been a public health concern worldwide (Grundmann et al. 2006). The mutation rate in bacteria is very high which helps bacteria to become resistant to antibiotics. In 1942, researchers have reported the occurrence of penicillin-resistant *S. aureus* within a few years of the discovery of penicillin (Barber et al. 1948; Kirby 1944). Most of the *S. aureus* strains produce penicillinases enzymes which hydrolyze the beta-lactam ring of penicillin and make it inactive. The infections resulted from penicillin-resistant *S. aureus* strains, primarily the phage-type 80/81, reached the community settings and became pandemic by the early 1960s (Bynoe et al. 1956; Rountree and Freeman 1955), and ended after the introduction of methicillin in 1959.

Methicillin-resistant *S. aureus* strains (MRSA) were first identified in 1962 from a hospital in the United Kingdom. The structural gene that is responsible for methicillin resistance, mecA, encodes penicillin-binding protein 2a (PBP2a), which has reduced affinity for β-lactam antibiotics. Acquisition of the *mecA* gene helps *S. aureus* to develop resistance against methicillin and all other β-lactam antibiotics. During the 1970s, MRSA strains were predominantly reported from European hospitals (Crisóstomo et al. 2001), whereas, very few cases were reported from hospital settings in the United States (Barrett et al. 1968; Bra et al. 1972). MRSA-associated outbreaks were reported in the hospitals of the United States in the late 1970s and became endemic by the 1980s (Crossley et al. 1979; James E. Peacock et al. 1980). The overall burden of infections caused by MRSA has been increasing in healthcare and community settings globally (Hersh et al. 2008; Hope et al. 2008; Kaplan et al. 2005; Laupland et al. 2008). A rapid spread of community-associated MRSA (CA-MRSA), causing skin and soft-tissue infections (Fridkin et al. 2005; Moran et al. 2006), has been observed in the United States. The first cases of CA-MRSA (USA400 clone) have been reported in children from the USA during the 1990s (CDC 1999; Herold et al. 1998). The death of children who had no exposure to the risk factors for hospital-associated MRSA suggested that the USA400 clone is a virulent strain of CA-MRSA. In addition to this, another epidemic clone, USA300, emerged during the early 2000s, are known to be responsible for CA-MRSA infections in the USA (Aiello et al. 2006; Kazakova et al. 2005; Miller and Diep 2008; Pannaraj et al. 2006). Later, *S. aureus* USA300 clone spread globally and became responsible for global epidemics in the skin and soft-tissue infections, as well as severe pneumonia in many countries around the world (Nimmo 2012). CA-MRSA has been reported from various populations regardless of the age of the subjects, including Native Americans (Sutcliffe et al. 2020) and Alaska natives (Baggett et al. 2004); Pacific Islanders (CDC 2004); athletes (Kazakova et al. 2005); prisoners (Aiello et al. 2006); military personnel (Aiello et al. 2006); and children in daycare facilities (Adcock et al. 1998). In addition to the United States, CA-MRSA strains have been reported in other countries including Canada, Australia, Sweden, Netherlands, Denmark, Finland, Switzerland, Argentina, Korea, India, and China (Conly and Johnston 2003; Fang et al. 2008; Francois et al. 2008; Gardella et al. 2008; Kanerva et al. 2009; Larsen et al. 2007; Nimmo and Coombs 2008; Park et al. 2009; Sangeeta Joshi 2013; Stam-Bolink et al. 2007; Wu et al. 2019a).

The overall increase in MRSA infections remains a significant threat to public health globally. This led to finding out an alternative treatment for MRSA infections. Vancomycin, a glycopeptide class of antibiotics, has been approved for humans in 1958 and was used to treat severe MRSA infections since the late 1980s (T C Sorrell 1982). *Staphylococcus* species with intermediate or complete resistance against vancomycin have been reported since the late 1990s (Hiramatsu et al. 1997) and have become a serious public health concern. In the United States, the first cases of vancomycin-resistant *S. aureus* (VRSA) infection were reported in Michigan in 2002 (CDC 2002; Chang et al. 2003). In addition to this, a second VRSA strain was isolated from Pennsylvania, the USA in 2002 (Tenover et al. 2004). A total of approximately 52 VRSA strains have been identified so far globally (Cong et al. 2020). With the reduced susceptibility to vancomycin, *S. aureus* has been classified into three groups based on the Clinical and Laboratory Standards Institute. These are vancomycin-susceptible *S. aureus* (VSSA) (MIC ≤ 2 µg/ml), vancomycin-intermediate *S. aureus* (VISA) (MIC of 4–8 µg/ml), and vancomycin-resistant *S. aureus* (VRSA) (MIC ≥ 16 µg/ml). VRSA is generally confirmed with the presence of approximately 11 *van* gene clusters, including VanA, VanB, VanD, Van F, VanI, VanM, VanC, VanE, VanG, VanL, and VanN, described to date (Boyd et al. 2008; Courvalin 2006; Kruse et al. 2014; Lebreton et al. 2011; Xu et al. 2010) with the help of molecular techniques.

Biofilm

Nowadays, the control of biofilm-producing candidate pathogens like *Staphylococcus* sp. is a challenging and serious threat to human health. Biofilm can be defined as the assemblage of the cell, covered by a hard layer that is made up of polysaccharides, lipids, proteins, and extracellular DNA. However, *Staphylococcus aureus* produces a slimy glycocalyx layer made up of protein (both host and agent) and teichoic acids (a polymer of alcohol phosphate and sugar) (Hussain et al. 1993). Later on, a specialized polysaccharide [poly-N-acetylglucosamine (PNAG) or polysaccharide intercellular adhesin (PIA)] has been identified, which is a linked polymer of non-N-acetylated D-glucosaminyl residues and β-1,6-linked N-acetylglucosamine residues (Nguyen et al. 2020). Interestingly, PIA is not present in *Staphylococcus* sp. but also have been detected in a wide range of biofilm-producing candidates like *Pseudomonas fluorescens, Yersinia pestis, Escherichia coli, Bordetella bronchiseptica, B. pertussis,* etc. (Nguyen et al. 2020). In the case of *Staphylococcus* sp., the operon called icaADBC is responsible for the production of enzymes required for the synthesis of PIA which is situated in ica locus. According to Hoang et al., the production of PIA among *Staphylococcus* sp. varies greatly; in the case of *S. epidermis*, it is controlled by two regulators called icaR and tcaR which can completely repress the icaADBC (Hoang et al. 2019). There is another regulator called staphylococcal accessory regulator (sarA) which produces a protein (14.5 kDa) with the help of sarA locus (1.2 kda region of DNA consisting of *arB, sarC,* and *sarA* units) which has the ability to bind the RNA and enhances the expression of agr gene cluster and RNAIII promoters, which ultimately reduces the production of biofilm (Morrison et al. 2012). On the other hand, sarA also plays a vital role in the agr-independent pathway (Valle et al. 2003) in which lower expression or mutation of sarA reduces the expression of ica operon. It is observed that there is a correlation between the ica gene cluster and the colonization on the surface of medical devices caused by *S. epidermidis* (Harris et al. 2016). Thus, it can be concluded that PIA plays a critical role in biofilm formation and this information might be useful in understanding the mechanism of biofilm formation in *Staphylococcus* sp.

Biofilm formation is a protective mechanism to avoid host immunity and antibiotics; however, it is negatively correlated with toxin production (Figure 1). The infection (mostly sepsis and bacteremia) caused by *Staphylococcus* sp. becomes notorious due to its ability to form resistant biofilm on several medical devices (mainly *S. epidermidis*) like joint prosthetics, pacemakers, and catheters (Zheng et al. 2018), which enhance the rate of mortality and morbidity. From a molecular biology perspective, biofilm formed by *Staphylococcus* sp. can be categorized into five major stages: adherence and attachment (stage1), division or multiplication (stage 2), exodus (stage 3), maturation (stage 4), and release or dispersal (stage 5) (Moormeier and Bayles 2017). A general diagram of biofilm formation in *Staphylococcus* sp. is presented in Figure 2.

The structure and architecture of the biofilm depend on several environmental factors and its thickness varies from a single-cell layer to multiple layers (Costerton et al. 1995). The cellular activities and metabolic states of the *Staphylococcus* cells inside the biofilm enclosure differ greatly. Bacterial cells on the upper surface of the matrix layer remain in an oxygen-enriched environment and become very active metabolically. Similarly, cell present near the attachment surface gets more nutrients compared to other (Rani et al. 2007). However, most of the cells in the biofilm are dormant or dead and remain in an anoxygenic environment. Thus, the expression profile of the biofilm gene varies from one cell to another (Archer et al. 2011). The biofilm formation in bacteria including *Staphylococcus* sp. is a multievent phenomenon (Table 2). In the case of *Staphylococcus* sp., the formation of biofilm varies greatly depending on the surface (biotic and abiotic) in terms of cell surface molecule profiling. During the formation of biofilm on the biotic surface, a variety of cell surface adhesin molecules (plasmin sensitive protein, serine-aspartate repeat family proteins, iron-regulated surface determinants, fibronectin-binding

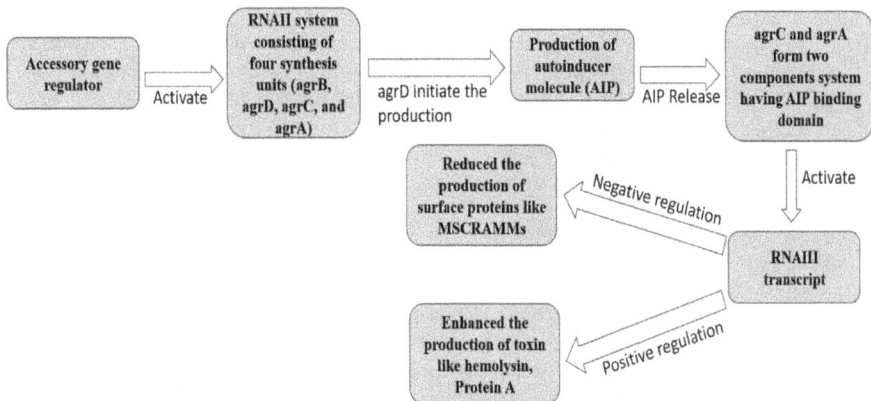

Figure 1. Biofilm formation is negatively regulated with toxin production and is a part of the quorum sensing network in *S. aureus*. RNAII and RNAIII are the main regulatory transcript that controls the expression of biofilm genes and toxin-producing genes in a density-dependent manner.

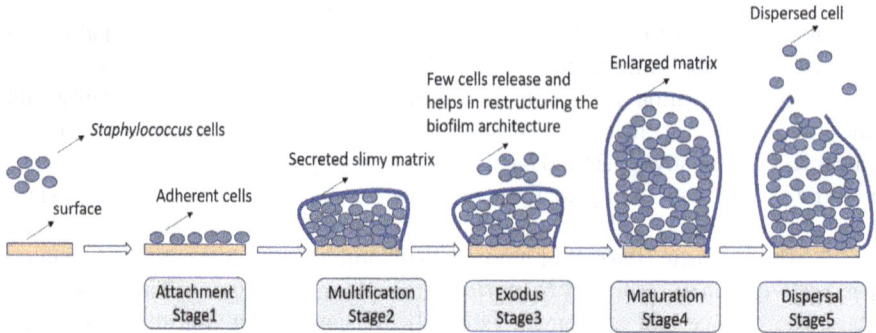

Figure 2. Different stages of biofilm formation from stage 1 (attachment) to stage 5 (dispersal) through sequential events.

Table 2. Different stages and their associated events during biofilm formation in *Staphylococcus* sp.

Stages	Names	Events	References
1	Attachment	a. Express surface proteins belong to the adhesin group (MSCRAMMs.) b. Adhesin can recognize and bind to glycoprotein receptors present on the surface host cell c. A wide diversity of this adhesin molecule helps bacteria to bind a variety of tissue surfaces like bone, nasopharynx, and vascular tissue	(Kot et al. 2018)
2	Multiplication	a. Cell start dividing, binding to each other and accumulating b. Cell-cell attachment is achieved by electrostatic interaction between PIA (positively charged) and teichoic acids (negatively charged) present in the cell wall c. Different types of CWA proteins help in the accumulation d. Formation of ECM starts with the help of cytoplasmic proteins e. ECM structure is stabilized by binding several proteins (GAPDH, enolase, Immunodominant surface antigen B, and beta-toxin) with eDNA	(Jan-Roblero et al. 2017) (Speziale et al. 2014)
3	Exodus	a. An early dispersal event helps in biofilm restructuring b. *Nuc* gene express in a small population of cells that produce nuclease and degrades the eDNA c. *Nuc* gene expression is controlled in a Sae-dependent manner	(Moormeier et al. 2017)
4	Maturation	a. Increase surface area for metabolic waste removal and nutrient exchange b. Fast-growing cells emerge from slow-growing basal layer cells c. The biofilm structure is shaped like a mushroom d. The emergence of metabolically distinct populations	(Nguyen et al. 2020; Moormeier et al. 2017)
5	Dispersal	a. PSMs play a critical role in dispersal by disrupting the non-covalent interaction between the matrix molecules b. It is controlled by *Agr* mediated quorum-sensing network c. Release of cells from the biofilm matrix	(Le et al. 2014)

MSCRAMMs: Microbial surface components recognizing adhesive matrix molecules
CWA: Cell wall-anchored
PSMs: Phenol soluble modulins

proteins, collagen adhesin, bone sialoprotein, etc.) are expressed and linked to the peptidoglycan layer with the help of another protein called sortase A (Moormeier et al. 2017). On the other hand, the formation of biofilm on abiotic surfaces like glass and polystyrene carries out with the help of another set of gene clusters like *AtlA* and *agr* (Jenul and Horswill 2019). However, the expression profile of these cell wall surface proteins varies among different species of *Staphylococcus* (Foster et al. 2014).

The ECM of biofilm in *Staphylococcus* sp. is composed of different types of polysaccharides, proteins, lipids, and eDNA. It is already established that ECM maintains the structural integrity of the biofilm and plays a very important role in protecting biofilm cells from antibiotics, host immune molecules, and environmental stress. The function of each ECM molecule and its interaction in forming the extraocular matrix is not elucidated clearly. In an experiment with biofilm produced by *Staphylococcus aureus* ATCC 25923, researchers have identified a total of 460 proteins that are associated with biofilm matrix formation (Hiltunen et al. 2019). However, protein concentration and abundance vary greatly from one surface to another at different time intervals. Furthermore, Hiltunen et al. (2019) also reported that 66 proteins are common in biofilm formed on all the different surfaces (borosilicate glass, hydroxyapatite, titanium, and plexiglass). Among several proteins, stress proteins, chaperones, hemolysins, leukocidins, and response regulators are considered to be the most important biofilm protein for infection (Hiltunen et al. 2019). However, other classes of proteins like surface protein C and G, the biofilm-associated protein, fibrinogen-binding proteins, clumping factor B, and serine aspartate repeat protein are important for attachment and host tissue interaction in *Staphylococcus aureus* (Speziale et al. 2014). In *Staphylococcus epidermidis*, there is another set of a protein called accumulation-associated protein which helps in attachment and host-cell interaction. The extracellular DNA (eDNA) is an important component of *Staphylococcus* biofilm matrix. Several researchers have tried to explore the importance of eDNA and its relationship with PIA using DNase I assay (Houston et al. 2011; Izano et al. 2008). However, their results are contradictory and did not provide a clear relation between PIA-eDNA-MRSA and MSSA. Similarly, a detailed experiment among 47 *S. aureus* strains was conducted to understand the dependency of eDNA on PIA-dependent and PIA-independent biofilm formation (Sugimoto et al. 2018). Their result indicated that PIA independent candidate poses a significantly higher eDNA level compared to PIA dependent produced. However, it is not always true as the eDNA level also depends on culture media and the genetic makeup of the strain.

Epidemiology

Different species of *Staphylococcus* are considered to be potent pathogens in clinical sectors and food industries. World Health Organization (WHO) has considered *Staphylococcus aureus* also to be a foodborne pathogen that causes serious illness in humans. Whereas biofilm forming (on implanted medical devices) *Staphylococcus epidermidis* is considered to be very dangerous in medical sectors. It was well known that *S. aureus* produces heat-stable toxins in the food matrix (like milk, meat, egg, and cream-filled bakery products), which cause a very common illness named SFP (e Silva et al. 2017). *Staphylococcus aureus* bacteremia (SAB) is the most common infection manifestation which causes 20% to 25% mortality (Monnier et al. 2020), and its prevalence and prognosis have been diagnosed mostly in the industrial region of the world. Several reports have been published regarding the rate of SAB in different time frames. In Denmark, the SAB rate during the period of 1957–1990 was increased from 3 per 100,000 people to 20 per 100,000 (Frimodt-Møller et al. 1997); however, it increased by 48% from 2008–2015 (Thorlacius-Ussing et al. 2019). The prevalence of healthcare-associated MRSA is recorded to be highest in Asian countries (28% in Indonesia and > 70% in Korea), whereas the occurrence of community-associated MRSA varies from 5% to 35% depending on geographical area and age structure (Chen and Huang 2014; Wu et al. 2019b). Along with the previous strains of MRSA, a new strain of *S. aureus* has emerged (in China) in a farmed animal that was diagnosed as a potent human pathogen (Wan et al. 2013). In the United States, healthcare-associated MRSA-causing bloodstream infection was more common, but community-associated infections affecting the skin and soft tissue also emerged in the early 1990s and started to increase rapidly thereby (Kourtis et al. 2019). However, according to the surveillance system (National Healthcare Safety Network and Emerging Infections Program), the MRSA infection rate might have decreased in recent years but still the United States is not on track to fulfill the goal taken by Healthcare-Associated Infection National Action Plan in 2020 (Health and Services 2011). According to Frimodt-Møller et al. (2019), both hospital and community-associated bloodstream infection (from 2005–2016) caused by MRSA has been reduced by 74% and 40%, respectively. In the last 60 years, *S. aureus* has become resistant to a wide range of antibiotics (like carbapenems, penicillin, monobactams, cephalosporins, and carbacephem), which is a big challenge for clinical management to control the SAB infection (Monnier et al. 2020). The emergence of MRSA varies greatly (1.2% to 50.5% in different European countries) depending on geographical location (Struyf et al. 2017). Due to the hard work of the surveillance committee, good clinical practice, and proper management, SAB infection reduced significantly in UK and France in recent years (Lawes et al. 2015).

The rate of SAB infection depends on several factors and age is considered to be the most important one. It was recorded that people in the first year of age and later age (> 70 years) are more susceptible to SAB compared to younger (Tong et al. 2015). Furthermore, the rate of infection is observed to be 4.0% (people below 80 years of age), 8.4% (people between 80–89 years of age), and 13.0% (people above 90 years of age) depending on geographical regions (Thorlacius-Ussing et al. 2019). Not only age, gender, and ethnicity are also considered to be important for SAB infection. Clinical data strongly supports that males are more susceptible than females, however, the logical argument is not clear (Landrum et al. 2012). In the United States, the black population is more susceptible to the SAB risk group compared to the white population (Kallen et al. 2010). Similarly, skin and soft-tissue infections caused by *S. aureus* vary significantly in people's lives in Australia and New Zealand based on indigenous and nonindigenous origin (Hill et al. 2001; Tong et al. 2009). Infection caused by the hospital and community-associated MRSA has

Table 3. MRSA outbreaks at different times in different locations.

Area/countries	Year of outbreak	Strains	Symptoms	References
Australia and Canada	1953	80/81 penicillin-resistant *S. aureus*	Pneumonia, skin infection, and sepsis in children and adults	(DeLeo et al. 2011)
Northamptonshire (England)	1991 to 1992	EMRSA-16	Throat infection	(Cox et al. 1995)
The Health Sciences Center, Manitoba, Canada	1996	NM	Skin, throat, and burn wound	(Embil et al. 2001)
Hospitals in Chicago and Los Angeles	2006	NM	Skin and soft-tissue infection in newborn	(Waknine 2006)
The University Hospital of Lausanne (Switzerland)	2008 to 2012	MRSA ST228-I	NM	(Senn et al. 2016)
Cambridge University Hospitals National Health Service (UK)	2011	MRSA spa-type t2068	Skin infection in infants	(Brown et al. 2019)
Affiliated People's Hospital of Jiangsu University, China	2012	MRSA T0131 and TW20	NM	(Kong et al. 2016)
Aga Khan University Hospital (AKUH), Pakistan	2013	NM	Urine infection in newborn	(Irfan et al. 2019)
Hvidovre Hospital, Denmark	2013 to 2017	MRSA spa type t267	NM	(Rubin et al. 2018)
Germany	2018	MRSA spa-type t020	Skin infection in adult	(Baier et al. 2020)
Tama district, Tokyo, Japan	2018 to 2019	MRSA ST8-IV	Pneumonia and Surgical site infection in adult and old people	(Kobayashi et al. 2020)

NM: not mentioned

been reported from different parts of the world (Table 3), and still it is one of the major threats for human being. Most of the time, MRSA cause infection on the skin, and thus it spread easily from one person to other in community transmitted manner. According to the Centers for Disease Control and Prevention, US (CDC) MRSA infection spread can be controlled through proper clinical management like washing hands with soap or alcohol, regular cleaning of hospital bed, cleaning of medical devices with detergent and alcohol, regular monitoring of wound site, sampling of germ from nose and skin, and wearing gloves during patient visit.

Identification of *S. aureus* Using Molecular Techniques

Identification of S. aureus Using Polymerase Chain Reaction (PCR)

S. aureus produces an extracellular thermostable protein, thermonuclease with a molecular mass of 17,000 Da (Tucker et al. 1978). Thermonuclease, an endonuclease that can degrade both DNA and RNA, can resist 1,000 C for 1 hour (Lachica et al. 1972). Thermonuclease is encoded by the *nuc* gene which is known as highly conserved in most *S. aureus* strains. Brakstad et al. (1992) developed primers Nuc1 and Nuc2 (Table 2) for PCR assay to identify the presence of the *nuc* genes in *S. aureus*. The primers (Nuc1 and Nuc2) were able to amplify the *nuc* genes of clinical *S. aureus* isolates, however, did not amplify the *nuc* genes of other *Staphylococcus* species and non-*Staphylococcus* species (Brakstad et al. 1992). The staphylococcal coagulase enzyme, encoded by the *coa* gene, is an important virulence factor *S. aureus*. Goh et al. (1992b) developed a typing method to identify *S. aureus* by amplifying a variable region of the coagulase genes in PCR followed by AluI restriction enzyme digestions and RFLP (Restriction fragment length polymorphism) analysis. The primers COAG1 and COAG4 (Table 4) were able to amplify the 1557-bp size of the *coa* gene. The nested primers COAG2 and COAG3 (Table 4) were used to amplify the variable tandem 3' region of the *coa* gene. This method was successfully used to identify the source of the MRSA outbreak that occurred at a hospital. In the year 1995, Saruta et al. developed a primer set (16SSAIII:5'-TATAGATGGATCCGCGCT-3', and 16SSAIV:5'-GATTAGGTACCGTCAAGAT-3') considering 16S rRNA gene as a target to identify the *S. aureus* (Saruta et al. 1995). They have considered 28 staphylococcal and non-staphylococcal strains, and the primer (273 bp amplicons) can effectively identify *S. aureus*. Later on, Martineau et al. developed another set of primers: Sa442-1 (5'-AATCTTTGTCGGTACACGATATTCTTC ACG-3'; positions 5 to 34), and Sa442-2 (5'-CGTAATGAGATTTCAGTAGATAATACAACA-3'; positions 83 to 112) based on the chromosomal sequence of 442 bp long to identify *S. aureus* (Martineau et al. 1998). Results of their investigation indicated that the primer set was very much specific to *S. aureus* but not other species of *Staphylococcus*. Similarly, Hookey et al. targeted the coagulase gene (*coa*) to develop a novel typing method for identifying *S. aureus* (Hookey et al. 1998). The primer set coa-F (5'-ATAGAGATGCTGGTACAGG-3'), and coa-R (5'-GCTTCCGATTGTTCGATGC-3') produces 875, 660, and 603 bp amplicon for coagulase-positive staphylococci, while 547 bp amplicon is produced in case of the MRSA isolates. Marcos et al. have developed an another developed another typing method to rapidly identify *S. aureus* targeting the *aroA* gene which encodes 5-enolpyruvylshikimate-3-phosphate synthase (Marcos et al. 1999). The primer Set FA1 and RA2 (Table 2) can effectively identify *S. aureus* by amplifying the 1,153 bp region of the *aroA* gene. Furthermore, Lange et al. have used three sets of primer (Table 4) to amplify the region of coagulase gene, protein A gene, and intermediate space between 16S and 23S rRNA to detect the *S. aureus* from the milk obtained from dairy cows diagnosed with subclinical mastitis (Lange et al. 1999). The results of their study depicted that all these PCR typing methods are sensitive. However, the spacer region of *16S rRNA* and *23S rRNA* was reported to be more

Table 4. List of primer used to detect and identify *Staphylococcus* species.

PCR Methods	Genes	Primers	Amplicon Size (bp)	Functions	Candidate Pathogens	References
Singleplex PCR	*nuc*	GCGATTGATGGTGATACGGTI AGCCAAGCCTTGACGAACTAAAGC	267	➢ It encodes a nuclease which acts as a virulence factor	*S. aureus*	(Brakstad et al. 1992)
	coa	ATACTCAACCGACGACACCG GATTTTGGATGAAGCGGATT	1,557	➢ Encodes the virulence factor coagulase	*S. aureus*	(Goh et al. 1992a)
	16S rRNA *Gap*	GGAATTCAAAGGAATTGACG GGGGC CGGGATC CCAGGCCCGGGAACGTATTCAC ATG GTTTTGGTAGAATTGGTCGTTTA GACATTTCGTTATCATACCAAGCTG	479 933	➢ Part of the 30S subunit ➢ Encodes glyceraldehyde-3-phosphate dehydrogenase	*S. aureus* and *Staphylococcus* sp.	(Ali, Al-Achkar et al. 2014)
	Sau-02	GTAAAAAGACGACATGCAGGAA CCATCATTTCAAAACTTTGACA	110	➢ Small RNA that does not code any protein	*S. aureus* and MRSA	(Soo Yean et al. 2016)
	mecA *PVL* *nuc*	AACCATCGTTACGGATTGCTTC AAATGCTGGACAAAACTTCTTGG TTTGCAGCGTTTGTTTTCG GGCATATGTATGGCAATTGTTTC CGTATTGCCCTTTCGAAACATT	107 108 73	➢ Produces PBP2a enzyme helps from methicillin ➢ Produces a cytotoxin ➢ Produces a thermonuclease	*S. aureus* (both MRSA and MSSA)	(Galia et al. 2019)
	aroA	AAGGGCGAAATAGAAGTGCCGGGC CACAAGCAACTGCAAGCAT	1,153	➢ Encodes 5-enolpyruvylshikimate-3-phosphate synthase	*S. aureus*	(Marcos et al 1999)
	coa *spa* *ribosomal gene*	CGAGACCAAGATTCAACAAG AAAGAAAACCACTCACATCA CAAGCACCAAAAGAGGAA CACCAGGTTTAACGACAT TTGTACACACCGCCCGTCA GGTACCTTAGATGTTTCAGTTC		➢ Encodes coagulase ➢ Encodes a surface protein that helps in virulence and survival ➢ Encodes the structural part of the ribosomal RNA	*S. aureus*	(Lange et al. 1999)
	23S rRNA *23S rRNA*	ACGGAGTTACAAAGGACGAC AGCTCAGCCTTAACGAGTAC ATCATCTGGAAAGATGAATCAA ATCGATTAAAAACGATTATAGGT		➢ Both primers amplified the structural region of 23S rRNA	*S. aureus*	(Straub et al. 1999)

Table 4 contd. ....

...Table 4 contd.

PCR Methods	Genes	Primers	Amplicon Size (bp)	Functions	Candidate Pathogens	References
Multiplex PCR	sea	GGTTATCAATGTGCGGGTGG CGGCACTTTTTCTCTTCTCGG	102	➤ Encodes enterotoxins A	S. aureus and MRSA	(Mehrotra et al. 2000)
	seb	GTATGGTGGTGTAACTGAGC CCAAATAGTGACGAGTTAGG	164	➤ Encodes enterotoxins B		
	sec	AGATGAAGTAGTTGATGTGTATGG CACACTTTTAGAATCAACCG	451	➤ Encodes enterotoxins C		
	sed	CCAATAATAGGAGAAAATAAAAG ATTGGTATTTTTTTCGTTC	278	➤ Encodes enterotoxins D		
	see	AGGTTTTTTCACAGGTCATCC CTTTTTTTTCTTCGGTCAATC	209	➤ Encodes enterotoxins E		
	femA	AAAAAGCACATAACAAGCG GATAAAGAAGAAACCAGCAG	132	➤ Helps in methicillin resistance		
	etaA	GCAGGTGTTGATTTAGCATT AGATGTCCCTATTTTTGCTG	93	➤ Encodes exfoliative toxins A		
	etaB	ACAAGCAAAAGAATACAGCG GTTTTTGGCTGCTTCTCTTG	226	➤ Encodes exfoliative toxins B		
	tst	ACCCCTGTTCCCTTATCATC TTTTCAGTATTTGTAACGCC	326	➤ Encodes TSS toxin 1		
	mecA	ACTGCTATCCACCCTCAAAC CTGGTGAAGTTGTAATCTGG	163	➤ Provides resistance to methicillin		
	etd	CCCGTTGATTAGTCATGCAG TCCAGAATTCCCGACTCAG	607	➤ Encodes exfoliative toxin D	CA-MRSA	(Strommenger et al. 2008)
	arcA	TTGCTCAAACTTTGAGAGATGAA TTACGTACGCCAGCCATGAT	215	➤ Encodes arginine deiminase		
	seh	CAACTGCTGATTTAGCTCAG GTCGAATGAGTAATCTCTAGG	358	➤ Encodes enterotoxin H		
	lukPV	ATCATTAGGTAAAATGTCTGGACATGATCCA GCATCAAGTGTATTGGATAGCAAAAGC	432	➤ Produces leukocidin		

Gene	Sequence	Size	Function	Target	Reference
Tuf	TACCAGCACATTAGTAGTATTCTTAAAACAAAGTTG TGCTGAACCAGCGATTACAG	143	➤ Encodes Tu protein that involves in translation	*S. aureus* and coagulase-negative staphylococci	(Okolie et al. 2015)
vanA	GCTGTGAGGTCGGTTGTG GCTCGACTTCCTGATGAATACG	111	➤ Helps bacteria from vancomycin		
cns	TATCCACGAAACTTCTAAAACAAACTGTTACT TCTTTAGATAAATACGTATACTTCAGCTTTGAATTT	204	➤ Bb		
16S rRNA	CTAGTAATCGCGGATCAGCAT GATACGGGCTACCTTGTTACGACTT	174	➤ It is a part of 30S unit		
mecA	TGGTATGTGGAAGTTAGATTGGGAT CTAATCTCATATGTGTTCCTGTATTGGC	155	➤ Helps from methicillin ➤ Encodes a virulence factor called protein A	CA-MRSA, *S. aureus* and coagulase-negative staphylococci	(Nakagawa et al. 2005; Okolie et al. 2015)
spa	CAGCAAACCATGCAGATGCTA CGCTAATGATATCCACCAAATACA	101			
pvl	TTACACAGTTAAATATGAAGTGAACTGGA CGCTAATGATAATCCACCAAATACA	118	➤ Produces a toxin called Panton-Valentine leucocidin		
mecA	TGGCTATCGTGTCACAATCG CTGGAACTTGTTGAGCAGAG		➤ Provides resistance to methicillin	MRSA and other staphylococci species	(Vannuffel et al. 1995)
femA	CTTACTTACTGGCTGTACCTG ATGT CGCTTGTTATGTGC		➤ Encodes protein which is responsible for enhancing methicillin resistance level		
mecA	CTTACTTACTGCTGTACCTG ATCTCGCTTGTTATGTGC	684	➤ Provides resistance to methicillin	Mupirocin resistance MRSA	(de Castro Nunes et al. 1999)
femA	TAGAAATGACTGAACGTCCG TTGCGATCAATGTTACCGTAG	154	➤ Encodes protein which is responsible for enhancing methicillin–resistance level		
ileS-2	GTTTATTCTTCTGATGCTGAG CCCCAGTTACACCGATATAA	237	➤ Provides resistance to mupirocin		
mecA	CGGTAACATTGATCGCAACGTTCA CTTTGGAACGATGCCTAATCTCAT	214	➤ Provides resistance to methicillin	MRSA	(Kearns et al. 1999)
coa	GTAGATTGGGCAATTACATTTTGGAGG CGCATCAGCTTTGTTATCCCATGTA	117	➤ Encodes coagulase		

Table 4 contd. ...

...*Table 4 contd.*

PCR Methods	Genes	Primers	Amplicon Size (bp)	Functions	Candidate Pathogens	References
	sea	ATGGTTATCAATGTGCGGGTGᴵᴵᴵᴵᴵCCAAACAAAAC TGAATACTGTCCTTGAGCACCAᴵᴵᴵᴵᴵATCGTAATTAAC	344	➤ All these genes (*sea–see* and *seg–sei*) encode different types of enterotoxins	*S. aureus*	(Hwang et al. 2007)
	seb	TGGTATGACATGATGCCTGCACᴵᴵᴵᴵᴵGATAAATTTGAC AGGTACTCTATAAGTGCCTGCCTᴵᴵᴵᴵᴵACTAACTCTT	196			
	sec	GATGAAGTAGTTGATGTGTATGGATCᴵᴵᴵᴵᴵACTATGTGTAAAC AGATTGGTCAAACTTATCGCCTGGᴵᴵᴵᴵᴵGCATCATATC	399			
	sed	CTGAATTAAGTAGTACCGCGCTᴵᴵᴵᴵᴵATATGAAAC TCCTTTTGCAAATAGCGCCTTGᴵᴵᴵᴵᴵGCATCTAATTC	451			
	see	CGGGGGTGTAACATTACAATGATᴵᴵᴵᴵᴵCCGATTGACC CCCTTGAGCATCAAACAAATCATAAᴵᴵᴵᴵᴵCGTGGACCCTTC	286			
	seg	ATAGACTGAATAAGTTAGAGGAGGTᴵᴵᴵᴵᴵGAAGAAATTATC TTAGTGAGCCAGTGTCTTGCᴵᴵᴵᴵᴵAATCTAGTTC	594			
	seh	CATTCACATCATATGCGAAAGCAGᴵᴵᴵᴵᴵTTACACG CTTCTGAGCTAAATCAGCAGTTGCᴵᴵᴵᴵᴵTTACTCTC	218			
	sei	AGGCGTCACAGATAAAAACCTACCᴵᴵᴵᴵᴵCAAATCAACTC ACAAGGACCATTATAATCAATGCCᴵᴵᴵᴵᴵTATCCAGTTTC	154			
	sej	TGTATGGTGGAGTAAACACTGCATGᴵᴵᴵᴵᴵAATCAACTTTATG CTAGCGGAACAACAGTTCTGATGCᴵᴵᴵᴵᴵATCCATAAAT	102	➤ All the genes (*sej-ser* and *seu*) encode SE-like toxins		
	sek	GTGTCTCTAATAATGCCAGCGCTᴵᴵᴵᴵᴵCGATATAGG CGTTAGTAGCTGTGACTCCACCᴵᴵᴵᴵᴵTGTATTTAG	282			
	sel	ATTCACCAGAATCACACGCTᴵᴵᴵᴵᴵTTACTCGTA GTGTAAAATAAATCATACGAGᴵᴵᴵᴵᴵAGAAACCATCATTC	469			
	sem	CGCAACCGCTGATGTCGGᴵᴵᴵᴵᴵTGAATCTTAGG CAGCTTGTCCTGTTCCAGTATCᴵᴵᴵᴵᴵAGTCATAAG	572			
	sen	TCATGCTTATACGGAGGAGTTACGᴵᴵᴵᴵᴵTGATGGAAATC AACCTTCTTGTTGGACACCATCᴵᴵᴵᴵᴵATACATTAACGC	103			
	seo	GTGGAATTTAGCTCATCAGCGATTTCᴵᴵᴵᴵᴵAATTTCTAGG GTACAGGCAGTATCCACTTGATGCᴵᴵᴵᴵᴵATGACAATGTGC	116			

sep	ATCATAACCAACCGAATCACCAGIIIIIGGGTGAAACTC GTCTGAATTGCAGGGAACTGCIIIIIGCAATCTTAG	547	
seq	GGTGGAATTACGTTGGCGAATCAIIIIITAGATAAACC CTCTGCTTGACCAGTTCCGGTGIIIIICAAATCGTATG	330	
ser	TTCAGTAAGTGCTAAACCAGATCCIIIIICTGGAGAATTG CTGTGGAGTGCATTGTAACGCCIIIIIATATGCAAACTCC	368	
seu	ATGGCTCTAAAATTGATGGTTCTAIIIIITTAAAAACAG GCCAGACTCATAAGGCGAACTAIIIIITTCATATAAA	410	
tst	GTTGCTTGCGACAACTGCTACACAGIIIIIACCCCTGTTC TCAAGCTGATGCTGCCATCTGTGIIIIITATACGCATAG	209	➤ Encodes TSS toxin-1
femA	ACAGCTAAAGAGTTTGGTGCCTIIIIIGATAGCATGC TTCATCAAAGTTGATATACGCTAAAGGTIIIIICACACGGTC	723	➤ Encodes proteins that influence the resistance to methicillin

sensitive, followed by coagulase and *protein A* gene. In the same year, Straub et al. also have developed a PCR protocol taking *23S rRNA* as a gene of interest (Straub et al. 1999). The amplicon size of 1,250 bp and 1,122 bp produced by the primer sets Staur4 and Staur6, and Staur5 and Staur9 (Table 4) were very much specific to *S. aureus* collected from various milk product. In 2016, Mukherjee et al. conducted a risk assessment study taking 32 samples collected from the surface of different exercise equipments from fitness centers located in Tennessee, USA, and have identified the community-associated MRSA amplifying the four target genes (*16S rRNA, mecA, femA,* and *femB*) (Mukherjee et al. 2016).

The use of multiplex PCR in the identification of bacterial species is another typing method that has more advantages (i. provides more information, ii. requires less sample, iii. cost-effective, and iv. time-saving) than conventional PCR. In 1995, Vannuffel et al. developed a multiplex PCR considering *mecA* and *femA* genes to identify MRSA (Vannuffel et al. 1995). The primer sets (Table 4) produce a 310-bp and a 686-bp amplicon for *mecA* and *femA*, respectively, and were very much sensitive to identifying all the tested species of staphylococci. In 1999, de Castro Nunes et al. used three sets of primers (Table 4) and to identify MRSA (de Castro Nunes et al. 1999). The primer sets *femA* (Vannuffel et al. 1995), *mecA* (Del Vecchio et al. 1995), and *ileS-2* (this work) amplified 684 bp, 154 bp, and 237 bp region, respectively, and have successfully identified the mupirocin resistance in MRSA. Later on, in 2000, a researcher group from Canada used the analytical power of multiplex PCR targeting five sets of enterotoxin-encoding genes, TSS toxin 1 encoding gene (*tst*), exfoliative toxins encoding genes (*etaA* and *etaB*), and methicillin resistance genes (*mecA*) (Table 4) to identify MRSA (Mehrotra et al. 2000). In 2007, Hwang et al. (2007) designed a novel multiplex strategy taking 19 genes (Table 4) for an investigation to identify the *S. aureus* contamination in raw meat. In recent years, Qin et al. developed a unique technique of multiplex PCR coupled with propidium monoazide and sodium dodecyl sulfate to detect and identify live food pathogens including *S. aureus* in milk (Qin et al. 2020). They have targeted *gltS* gene (primer F-TTCTTCACGACTAAATAAACGCTCA and primer R-GGTACTACTAAAGATTATCAAGACGGCT), which encodes glutamate transporter for the identification of *S. aureus*. To date, several primers set for singleplex and multiplex PCR were developed by different research groups which are tabulated in Table 4.

Identification of S. aureus Using Mass Spectrometry and VITEK 2

Like PCR, MALDI-TOF-MS is another gold standard technique for the identification of a wide spectrum of bacterial candidates. Compared to PCR, it is a rapid method and does not require any DNA extraction process. The detail of sample preparation is given in Figure 3.

The use of MALDI-TOF-MS has been started several years ago and different research groups have successfully used this technique for the molecular identification of bacterial candidates (Bernardo et al. 2002; Edwards-Jones et al. 2000). Singhal et al. (2015) have stated that MALDI-TOF-MS is gaining popularity in the microbiology method developments process due to its strength in detecting and

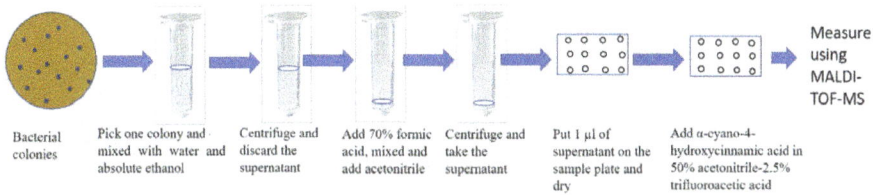

Bacterial colonies → Pick one colony and mixed with water and absolute ethanol → Centrifuge and discard the supernatant → Add 70% formic acid, mixed and add acetonitrile → Centrifuge and take the supernatant → Put 1 µl of supernatant on the sample plate and dry → Add α-cyano-4-hydroxycinnamic acid in 50% acetonitrile-2.5% trifluoroacetic acid → Measure using MALDI-TOF-MS

Figure 3. Sample preparation for MALDI-TOF analysis. Usually, a single colony is used for identification.

identifying bacterial pathogens from different sources, like food, blood samples, urine, water, and many others (Singhal et al. 2015). The identification of *Staphylococcus* sp. using MALDI-TOF-MS is a very common method nowadays, and its accuracy depends on the algorithm and the bioinformatics pipeline (Table 5). In the year 2020, a research group from the UK tested the efficiency of MALDI-TOF in identifying *Staphylococcus* sp. from blood culture samples (Hamilton et al. 2021). According to their observation, MALDI-TOF is very accurate in identifying *S. aureus* (93.6%), however, sometimes it fails to identify coagulase-negative staphylococci. There are several advantages (rapid, cost-effective, accuracy, etc.) of using MALDI-TOF-MS in bacterial candidate identifications. However, due to the limited database, sometimes it is not able to identify bacterial candidates (Tonamo et al. 2021). Furthermore, for identifying *Staphylococcus* sp. MALDI-TOF-MS works well but there are a few bacterial candidates which cannot be discriminated against using MALDI-TOF-MS (for example *E. coli* from Shigella) (Rychert 2019).

VITEK MS is an automated system that uses the technology of MALDI-TOF and is very popular for the identification of bacterial isolates. In 2016, Shan et al. (2016) developed a novel approach to distinguish between MRSA and MSSA from clinical samples using the VITEK MS system based on the MS peak (2,305.6 and 3,007.3 Da for MRSA, and 6,816.7 Da for MSSA). Later on, in 2017, Sulaiman et al. used the VITEK MS system to identify 47 staphylococci isolated from different sources like the cosmetic product, environment, clinical samples, food, etc. (Sulaiman et al. 2018). Guo et al. have compared the efficacy of MALDI-TOF-MS and VITEK 2 taking 1,025 isolates that belong to 25 genera (Guo et al. 2014). The result of their investigation indicated that MALDI-TOF-MS offers more resolution in species identification compared to VITEK 2 in terms of accuracy and lower error rates. Similarly, Martins et al. (2018) also compared the resolution depth of MALDI-TOF-MS and VITEK 2 in the identification of coagulase-negative staphylococci isolates. Their study clearly stated the superiority of MALDI-TOF-MS over VITEK 2.

Outbreak Detection and Identification of S. aureus Using Whole Genome Sequencing

Hospital-acquired and community-associated infection caused by *S. aureus* is a worldwide burden and whole genome sequencing (WGS) plays an important role in identifying such entities for better health management. In recent year scientist have used the enormous power of WGS in revolutionizing the infection biology of

Table 5. Identification of *Staphylococcus* sp. using MALDI-TOF-MS.

Model	Bacterial candidates	M/Z range (Da)	References
Autoflex (Bruker Daltonics)	Coagulase-negative staphylococci	1,000 to 11,000	(Carbonnelle et al. 2007)
Ultraflex (Bruker Daltonics)	22 species of *Staphylococcus*	2,000 to 20,000	(Dubois et al. 2010)
Microflex LT (Bruker Daltonics)	Different species of *Staphylococcus* including *S. aureus, S. epidermis, S. xylosus, S. lentus*, etc.	3,000–15,000	(Szabados 2010)
MALDI Biotyper (Bruker Daltonics)	MRSA	NM	(Nix et al. 2020)
4800 Proteomics Analyzer (with TOF-TOF Optics, Applied Biosystems)	MRSA and MSSA	399 to 4,012	(Tang et al. 2019)
Autoflex (Bruker Daltonics)	Coagulase-negative staphylococci	2,000–20,000	(Dupont et al. 2010)
UltrafleXtreme device (Bruker Daltonik)	*S. aureus* subsp. *aureus* and *anaerobius*	2,000–20,000	(Pérez-Sancho et al. 2018)
Biotyper Microflex (Bruker Daltonik)	Different species of *Staphylococcus*	2,000–20,000	(Zhu et al. 2015)
Microflex LT (Bruker Daltonics)	MRSA	1,960 to 20,000	(Kim et al. 2019)

MRSA: Methicillin-resistant *Staphylococcus aureus*
MSSA: Methicillin sensitive *Staphylococcus aureus*
NM: Not mentioned

S. aureus like virulence determination, antibiotic resistance, lineage, and the emergence of disease and genetic variation (Humphreys and Coleman 2019; Price et al. 2013). In a study, researchers from a semirural health board in Scotland used WGS in testing different areas in a hospital (surfaces, staff hands, and air in the intensive care unit) to address the load and transmission efficiency of hospital-acquired *S. aureus* (Adams et al. 2020). The result of their investigation clearly indicated that transmission of *S. aureus* was least contributed from surface and staff hand, while the infection in the ICU care unit was autogenous. A similar WGS study was also conducted in a territory care hospital in Tanzania to understand the epidemiology and antibiotic resistance of *S. aureus* (Kumburu et al. 2018). The result of their investigation indicated that the population structure of *S. aureus* in a hospital environment is diverse and unrelated. Due to high resolution, WGS can detect a single nucleotide mutation in the genome and predict the genetic basis of antimicrobial resistance in bacteria which provides a new mechanistic insight into antibiotic resistance.

Due to low cost, accuracy, and rapidity, WGS is used for source tracking and outbreak detection, and many government authorities including the U.S. Food and Drug Administration (FDA) is sequencing it is all collected pathogens spanning 20 years for better health management practices (Allard et al. 2018). In this direction, several investigations have been done on *S. aureus* to explore antibiotic resistance,

Table 6. Use of WGS in characterizing *S. aureus* strains.

Platform used	Findings	References
Illumina HiSeq 2000	501 isolates were sequenced to predict the antimicrobial resistance profile and compared the result with phenotypic characteristics using 12 antibiotics	(Gordon et al. 2014)
Illumina HiSeq or MiSeq	20 outbreak isolates of MRSA and MSSA were used to understand the genetic variation and to predict the most recent ancestor	(Gordon et al. 2017)
Illumina MiSeq	42 outbreak and 12 reference strains of *S. aureus* were used to compare the WGS and PAGE result	(Cunningham et al. 2017)
Illumina HiSeq	They have identified 131 isolates of MRSA and MSSA belongs to different clonal complexes like CC1, CC5, CC8, CC30	(Durand et al. 2018)
Illumina MiSeq	152 isolates of MRSA and MSSA were sequenced to determine the virulence genes responsible for adhesion and immune evasion	(Park et al. 2019)
Illumina MiSeq	100 isolates from prosthetic joint infection and 101 isolates from nares were sequenced and compared to check the relatedness. The result indicated no significant difference in the genomic constitution	(Wildeman et al. 2020)
Illumina MiSeq	69 outbreak isolates have been sequenced to understand the virulence factors and evolutionary relationship	(Hait et al. 2021)

PFGE: pulsed-field gel electrophoresis

virulence factor determination, genomic comparison, identification of the strain, and outbreak detection (Giulieri et al. 2016; Kashif et al. 2019; Price et al. 2013; Sullivan et al. 2019) (Table 6). The understanding of WGS data is quite tricky and the prediction depends on downstream analysis. Until now, a series of bioinformatics software is available, and the result varies slightly from one pipeline to another. In this direction, the Swiss clinical bacteriology community conducted a ring trial (nine clinical laboratories) to evaluate the variation in the data analysis pipeline considering three parameters (sample preparation, SNP calling, and phylogenetic tree construction) for better characterization of the outbreak strain of *S. aureus* (Dylus et al. 2020). The result indicated that there are differences in SNP calling using different bioinformatics software, however, the construction of the tree and identification of cluster based on SNP from the core genome (defined by cgMLST scheme) of *S. aureus* was almost similar.

Conclusion

Staphylococcus aureus has a widespread occurrence in humans and animals. Different sources like foodborne, animal origin, environment, food handlers, and companion animals cause transmission of the bacteria. Identification methods like PCR and more recently MALDI-TOF-MS and WGS are widely used in laboratory and industrial settings. There are certain advantages and disadvantages associated with each of the methods. The biofilm formation property of the *Staphylococcus epidermidis* is a crucial factor in medical settings whereas *Staphylococcus aureus* is more of a foodborne pathogen. Drug resistance has caused the increased occurrence of MRSA, attempts are being made to develop drugs to combat community-associated as well as hospital-associated MRSA.

References

Abdurrahman, G., F. Schmiedeke, C. Bachert, B.M. Bröker and S. Holtfreter. 2020. Allergy—a new role for T cell superantigens of *Staphylococcus aureus*? Toxins 12: 176.

Abolghait, S.K., A.G. Fathi, F.M. Youssef and A.M. Algammal. 2020. Methicillin-resistant *Staphylococcus aureus* (MRSA) isolated from chicken meat and giblets often produces staphylococcal enterotoxin B (SEB) in non-refrigerated raw chicken livers. International Journal of Food Microbiology 328: 108669.

Adams, C.E. and S.J. Dancer. 2020. Dynamic transmission of *Staphylococcus aureus* in the intensive care unit. International Journal of Environmental Research and Public Health 17: 2109.

Adcock, P.M., P. Pastor, F. Medley, J.E. Patterson and T.V. Murphy. 1998. Methicillin-Resistant *Staphylococcus aureus* in two child care centers. The Journal of Infectious Diseases 178: 577–580.

Aiello, A.E., F.D. Lowy, L.N. Wright and E.L. Larson. 2006. Methicillin-resistant *Staphylococcus aureus* among US prisoners and military personnel: review and recommendations for future studies. The Lancet Infectious Diseases 6: 335–341.

Ali, R., K. Al-Achkar, A. Al-Mariri and M. Safi. 2014. Role of Polymerase Chain Reaction (PCR) in the detection of antibiotic-resistant *Staphylococcus aureus*. Egyptian Journal of Medical Human Genetics 15: 293–298.

Allard, M.W., R. Bell, C.M. Ferreira, N. Gonzalez-Escalona, M. Hoffmann, T. Muruvanda, A. Ottesen, P. Ramachandran, E. Reed and S. Sharma. 2018. Genomics of foodborne pathogens for microbial food safety. Current Opinion in Biotechnology 49: 224–229.

Allen, K., J. Anson, L. Parsons and N. Frost. 1997. Staff carriage of methicillin-resistant *Staphylococcus aureus* (EMRSA 15) and the home environment: a case report. Journal of Hospital Infection 35: 307–311.

Aman, M.J., H. Karauzum, M.G. Bowden and Nguyen, T.L. 2010. Structural model of the pre-pore ring-like structure of Panton-Valentine leukocidin: providing dimensionality to biophysical and mutational data. Journal of Biomolecular Structure and Dynamics 28: 1–12.

Archer, N.K., M.J. Mazaitis, J.W. Costerton, J.G. Leid, M.E. Powers and M.E. Shirtliff. 2011. *Staphylococcus aureus* biofilms: properties, regulation, and roles in human disease. Virulence 2: 445–459.

Asao, T., Y. Kumeda, T. Kawai, T. Shibata, H. Oda, K. Haruki, H. Nakazawa and S. Kozaki. 2003. An extensive outbreak of staphylococcal food poisoning due to low-fat milk in Japan: estimation of enterotoxin A in the incriminated milk and powdered skim milk. Epidemiology and Infection 130: 33.

Baggett, H.C., T.W. Hennessy, K. Rudolph, D. Bruden, A. Reasonover, A. Parkinson, R. Sparks, R.M. Donlan, P. Martinez, K. Mongkolrattanothai and J.C. Butler. 2004. Community-onset Methicillin-Resistant *Staphylococcus aureus* associated with antibiotic use and the cytotoxin panton-valentine leukocidin during a furunculosis outbreak in rural Alaska. The Journal of Infectious Diseases 189: 1565–1573.

Baier, C., E. Ebadi, T.R. Mett, M. Stoll, G. Küther, P.M. Vogt and F.-C. Bange. 2020. Epidemiologic and molecular investigation of a MRSA outbreak caused by a contaminated bathtub for carbon dioxide hydrotherapy and review of the literature. Canadian Journal of Infectious Diseases and Medical Microbiology 2020.

Barber, M. and M. Rozwadowska-dowzenko. 1948. Infection by Penicillin-Resistant Staphyloeoeci. Lancet, 641–644.

Barrett, F.F., R.F. McGehee and M. Finland. 1968. Methicillin-Resistant *Staphylococcus aureus* at Boston City Hospital. New England Journal of Medicine 279: 441–448.

Beard-Pegler, M.A., E. Stubbs and A.M. Vickery. 1988. Observations on the resistance to drying of staphylococcal strains. Journal of Medical Microbiology 26: 251–255.

Becker, K., C. Heilmann and G. Peters. 2014. Coagulase-Negative Staphylococci. Clinical Microbiology Reviews 27: 870–926.

Benkerroum, N. 2018. Staphylococcal enterotoxins and enterotoxin-like toxins with special reference to dairy products: An overview. Critical Reviews in Food Science and Nutrition 58: 1943–1970.

Bernardo, K., N. Pakulat, M. Macht, O. Krut, H. Seifert, S. Fleer, F. Hünger and M. Krönke. 2002. Identification and discrimination of *Staphylococcus aureus* strains using matrix-assisted laser desorption/ionization-time of flight mass spectrometry. Proteomics 2: 747–753.

Böcher, S., B. Tønning, R.L. Skov and J. Prag. 2009. *Staphylococcus lugdunensis*, a common cause of skin and soft tissue infections in the community. Journal of Clinical Microbiology 47: 946–950.

Boyd, D.A., B.M. Willey, D. Fawcett, N. Gillani and M.R. Mulvey. 2008. Molecular characterization of *Enterococcus faecalis* N06-0364 with low-level vancomycin resistance harboring a novel d-Ala-d-Ser Gene Cluster, *vanL*. Antimicrobial Agents and Chemotherapy 52: 2667–2672.

Brakstad, O.G., K. Aasbakk and J.A. Maeland. 1992. Detection of *Staphylococcus aureus* by polymerase chain reaction amplification of the nuc gene. Journal of Clinical Microbiology 30: 1654–1660.

Bran, J.L., M.E. Levison and D. Kaye. 1972. Survey for Methicillin-Resistant Staphylococci. Antimicrobial Agents and Chemotherapy 1: 235–236.

Brown, N., M. Reacher, W. Rice, I. Roddick, L. Reeve, N. Verlander, S. Broster, A. Ogilvy-Stuart, A. D'Amore and J. Ahluwalia. 2019. An outbreak of meticillin-resistant *Staphylococcus aureus* colonization in a neonatal intensive care unit: use of a case–control study to investigate and control it and lessons learnt. Journal of Hospital Infection 103: 35–43.

Bukowski, M., B. Wladyka and G. Dubin. 2010. Exfoliative toxins of *Staphylococcus aureus*. Toxins 2: 1148–1165.

Bynoe, E.T., R.H. Elder and R.D. Comtois. 1956. Phage-typing and antibiotic-resistance of staphylococci isolated in a general hospital. Canadian Journal of Microbiology 2: 346–358.

Cao, R., N. Zeaki, N. Wallin-Carlquist, P.N. Skandamis, J. Schelin and P. Rådström. 2012. Elevated enterotoxin A expression and formation in *Staphylococcus aureus* and its association with prophage induction. Applied and Environmental Microbiology 78: 4942–4948.

Capurro, A., A. Aspán, H.E. Unnerstad, K.P. Waller and K. Artursson. 2010. Identification of potential sources of *Staphylococcus aureus* in herds with mastitis problems. Journal of Dairy Science 93: 180–191.

Carbonnelle, E., J.-L. Beretti, S. Cottyn, G. Quesne, P. Berche, X. Nassif and A. Ferroni. 2007. Rapid identification of Staphylococci isolated in clinical microbiology laboratories by matrix-assisted laser desorption ionization-time of flight mass spectrometry. Journal of Clinical Microbiology 45: 2156–2161.

Castro, A., C. Santos, H. Meireles, J. Silva and P. Teixeira. 2016. Food handlers as potential sources of dissemination of virulent strains of *Staphylococcus aureus* in the community. Journal of Infection and Public Health 9: 153–160.

Castro, R., S. Pedroso, S. Sandes, G. Silva, K. Luiz, R. Dias, H. Figueiredo, S. Santos, A. Nunes and M. Souza. 2020. Virulence factors and antimicrobial resistance of *Staphylococcus aureus* isolated from the production process of Minas artisanal cheese from the region of Campo das Vertentes, Brazil. Journal of Dairy Science 103: 2098–2110.

CDC. 1999. Four pediatric deaths from community-acquired methicillin-resistant *Staphylococcus aureus*—Minnesota and North Dakota, 1997–1999. Morbidity and Mortality Weekly Report 48: 707–710.

CDC. 2002. *Staphylococcus aureus* resistant to vanomycin—United States, 2002. Infectious Diseases in Clinical Practice 11: 163.

CDC. 2004. Community-associated methicillin-resistant *Staphylococcus aureus* infections in Pacific Islanders—Hawaii, 2001–2003. Morbidity and Mortality Weekly Report 53: 767–770.

Chang, S., D.M. Sievert, J.C. Hageman, M.L. Boulton, F.C. Tenover, F.P. Downes, S. Shah, J.T. Rudrik, G.R. Pupp, W.J. Brown, D. Cardo and S.K. Fridkin. 2003. Infection with vancomycin-resistant *Staphylococcus aureus* containing the vanA resistance gene. New England Journal of Medicine 348: 1342–1347.

Chen, C.-J. and Y.-C. Huang. 2014. New epidemiology of *Staphylococcus aureus* infection in Asia. Clinical Microbiology and Infection 20: 605–623.

Cong, Y., S. Yang and X. Rao. 2020. Vancomycin resistant *Staphylococcus aureus* infections: A review of case updating and clinical features. Journal of Advanced Research 21: 169–176.

Conly, J.M. and B.L. Johnston. 2003. The emergence of methicillin-resistant *Staphylococcus aureus* as a community-acquired pathogen in Canada. Canadian Journal of Infectious Diseases 14: 197126.

Costerton, J.W., Z. Lewandowski, D.E. Caldwell, D.R. Korber and H.M. Lappin-Scott. 1995. Microbial biofilms. Annual Review of Microbiology 49: 711–745.

Courvalin, P. 2006. Vancomycin resistance in gram-positive cocci. Clinical Infectious Diseases 42: S25–S34.

Cox, R., C. Conquest, C. Mallaghan and R. Marples. 1995. A major outbreak of methicillin-resistant *Staphylococcus aureus* caused by a new phage-type (EMRSA-16). Journal of Hospital Infection 29: 87–106.

Crisóstomo, M.I., H. Westh, A. Tomasz, M. Chung, D.C. Oliveira and H. de Lencastre. 2001. The evolution of methicillin resistance in *Staphylococcus aureus*: Similarity of genetic backgrounds in historically early methicillin-susceptible and -resistant isolates and contemporary epidemic clones. Proceedings of the National Academy of Sciences 98: 9865–9870.

Crossley, K., B. Landesman and D. Zaske. 1979. An outbreak of infections caused by strains of *Staphylococcus aureus* resistant to methicillin and aminoglycosides. II. Epidemiologic studies. The Journal of Infectious Diseases 139: 280–287.

Cunningham, S.A., N. Chia, P.R. Jeraldo, D.J. Quest, J.A. Johnson, D.J. Boxrud, A.J. Taylor, J. Chen, G.D. Jenkins and T.M. Drucker. 2017. Comparison of whole-genome sequencing methods for analysis of three methicillin-resistant *Staphylococcus aureus* outbreaks. Journal of Clinical Microbiology 55: 1946–1953.

Davis, M.F., P. Baron, L.B. Price, L.W. D'Ann, S. Jeyaseelan, I.R. Hambleton, G.B. Diette, P.N. Breysse and M.C. McCormack. 2012. Dry collection and culture methods for recovery of methicillin-susceptible and methicillin-resistant *Staphylococcus aureus* strains from indoor home environments. Applied and Environmental Microbiology 78: 2474–2476.

De Boer, E., J. Zwartkruis-Nahuis, B. Wit, X. Huijsdens, A. De Neeling, T. Bosch, R. Van Oosterom, A. Vila and A. Heuvelink. 2009. Prevalence of methicillin-resistant *Staphylococcus aureus* in meat. International Journal of Food Microbiology 134: 52–56.

de Castro Nunes, E.L., K.R.N. dos Santos, P.J.J. Mondino, M.d.C. de Freire Bastos and M. Giambiagi-deMarval. 1999. Detection of ileS-2 gene encoding mupirocin resistance in methicillin-resistant *Staphylococcus aureus* by multiplex PCR. Diagnostic Microbiology and Infectious Disease 34: 77–81.

De Neeling, A., M. Van den Broek, E. Spalburg, M. van Santen-Verheuvel, W. Dam-Deisz, H. Boshuizen, A. Van De Giessen, E. Van Duijkeren and X. Huijsdens. 2007. High prevalence of methicillin-resistant *Staphylococcus aureus* in pigs. Veterinary Microbiology 122: 366–372.

Del Vecchio, V.G., J.M. Petroziello, M.J. Gress, F.K. McCleskey, G.P. Melcher, H.K. Crouch and J.R. Lupski. 1995. Molecular genotyping of methicillin-resistant *Staphylococcus aureus* via fluorophore-enhanced repetitive-sequence PCR. Journal of Clinical Microbiology 33: 2141–2144.

DeLeo, F.R., A.D. Kennedy, L. Chen, J.B. Wardenburg, S.D. Kobayashi, B. Mathema, K.R. Braughton, A.R. Whitney, A.E. Villaruz and C.A. Martens. 2011. Molecular differentiation of historic phage-type 80/81 and contemporary epidemic *Staphylococcus aureus*. Proceedings of the National Academy of Sciences 108: 18091–18096.

Dietze, B., A. Rath, C. Wendt and H. Martiny. 2001. Survival of MRSA on sterile goods packaging. Journal of Hospital Infection 49: 255–261.

Dubois, D., D. Leyssene, J.P. Chacornac, M. Kostrzewa, P.O. Schmit, R. Talon, R. Bonnet and J. Delmas. 2010. Identification of a variety of *Staphylococcus* species by matrix-assisted laser desorption ionization-time of flight mass spectrometry. Journal of Clinical Microbiology 48: 941–945.

DuMont, A.L., T.K. Nygaard, R.L. Watkins, A. Smith, L. Kozhaya, B.N. Kreiswirth, B. Shopsin, D. Unutmaz, J.M. Voyich and V.J. Torres. 2011. Characterization of a new cytotoxin that contributes to *Staphylococcus aureus* pathogenesis. Molecular Microbiology 79: 814–825.

Dupont, C., V. Sivadon-Tardy, E. Bille, B. Dauphin, J. Beretti, A. Alvarez, N. Degand, A. Ferroni, M. Rottman and J. Herrmann. 2010. Identification of clinical coagulase-negative staphylococci, isolated in microbiology laboratories, by matrix-assisted laser desorption/ionization-time of flight mass spectrometry and two automated systems. Clinical Microbiology and Infection 16: 998–1004.

Durand, G., F. Javerliat, M. Bes, J.-B. Veyrieras, G. Guigon, N. Mugnier, S. Schicklin, G. Kaneko, E. Santiago-Allexant and C. Bouchiat. 2018. Routine whole-genome sequencing for outbreak investigations of *Staphylococcus aureus* in a national reference center. Frontiers in Microbiology 9: 511.

Dylus, D., T. Pillonel, O. Opota, D. Wüthrich, H. Seth-Smith, A. Egli, S. Leo, V. Lazarevic, J. Schrenzel and S. Laurent. 2020. NGS-based *S. aureus* typing and outbreak analysis in clinical microbiology laboratories: lessons learned from a Swiss-wide proficiency test. Frontiers in Microbiology 11: 2822.

e Silva, S., J. Carvalho, C. Aires and M. Nitschke. 2017. Disruption of *Staphylococcus aureus* biofilms using rhamnolipid biosurfactants. Journal of Dairy Science 100: 7864–7873.

Edwards-Jones, V., M.A. Claydon, D.J. Evason, J. Walker, A. Fox and D. Gordon. 2000. Rapid discrimination between methicillin-sensitive and methicillin-resistant *Staphylococcus aureus* by intact cell mass spectrometry. Journal of Medical Microbiology 49: 295–300.

Eells, S.J., M.Z. David, A. Taylor, N. Ortiz, N. Kumar, J. Sieth, S. Boyle-Vavra, R.S. Daum and L.G. Miller. 2014. Persistent environmental contamination with USA300 methicillin-resistant *Staphylococcus aureus* and other pathogenic strain types in households with *S. aureus* skin infections. Infection Control & Hospital Epidemiology 35: 1373–1382.

Embil, J.M., J.A. McLeod, A.M. Al-Barrak, G.M. Thompson, F.Y. Aoki, E.J. Witwicki, M.F. Stranc, A.M. Kabani, D.R. Nicoll and L.E. Nicolle. 2001. An outbreak of methicillin resistant *Staphylococcus aureus* on a burn unit: potential role of contaminated hydrotherapy equipment. Burns 27: 681–688.

Fang, H., G. Hedin, G. Li and C.E. Nord. 2008. Genetic diversity of community-associated methicillin-resistant *Staphylococcus aureus* in southern Stockholm, 2000–2005. Clinical Microbiology and Infection 14: 370–376.

Fertner, M., K. Pedersen, V.F. Jensen, G. Larsen, M. Lindegaard, J.E. Hansen and M. Chriél. 2019. Within-farm prevalence and environmental distribution of livestock-associated methicillin-resistant *Staphylococcus aureus* in farmed mink (Neovison vison). Veterinary Microbiology 231: 80–86.

Fetsch, A., M. Contzen, K. Hartelt, A. Kleiser, S. Maassen, J. Rau, B. Kraushaar, F. Layer and B. Strommenger. 2014. *Staphylococcus aureus* food-poisoning outbreak associated with the consumption of ice-cream. International Journal of Food Microbiology 187: 1–6.

Fisher, E.L., M. Otto and G.Y. Cheung. 2018. Basis of virulence in enterotoxin-mediated staphylococcal food poisoning. Frontiers in Microbiology 9: 436.

Foster, T.J., J.A. Geoghegan, V.K. Ganesh and M. Höök. 2014. Adhesion, invasion and evasion: the many functions of the surface proteins of *Staphylococcus aureus*. Nature Reviews Microbiology 12: 49–62.

Francois, P., S. Harbarth, A. Huyghe, G. Renzi, M. Bento, A. Gervaix, D. Pittet and J. Schrenzel. 2008. Methicillin-resistant *Staphylococcus aureus*, Geneva, Switzerland, 1993–2005. Emerging Infectious Diseases 14: 304–307.

Fridkin, S.K., J.C. Hageman, M. Morrison, L.T. Sanza, K. Como-Sabetti, J.A. Jernigan, K. Harriman, L.H. Harrison, R. Lynfield and M.M. Farley. 2005. Methicillin-Resistant *Staphylococcus aureus* disease in three communities. New England Journal of Medicine 352: 1436–1444.

Frimodt-Møller, N., F. Espersen, P. Skinhøj and V.T. Rosdahl. 1997. Epidemiology of *Staphylococcus aureus* bacteremia in Denmark from 1957 to 1990. Clinical Microbiology and Infection 3: 297–305.

Fudaba, Y., K. Nishifuji, L.O. Andresen, T. Yamaguchi, H. Komatsuzawa, M. Amagai and M. Sugai. 2005. *Staphylococcus hyicus* exfoliative toxins selectively digest porcine desmoglein 1. Microbial Pathogenesis 39: 171–176.

Galia, L., M. Ligozzi, A. Bertoncelli and A. Mazzariol. 2019. Real-time PCR assay for detection of *Staphylococcus aureus*, Panton-Valentine Leucocidin and Methicillin Resistance directly from clinical samples. AIMS Microbiology 5: 138.

Gardella, N., M. von Specht, A. Cuirolo, A. Rosato, G. Gutkind and M. Mollerach. 2008. Community-associated methicillin-resistant *Staphylococcus aureus*, eastern Argentina. Diagnostic Microbiology and Infectious Disease 62: 343–347.

Giulieri, S.G., N.E. Holmes, T.P. Stinear and B.P. Howden. 2016. Use of bacterial whole-genome sequencing to understand and improve the management of invasive *Staphylococcus aureus* infections. Expert Review of Anti-infective Therapy 14: 1023–1036.

Goh, S.-H., S. Byrne, J. Zhang and A. Chow. 1992a. Molecular typing of *Staphylococcus aureus* on the basis of coagulase gene polymorphisms. Journal of Clinical Microbiology 30: 1642–1645.

Goh, S.H., S.K. Byrne, J.L. Zhang and A.W. Chow. 1992b. Molecular typing of *Staphylococcus aureus* on the basis of coagulase gene polymorphisms. Journal of Clinical Microbiology 30: 1642–1645.

Goldstein, R.E.R., S.A. Micallef, S.G. Gibbs, J.A. Davis, X. He, A. George, L.M. Kleinfelter, N.A. Schreiber, S. Mukherjee and A. Sapkota. 2012. Methicillin-resistant *Staphylococcus aureus* (MRSA) detected at four US wastewater treatment plants. Environmental Health Perspectives 120: 1551–1558.

Gordon, N., J. Price, K. Cole, R. Everitt, M. Morgan, J. Finney, A. Kearns, B. Pichon, B. Young and D. Wilson. 2014. Prediction of *Staphylococcus aureus* antimicrobial resistance by whole-genome sequencing. Journal of Clinical Microbiology 52: 1182–1191.

Gordon, N., B. Pichon, T. Golubchik, D. Wilson, J. Paul, D. Blanc, K. Cole, J. Collins, N. Cortes and M. Cubbon. 2017. Whole-genome sequencing reveals the contribution of long-term carriers in *Staphylococcus aureus* outbreak investigation. Journal of Clinical Microbiology 55: 2188–2197.

Grundmann, H., M. Aires-de-Sousa, J. Boyce and E. Tiemersma. 2006. Emergence and resurgence of meticillin-resistant *Staphylococcus aureus* as a public-health threat. The Lancet 368: 874–885.

Guo, L., L. Ye, Q. Zhao, Y. Ma, J. Yang and Y. Luo. 2014. Comparative study of MALDI-TOF MS and VITEK 2 in bacteria identification. Journal of Thoracic Disease 6: 534.

Hait, J.M., G. Cao, G. Kastanis, L. Yin, J.B. Pettengill and S.M. Tallent. 2021. Evaluation of virulence determinants using whole-genome sequencing and phenotypic biofilm analysis of outbreak-linked *Staphylococcus aureus* isolates. Frontiers in Microbiology 12.

Hamilton, F., R. Evans and A. MacGowan. 2021. The value of MALDI-TOF failure to provide an identification of Staphylococcal species direct from blood cultures and rule out *Staphylococcus aureus* bacteraemia: a post-hoc analysis of the RAPIDO trial. Access Microbiology 3: 000192.

Harris, L.G., S. Murray, B. Pascoe, J. Bray, G. Meric, L. Magerios, T.S. Wilkinson, R. Jeeves, H. Rohde and S. Schwarz. 2016. Biofilm morphotypes and population structure among *Staphylococcus epidermidis* from commensal and clinical samples. PLoS One 11: e0151240.

Haveri, M., M. Hovinen, A. Roslöf and S. Pyörälä. 2008. Molecular types and genetic profiles of *Staphylococcus aureus* strains isolated from bovine intramammary infections and extramammary sites. Journal of Clinical Microbiology 46: 3728–3735.

Health, U.D.o. and H. Services. 2011. National action plan to reduce healthcare-associated infections. Washington, DC: US Department of Health and Human Services; 2010.

Hébert, G.A. and G.A. Hancock. 1985. Synergistic hemolysis exhibited by species of staphylococci. Journal of Clinical Microbiology 22: 409–415.

Hennekinne, J.-A., M.-L. De Buyser and S. Dragacci. 2012. *Staphylococcus aureus* and its food poisoning toxins: characterization and outbreak investigation. FEMS Microbiology Reviews 36: 815–836.

Herold, B.C., L.C. Immergluck, M.C. Maranan, D.S. Lauderdale, R.E. Gaskin, S. Boyle-Vavra, C.D. Leitch and R.S. Daum. 1998. Community-acquired methicillin-resistant *Staphylococcus aureus* in children with no identified predisposing risk. JAMA 279: 593–598.

Hersh, A.L., H.F. Chambers, J.H. Maselli and R. Gonzales. 2008. National trends in ambulatory visits and antibiotic prescribing for skin and soft-tissue infections. Archives of Internal Medicine 168: 1585–1591.

Hill, P.C., C.G. Wong, L.M. Voss, S.L. Taylor, S. Pottumarthy, D. Drinkovic and A.J. Morris. 2001. Prospective study of 125 cases of *Staphylococcus aureus* bacteremia in children in New Zealand. The Pediatric Infectious Disease Journal 20: 868–873.

Hiltunen, A.K., K. Savijoki, T.A. Nyman, I. Miettinen, P. Ihalainen, J. Peltonen and A. Fallarero. 2019. Structural and functional dynamics of *Staphylococcus aureus* biofilms and biofilm matrix proteins on different clinical materials. Microorganisms 7: 584.

Hiramatsu, K., N. Aritaka, H. Hanaki, S. Kawasaki, Y. Hosoda, S. Hori, Y. Fukuchi and I. Kobayashi. 1997. Dissemination in Japanese hospitals of strains of *Staphylococcus aureus* heterogeneously resistant to vancomycin. The Lancet 350: 1670–1673.

Hoang, T.-M., C. Zhou, J. Lindgren, M. Galac, B. Corey, J. Endres, M. Olson and P. Fey. 2019. Transcriptional regulation of icaADBC by both IcaR and TcaR in *Staphylococcus epidermidis*. Journal of Bacteriology 201.

Hookey, J.V., J.F. Richardson and B.D. Cookson. 1998. Molecular typing of *Staphylococcus aureus* based on PCR restriction fragment length polymorphism and DNA sequence analysis of the coagulase gene. Journal of Clinical Microbiology 36: 1083–1089.

Hope, R., D.M. Livermore, G. Brick, M. Lillie, R. Reynolds and o.b.o.t.B.W.P.o.R. Surveillance. 2008. Non-susceptibility trends among staphylococci from bacteraemias in the UK and Ireland, 2001–06. Journal of Antimicrobial Chemotherapy 62: ii65–ii74.

Houston, P., S.E. Rowe, C. Pozzi, E.M. Waters and J.P. O'Gara. 2011. Essential role for the major autolysin in the fibronectin-binding protein-mediated *Staphylococcus aureus* biofilm phenotype. Infection and Immunity 79: 1153–1165.

Humphreys, H. and D. Coleman. 2019. Contribution of whole-genome sequencing to understanding of the epidemiology and control of meticillin-resistant *Staphylococcus aureus*. Journal of Hospital Infection 102: 189–199.

Hussain, M., M. Wilcox and P. White. 1993. The slime of coagulase-negative staphylococci: biochemistry and relation to adherence. FEMS Microbiology Reviews 10: 191–208.

Hwang, S.Y., S.H. Kim, E.J. Jang, N.H. Kwon, Y.K. Park, H.C. Koo, W.K. Jung, J.M. Kim and Y.H. Park. 2007. Novel multiplex PCR for the detection of the *Staphylococcus aureus* superantigen and its application to raw meat isolates in Korea. International Journal of Food Microbiology 117: 99–105.

Imanishi, I., A. Nicolas, A.-C.B. Caetano, T.L. de Paula Castro, N.R. Tartaglia, R. Mariutti, E. Guédon, S. Even, N. Berkova and R.K. Arni. 2019. Exfoliative toxin E, a new *Staphylococcus aureus* virulence factor with host-specific activity. Scientific Reports 9: 1–12.

Irfan, S., I. Ahmed, F. Lalani, N. Anjum, N. Mohammad, M. Owais and A. Zafar. 2019. Methicillin resistant *Staphylococcus aureus* outbreak in a neonatal intensive care unit. East Mediterr. Health J. 25(7): 514–518. doi: 10.26719/emhj.18.058.

Izano, E.A., M.A. Amarante, W.B. Kher and J.B. Kaplan. 2008. Differential roles of poly-N-acetylglucosamine surface polysaccharide and extracellular DNA in *Staphylococcus aureus* and *Staphylococcus epidermidis* biofilms. Applied and Environmental Microbiology 74: 470–476.

James E. Peacock, F.J.M. and Richard P. Wenzel. 1980. Methicillin-resistant *Staphylococcus aureus*: Introduction and spread within a hospital. Annals of Internal Medicine 93: 526–532.

Jan-Roblero, J., E. García-Gómez, S. Rodríguez-Martínez, M.E. Cancino-Diaz and J.C. Cancino-Diaz. 2017. Surface proteins of *Staphylococcus aureus*. The Rise of Virulence and Antibiotic Resistance in *Staphylococcus aureus*, 169.

Jenul, C. and A.R. Horswill. 2019. Regulation of *Staphylococcus aureus* virulence. Gram-Positive Pathogens, 669–686.

Kallen, A.J., Y. Mu, S. Bulens, A. Reingold, S. Petit, K. Gershman, S.M. Ray, L.H. Harrison, R. Lynfield and G. Dumyati. 2010. Health care–associated invasive MRSA infections, 2005–2008. JAMA 304: 641–647.

Kaneko, J. and Y. Kamio. 2004. Bacterial two-component and hetero-heptameric pore-forming cytolytic toxins: structures, pore-forming mechanism, and organization of the genes. Bioscience, Biotechnology, and Biochemistry 68: 981–1003.

Kanerva, M., S. Salmenlinna, J. Vuopio-Varkila, P. Lehtinen, T. Möttönen, M.J. Virtanen and O. Lyytikäinen. 2009. Community-associated methicillin-resistant *Staphylococcus aureus* isolated in Finland in 2004 to 2006. Journal of Clinical Microbiology 47: 2655–2657.

Kaplan, S.L., K.G. Hulten, B.E. Gonzalez, W.A. Hammerman, L. Lamberth, J. Versalovic and E.O. Mason, Jr. 2005. Three-year surveillance of community-acquired *Staphylococcus aureus* infections in children. Clinical Infectious Diseases 40: 1785–1791.

Kashif, A., J.-A. McClure, S. Lakhundi, M. Pham, S. Chen, J.M. Conly and K. Zhang. 2019. *Staphylococcus aureus* ST398 virulence is associated with factors carried on prophage φSa3. Frontiers in Microbiology 10: 2219.

Kazakova, S.V., J.C. Hageman, M. Matava, A. Srinivasan, L. Phelan, B. Garfinkel, T. Boo, S. McAllister, J. Anderson, B. Jensen, D. Dodson, D. Lonsway, L.K. McDougal, M. Arduino, V.J. Fraser, G. Killgore, F.C. Tenover, S. Cody and D.B. Jernigan. 2005. A clone of methicillin-resistant *Staphylococcus aureus* among professional football players. New England Journal of Medicine 352: 468–475.

Kearns, A., P. Seiders, J. Wheeler, R. Freeman and M. Steward. 1999. Rapid detection of methicillin-resistant staphylococci by multiplex PCR. Journal of Hospital Infection 43: 33–37.

Khanna, T., R. Friendship, C. Dewey and J. Weese. 2008. Methicillin resistant *Staphylococcus aureus* colonization in pigs and pig farmers. Veterinary Microbiology 128: 298–303.

Kim, J.-M., I. Kim, S.H. Chung, Y. Chung, M. Han and J.-S. Kim. 2019. Rapid discrimination of methicillin-resistant *Staphylococcus aureus* by MALDI-TOF MS. Pathogens 8: 214.

Kirby, W.M.M. 1944. Extraction of a highly potent penicillin inactivator from penicillin-resistant staphylococci. Science 99: 452–453.

Kobayashi, T., H. Nakaminami, H. Ohtani, K. Yamada, Y. Nasu, S. Takadama, N. Noguchi, T. Fujii and T. Matsumoto. 2020. An outbreak of severe infectious diseases caused by methicillin-resistant *Staphylococcus aureus* USA300 clone among hospitalized patients and nursing staff in a tertiary care university hospital. Journal of Infection and Chemotherapy 26: 76–81.

Kong, Z., P. Zhao, H. Liu, X. Yu, Y. Qin, Z. Su, S. Wang, H. Xu and J. Chen. 2016. Whole-genome sequencing for the investigation of a hospital outbreak of MRSA in China. PLoS One 11: e0149844.

Kot, B., H. Sytykiewicz and I. Sprawka. 2018. Expression of the biofilm-associated genes in methicillin-resistant *Staphylococcus aureus* in biofilm and planktonic conditions. International Journal of Molecular Sciences 19: 3487.

Kourtis, A.P., K. Hatfield, J. Baggs, Y. Mu, I. See, E. Epson, J. Nadle, M.A. Kainer, G. Dumyati and S. Petit. 2019. Vital signs: epidemiology and recent trends in methicillin-resistant and in methicillin-susceptible *Staphylococcus aureus* bloodstream infections—United States. Morbidity and Mortality Weekly Report 68: 214.

Kruse, T., M. Levisson, W.M. de Vos and H. Smidt. 2014. vanI: a novel d-Ala-d-Lac vancomycin resistance gene cluster found in Desulfitobacterium hafniense. Microbial Biotechnology 7: 456–466.

Kumburu, H.H., T. Sonda, P. Leekitcharoenphon, M. van Zwetselaar, O. Lukjancenko, M. Alifrangis, O. Lund, B.T. Mmbaga, G. Kibiki and F.M. Aarestrup. 2018. Hospital epidemiology of methicillin-resistant *Staphylococcus aureus* in a tertiary care hospital in Moshi, Tanzania, as determined by whole genome sequencing. BioMed Research International 2018.

Kwon, N.H., K.T. Park, J.S. Moon, W.K. Jung, S.H. Kim, J.M. Kim, S.K. Hong, H.C. Koo, Y.S. Joo and Y.H. Park. 2005. Staphylococcal cassette chromosome mec (SCC mec) characterization and molecular analysis for methicillin-resistant *Staphylococcus aureus* and novel SCC mec subtype IVg isolated from bovine milk in Korea. Journal of Antimicrobial Chemotherapy 56: 624–632.

Kwon, N.H., K.T. Park, W.K. Jung, H.Y. Youn, Y. Lee, S.H. Kim, W. Bae, J.Y. Lim, J.Y. Kim and J.M. Kim. 2006. Characteristics of methicillin resistant *Staphylococcus aureus* isolated from chicken meat and hospitalized dogs in Korea and their epidemiological relatedness. Veterinary Microbiology 117: 304–312.

Lachica, R.V., P.D. Hoeprich and H.P. Riemann. 1972. Tolerance of staphylococcal thermonuclease to stress. Applied Microbiology 23: 994–997.

Landrum, M.L., C. Neumann, C. Cook, U. Chukwuma, M.W. Ellis, D.R. Hospenthal and C.K. Murray. 2012. Epidemiology of *Staphylococcus aureus* blood and skin and soft tissue infections in the US military health system, 2005–2010. JAMA 308: 50–59.

Lange, C., M. Cardoso, D. Senczek and S. Schwarz. 1999. Molecular subtyping of *Staphylococcus aureus* isolates from cases of bovine mastitis in Brazil. Veterinary Microbiology 67: 127–141.

Larsen, A.R., M. Stegger, R.V. Goering, M. Sørum and R. Skov. 2007. Emergence and dissemination of the methicillin resistant *Staphylococcus aureus* USA300 clone in Denmark (2000–2005). Eurosurveillance 12: 3–4%P 682.

Latorre, A.A., P.A. Pachá, G. González-Rocha, I. San Martín, M. Quezada-Aguiluz, A. Aguayo-Reyes, H. Bello-Toledo, R. Oliva, A. Estay and J. Pugin. 2020. On-farm surfaces in contact with milk: The role of *Staphylococcus aureus*-containing biofilms for udder health and milk quality. Foodborne Pathogens and Disease 17: 44–51.

Laupland, K.B., T. Ross and D.B. Gregson. 2008. *Staphylococcus aureus* bloodstream infections: risk factors, outcomes, and the influence of methicillin resistance in Calgary, Canada, 2000–2006. The Journal of Infectious Diseases 198: 336–343.

Lawes, T., J.-M. López-Lozano, C. Nebot, G. Macartney, R. Subbarao-Sharma, C.R. Dare, G.F. Edwards and I.M. Gould. 2015. Turning the tide or riding the waves? Impacts of antibiotic stewardship and infection control on MRSA strain dynamics in a Scottish region over 16 years: non-linear time series analysis. BMJ Open 5.

Le, K.Y., S. Dastgheyb, T.V. Ho and M. Otto. 2014. Molecular determinants of staphylococcal biofilm dispersal and structuring. Frontiers in Cellular and Infection Microbiology 4: 167.

Lebreton, F., F. Depardieu, N. Bourdon, M. Fines-Guyon, P. Berger, S. Camiade, R. Leclercq, P. Courvalin and V. Cattoir. 2011. d-Ala-d-Ser VanN-Type transferable vancomycin resistance in *Enterococcus faecium*. Antimicrobial Agents and Chemotherapy 55: 4606–4612.

Lee, J.H. 2003. Methicillin (Oxacillin)-resistant *Staphylococcus aureus* strains isolated from major food animals and their potential transmission to humans. Applied and Environmental Microbiology 69: 6489–6494.

Löfblom, J., R. Rosenstein, M.-T. Nguyen, S. Ståhl and F. Götz. 2017. *Staphylococcus carnosus*: from starter culture to protein engineering platform. Applied Microbiology and Biotechnology 101: 8293–8307.

Marcos, J.Y., A.C. Soriano, M.S. Salazar, C.H. Moral, S.S. Ramos, M.S. Smeltzer and G.N. Carrasco. 1999. Rapid identification and typing of *Staphylococcus aureus* by PCR-restriction fragment length polymorphism analysis of the aroA gene. Journal of Clinical Microbiology 37: 570–574.

Martineau, F., F.J. Picard, P.H. Roy, M. Ouellette and M.G. Bergeron. 1998. Species-specific and ubiquitous-DNA-based assays for rapid identification of *Staphylococcus aureus*. Journal of Clinical Microbiology 36: 618–623.

Martins, K.B., A.M. Ferreira, A.L. Mondelli, T.T. Rocchetti and M.d. LR de S da Cunha. 2018. Evaluation of MALDI-TOF VITEK® MS and VITEK® 2 system for the identification of *Staphylococcus saprophyticus*. Future Microbiology 13: 1603–1609.

Masterton, R., J. Coia, A. Notman, L. Kempton-Smith and B. Cookson. 1995. Refractory methicillin-resistant *Staphylococcus aureus* carriage associated with contamination of the home environment. Journal of Hospital Infection 29: 318–319.

Medrano-Félix, A., C. Martínez, N. Castro-del Campo, J. León-Félix, F. Peraza-Garay, C.P. Gerba and C. Chaidez. 2011. Impact of prescribed cleaning and disinfectant use on microbial contamination in the home. Journal of Applied Microbiology 110: 463–471.

Mehrotra, M., G. Wang and W.M. Johnson. 2000. Multiplex PCR for detection of genes for *Staphylococcus aureus* enterotoxins, exfoliative toxins, toxic shock syndrome toxin 1, and methicillin resistance. Journal of Clinical Microbiology 38: 1032–1035.

Miller, L.G. and B.A. Diep. 2008. Colonization, fomites, and virulence: rethinking the pathogenesis of community-associated methicillin-resistant *Staphylococcus aureus* infection. Clinical Infectious Diseases 46: 752–760.

Monnier, A.A., E. Tacconelli, C. Årdal, M. Cavaleri and I.C. Gyssens. 2020. A case study on *Staphylococcus aureus* bacteraemia: available treatment options, antibiotic R&D and responsible antibiotic-use strategies. JAC-Antimicrobial Resistance 2: dlaa034.

Moormeier, D.E. and K.W. Bayles. 2017. *Staphylococcus aureus* biofilm: a complex developmental organism. Molecular Microbiology 104: 365–376.

Moran, G.J., A. Krishnadasan, R.J. Gorwitz, G.E. Fosheim, L.K. McDougal, R.B. Carey and D.A. Talan. 2006. Methicillin-resistant *S. aureus* infections among patients in the emergency department. New England Journal of Medicine 355: 666–674.

Morrison, J.M., K.L. Anderson, K.E. Beenken, M.S. Smeltzer and P.M. Dunman. 2012. The staphylococcal accessory regulator, SarA, is an RNA-binding protein that modulates the mRNA turnover properties of late-exponential and stationary phase *Staphylococcus aureus* cells. Frontiers in Cellular and Infection Microbiology 2: 26.

Moyle, T., K. Drake, V. Gole, K. Chousalkar and S. Hazel. 2016. Bacterial contamination of eggs and behaviour of poultry flocks in the free range environment. Comparative Immunology, Microbiology and Infectious Diseases 49: 88–94.

Mukherjee, N., S.E. Dowd, A. Wise, S. Kedia, V. Vohra and P. Banerjee. 2014. Diversity of bacterial communities of fitness center surfaces in a US metropolitan area. International Journal of Environmental Research and Public Health 11: 12544–12561.

Mukherjee, N., I.M. Sulaiman and P. Banerjee. 2016. Characterization of methicillin-resistant *Staphylococcus aureus* isolates from fitness centers in the Memphis metropolitan area, Tennessee. American Journal of Infection Control 44: 1681–1683.

Munoz, A., N. Dominguez-Gasca, C. Jimenez-Lopez and A.B. Rodriguez-Navarro. 2015. Importance of eggshell cuticle composition and maturity for avoiding trans-shell Salmonella contamination in chicken eggs. Food Control 55: 31–38.

Nakagawa, S., I. Taneike, D. Mimura, N. Iwakura, T. Nakayama, T. Emura, M. Kitatsuji, A. Fujimoto and T. Yamamoto. 2005. Gene sequences and specific detection for Panton-Valentine leukocidin. Biochemical and Biophysical Research Communications 328: 995–1002.

Nemati, M., K. Hermans, U. Lipinska, O. Denis, A. Deplano, M. Struelens, L.A. Devriese, F. Pasmans and F. Haesebrouck. 2008. Antimicrobial resistance of old and recent *Staphylococcus aureus* isolates from poultry: first detection of livestock-associated methicillin-resistant strain ST398. Antimicrobial Agents and Chemotherapy 52: 3817–3819.

Nguyen, H.T., T.H. Nguyen and M. Otto. 2020. The staphylococcal exopolysaccharide PIA–Biosynthesis and role in biofilm formation, colonization, and infection. Computational and Structural Biotechnology Journal.

Nimmo, G.R. and G.W. Coombs. 2008. Community-associated methicillin-resistant *Staphylococcus aureus* (MRSA) in Australia. International Journal of Antimicrobial Agents 31: 401–410.

Nimmo, G.R. 2012. USA300 abroad: global spread of a virulent strain of community-associated methicillin-resistant *Staphylococcus aureus*. Clinical Microbiology and Infection 18: 725–734.

Nix, I.D., E.A. Idelevich, L.M. Storck, K. Sparbier, O. Drews, M. Kostrzewa and K. Becker. 2020. Detection of methicillin resistance in *Staphylococcus aureus* from agar cultures and directly from positive blood cultures using MALDI-TOF mass spectrometry-based direct-on-target microdroplet growth assay. Frontiers in Microbiology 11: 232.

Okolie, C.E., K.G. Wooldridge, D.P. Turner, A. Cockayne and R. James. 2015. Development of a heptaplex PCR assay for identification of *Staphylococcus aureus* and CoNS with simultaneous detection of virulence and antibiotic resistance genes. BMC Microbiology 15: 1–7.

Pacha, P., M. Munoz, E. Paredes-Osses and A. Latorre. 2020. Virulence profiles of *Staphylococcus aureus* isolated from bulk tank milk and adherences on milking equipment on Chilean dairy farms. Journal of Dairy Science.

Pannaraj, P.S., K.G. Hulten, B.E. Gonzalez, E.O. Mason, Jr. and S.L. Kaplan. 2006. Infective pyomyositis and myositis in children in the era of community-acquired, methicillin-resistant *Staphylococcus aureus* infection. Clinical Infectious Diseases 43: 953–960.

Park, K.-H., K.E. Greenwood-Quaintance, S.A. Cunningham, G. Rajagopalan, N. Chia, P.R. Jeraldo, J. Mandrekar and R. Patel. 2019. Lack of correlation of virulence gene profiles of *Staphylococcus aureus* bacteremia isolates with mortality. Microbial Pathogenesis 133: 103543.

Park, S.H., C. Park, J.-H. Yoo, S.-M. Choi, J.-H. Choi, H.-H. Shin, D.-G. Lee, S. Lee, J. Kim, S.E. Choi, Y.-M. Kwon and W.-S. Shin. 2009. Emergence of community-associated methicillin-resistant *Staphylococcus aureus* strains as a cause of healthcare-associated bloodstream infections in Korea. Infection Control & Hospital Epidemiology 30: 146–155.

Pérez-Sancho, M., A.I. Vela, P. Horcajo, M. Ugarte-Ruiz, L. Domínguez, J.F. Fernández-Garayzábal and R. de la Fuente. 2018. Rapid differentiation of *Staphylococcus aureus* subspecies based on MALDI-TOF MS profiles. Journal of Veterinary Diagnostic Investigation 30: 813–820.

Perret, M., C. Badiou, G. Lina, S. Burbaud, Y. Benito, M. Bes, V. Cottin, F. Couzon, C. Juruj and O. Dauwalder. 2012. Cross-talk between *Staphylococcus aureus* leukocidins-intoxicated macrophages and lung epithelial cells triggers chemokine secretion in an inflammasome-dependent manner. Cellular Microbiology 14: 1019–1036.

Piccinini, R., L. Cesaris, V. Dapra, V. Borromeo, C. Picozzi, C. Secchi and A. Zecconi. 2009. The role of teat skin contamination in the epidemiology of *Staphylococcus aureus* intramammary infections. The Journal of Dairy Research 76: 36.

Pillsbury, A., M. Chiew, J. Bates and V. Sheppeard. 2013. An outbreak of staphylococcal food poisoning in a commercially catered buffet. Commun Dis. Intell. 37: E144–E148.

Pinheiro, L., C.I. Brito, A. de Oliveira, P.Y.F. Martins, V.C. Pereira and M.d.L.R.d.S. da Cunha. 2015. *Staphylococcus epidermidis* and *Staphylococcus haemolyticus*: Molecular detection of cytotoxin and enterotoxin genes. Toxins 7: 3688–3699.

Price, J., N.C. Gordon, D. Crook, M. Llewelyn and J. Paul. 2013. The usefulness of whole genome sequencing in the management of *Staphylococcus aureus* infections. Clinical Microbiology and Infection 19: 784–789.

Pu, S., F. Han and B. Ge. 2009. Isolation and characterization of methicillin-resistant *Staphylococcus aureus* strains from Louisiana retail meats. Applied and Environmental Microbiology 75: 265–267.

Qin, H., X. Shi, L. Yu, K. Li, J. Wang, J. Chen, F. Yang, H. Xu and H. Xu. 2020. Multiplex real-time PCR coupled with sodium dodecyl sulphate and propidium monoazide for the simultaneous detection of viable *Listeria monocytogenes*, *Cronobacter sakazakii*, *Staphylococcus aureus* and *Salmonella* spp. in milk. International Dairy Journal 108: 104739.

Que, Y.-A. and P. Moreillon. 2015. 196 - *Staphylococcus aureus* (Including staphylococcal toxic shock syndrome). *In*: Bennett, J.E., R. Dolin and M.J. Blaser (eds.). Mandell, Douglas, and Bennett's Principles and Practice of Infectious Diseases (Eighth Edition). W.B. Saunders, Philadelphia. 2237–2271.e2235.

Rani, S.A., B. Pitts, H. Beyenal, R.A. Veluchamy, Z. Lewandowski, W.M. Davison, K. Buckingham-Meyer and P.S. Stewart. 2007. Spatial patterns of DNA replication, protein synthesis, and oxygen concentration within bacterial biofilms reveal diverse physiological states. Journal of Bacteriology 189: 4223–4233.

Rankin, S., S. Roberts, K. O'Shea, D. Maloney, M. Lorenzo and C.E. Benson. 2005. Panton valentine leukocidin (PVL) toxin positive MRSA strains isolated from companion animals. Veterinary Microbiology 108: 145–148.

Raz, R., R. Colodner and C.M. Kunin. 2005. Who Are You—*Staphylococcus saprophyticus*? Clinical Infectious Diseases 40: 896–898.

Roberts, M.C., O.O. Soge and D. No. 2013. Comparison of multi-drug resistant environmental methicillin-resistant *Staphylococcus aureus* isolated from recreational beaches and high touch surfaces in built environments. Frontiers in Microbiology 4: 74.

Rödström, K.E., P. Regenthal and K. Lindkvist-Petersson. 2015. Structure of staphylococcal enterotoxin E in complex with TCR defines the role of TCR loop positioning in superantigen recognition. PLoS One 10: e0131988.

Rountree, P.M. and B.M. Freeman. 1955. Infections caused by a particular phage type of *Staphylococcus aureus*. Medical Journal of Australia 2: 157–161.

Rubin, I.M., T.A. Hansen, A.M. Klingenberg, A.M. Petersen, P. Worning, H. Westh and M.D. Bartels. 2018. A sporadic four-year hospital outbreak of a ST97-IVa MRSA with half of the patients first identified in the community. Frontiers in Microbiology 9: 1494.

Rychert, J. 2019. Benefits and limitations of MALDI-TOF mass spectrometry for the identification of microorganisms. Journal of Infectiology 2.

Sánchez, M., C. Neira, A. Laca, A. Laca and M. Díaz. 2019. Survival and development of *Staphylococcus* in egg products. LWT 101: 685–693.

Sangeeta Joshi, P.R., Vikas Manchanda, Jyoti Bajaj, D.S. Chitnis, Vikas Gautam, Parijath Goswami, Varsha Gupta, B.N. Harish, Anju Kagal, Arti Kapil, Ratna Rao, Camilla Rodrigues, Raman Sardana, Kh Sulochana Devi, Anita Sharma and Veeragaghavan Balaji. 2013. Indian Network for Surveillance of Antimicrobial Resistance group, India. Methicillin resistant *Staphylococcus aureus* (MRSA) in India: prevalence & susceptibility pattern. The Indian Journal of Medical Research 137: 363–369.

Saruta, K., S. Hoshina and K. Machida. 1995. Genetic identification of *Staphylococcus aureus* by polymerase chain reaction using single-base-pair mismatch in 16S ribosomal RNA gene. Microbiology and Immunology 39: 839–844.

Schelin, J., N. Wallin-Carlquist, M. Thorup Cohn, R. Lindqvist and G.C. Barker. 2011. The formation of *Staphylococcus aureus* enterotoxin in food environments and advances in risk assessment. Virulence 2: 580–592.

Scott, E., S. Duty and M. Callahan. 2008. A pilot study to isolate *Staphylococcus aureus* and methicillin-resistant *S. aureus* from environmental surfaces in the home. American Journal of Infection Control 36: 458–460.

Senn, L., O. Clerc, G. Zanetti, P. Basset, G. Prod'hom, N.C. Gordon, A.E. Sheppard, D.W. Crook, R. James and H.A. Thorpe. 2016. The stealthy superbug: the role of asymptomatic enteric carriage in maintaining a long-term hospital outbreak of ST228 methicillin-resistant *Staphylococcus aureus*. MBio 7.

Sergelidis, D. and A. Angelidis. 2017. Methicillin-resistant *Staphylococcus aureus*: a controversial food-borne pathogen. Letters in Applied Microbiology 64: 409–418.

Shan, W., J. Li, Y. Fang, X. Wang, D. Gu and R. Zhang. 2016. Rapid identification of methicillin-resistant *Staphylococcus aureus* (MRSA) by the vitek MS saramis system. Current Microbiology 72: 29–32.

Sharma-Kuinkel, B.K., S.H. Ahn, T.H. Rude, Y. Zhang, S.Y. Tong, F. Ruffin, F.C. Genter, K.R. Braughton, F.R. DeLeo and S.L. Barriere. 2012. Presence of genes encoding Panton-Valentine leukocidin is not the primary determinant of outcome in patients with hospital-acquired pneumonia due to *Staphylococcus aureus*. Journal of Clinical Microbiology 50: 848–856.

Singhal, N., M. Kumar, P.K. Kanaujia and J.S. Virdi. 2015. MALDI-TOF mass spectrometry: an emerging technology for microbial identification and diagnosis. Frontiers in Microbiology 6: 791.

Soo Yean, C.Y., K. Selva Raju, R. Xavier, S. Subramaniam, S.C. Gopinath and S.V. Chinni. 2016. Molecular detection of methicillin-resistant *Staphylococcus aureus* by non-protein coding RNA-mediated monoplex polymerase chain reaction. PLoS One 11: e0158736.

Speziale, P., G. Pietrocola, T.J. Foster and J.A. Geoghegan. 2014. Protein-based biofilm matrices in Staphylococci. Frontiers in Cellular and Infection Microbiology 4: 171,

Stam-Bolink, E.M., D. Mithoe, W.H. Baas, J.P. Arends and A.V.M. Möller. 2007. Spread of a methicillin-resistant *Staphylococcus aureus* ST80 strain in the community of the northern Netherlands. European Journal of Clinical Microbiology & Infectious Diseases 26: 723.

Stepanović, S., D. Vuković, V. Trajković, T. Samardžić, M. Ćupić and M. Švabić-Vlahović. 2001. Possible virulence factors of *Staphylococcus sciuri*. FEMS Microbiology Letters 199: 47–53.

Stevens, D.L. and A.E. Bryant. 2011. CHAPTER 30 - Group A streptococcal and staphylococcal infections. pp. 203–211. *In*: Guerrant, R.L., D.H. Walker and P.F. Weller (eds.). Tropical Infectious Diseases: Principles, Pathogens and Practice (Third Edition). W.B. Saunders, Edinburgh.

Stevens, F.A. 1927. The occurrence of *Staphylococcus aureus* infection with a scarlatiniform rash. Journal of the American Medical Association 88: 1957–1958.

Straub, J.A., C. Hertel and W.P. Hammes. 1999. A 23S rDNA-targeted polymerase chain reaction–based system for detection of *Staphylococcus aureus* in meat starter cultures and dairy products. Journal of Food Protection 62: 1150–1156.

Strommenger, B., C. Braulke, B. Pasemann, C. Schmidt and W. Witte. 2008. Multiplex PCR for rapid detection of *Staphylococcus aureus* isolates suspected to represent community-acquired strains. Journal of Clinical Microbiology 46: 582–587.

Struyf, T. and K. Mertens. 2017. European Antimicrobial Resistance Surveillance Network (EARS-Net Belgium) Report 2017.

Sugimoto, S., F. Sato, R. Miyakawa, A. Chiba, S. Onodera, S. Hori and Y. Mizunoe. 2018. Broad impact of extracellular DNA on biofilm formation by clinically isolated Methicillin-resistant and-sensitive strains of *Staphylococcus aureus*. Scientific Reports 8: 1–11.

Sulaiman, I.M., P. Banerjee, Y.-H. Hsieh, N. Miranda, S. Simpson and K. Kerdahi. 2018. Rapid detection of *Staphylococcus aureus* and related species isolated from food, environment, cosmetics, a medical device, and clinical samples using the VITEK MS microbial identification system. Journal of AOAC International 101: 1135–1143.

Sullivan, M.J., D.R. Altman, K.I. Chacko, B. Ciferri, E. Webster, T.R. Pak, G. Deikus, M. Lewis-Sandari, Z. Khan and C. Beckford. 2019. A complete genome screening program of clinical methicillin-resistant *Staphylococcus aureus* isolates identifies the origin and progression of a neonatal intensive care unit outbreak. Journal of Clinical Microbiology 57: e01261–01219.

Sutcliffe, C.G., L.R. Grant, A. Reid, G. Douglass, L.B. Brown, K. Kellywood, R.C. Weatherholtz, R. Hubler, A. Quintana, R. Close, J.B. McAuley, M. Santosham, K.L. O'Brien and L.L. Hammitt. 2020. High burden of *Staphylococcus aureus* among native American individuals on the white mountain apache tribal lands. Open Forum Infectious Diseases 7.

Szabados, F., J. Woloszyn, C. Richter, M. Kaase and S. Gatermann. 2010. Identification of molecularly defined *Staphylococcus aureus* strains using matrix-assisted laser desorption/ionization time of flight mass spectrometry and the Biotyper 2.0 database. Journal of Medical Microbiology 59: 787–790.

T C Sorrell, D.R.P., S. Shanker, M. Foldes and R. Munro. 1982. Vancomycin therapy for methicillin-resistant *Staphylococcus aureus*. Annals of Internal Medicine 97: 344–350.

Tang, W., N. Ranganathan, V. Shahrezaei and G. Larrouy-Maumus. 2019. MALDI-TOF mass spectrometry on intact bacteria combined with a refined analysis framework allows accurate classification of MSSA and MRSA. PLoS One 14: e0218951.

Tenover, F.C., L.M. Weigel, P.C. Appelbaum, L.K. McDougal, J. Chaitram, S. McAllister, N. Clark, G. Killgore, C.M. O'Hara, L. Jevitt, J.B. Patel and B. Bozdogan. 2004. Vancomycin-resistant *Staphylococcus aureus* isolate from a patient in Pennsylvania. Antimicrobial Agents and Chemotherapy 48: 275–280.

Thorlacius-Ussing, L., H. Sandholdt, A.R. Larsen, A. Petersen and T. Benfield. 2019. Age-dependent increase in incidence of *Staphylococcus aureus* bacteremia, Denmark, 2008–2015. Emerging Infectious Diseases 25: 875.

Tolba, O., A. Loughrey, C.E. Goldsmith, B.C. Millar, P.J. Rooney and J.E. Moore. 2008. Survival of epidemic strains of healthcare (HA-MRSA) and community-associated (CA-MRSA) meticillin-resistant *Staphylococcus aureus* (MRSA) in river-, sea-and swimming pool water. International Journal of Hygiene and Environmental Health 211: 398–402.

Tonamo, A., I. Komlósi, L. Varga, M. Kačániová and F. Peles. 2021. Identification of ovine-associated staphylococci by MALDI-TOF mass spectrometry. Acta Alimentaria.

Tong, S.Y., E.J. Bishop, R.A. Lilliebridge, A.C. Cheng, Z. Spasova-Penkova, D.C. Holt, P.M. Giffard, M.I. McDonald, B.J. Currie and C.S. Boutlis. 2009. Community-associated strains of methicillin-resistant *Staphylococcus aureus* and methicillin-susceptible *S. aureus* in indigenous Northern Australia: epidemiology and outcomes. The Journal of Infectious Diseases 199: 1461–1470.

Tong, S.Y., J.S. Davis, E. Eichenberger, T.L. Holland and V.G. Fowler. 2015. *Staphylococcus aureus* infections: epidemiology, pathophysiology, clinical manifestations, and management. Clinical Microbiology Reviews 28: 603–661.

Tsai, H.-C., C.-W. Tao, B.-M. Hsu, Y.-Y. Yang, Y.-C. Tseng, T.-Y. Huang, S.-W. Huang, Y.-J. Kuo and J.-S. Chen. 2020. Multidrug-resistance in methicillin-resistant *Staphylococcus aureus* (MRSA) isolated from a subtropical river contaminated by nearby livestock industries. Ecotoxicology and Environmental Safety 200: 110724.

Tucker, P.W., E.E. Hazen and F.A. Cotton. 1978. Staphylococcal nuclease reviewed: A prototypic study in contemporary enzymology. I isolation; physical and enzymatic properties. Molecular and Cellular Biochemistry 22: 67–78.

Udo, E.E., S. Al-Mufti and M.J. Albert. 2009. The prevalence of antimicrobial resistance and carriage of virulence genes in *Staphylococcus aureus* isolated from food handlers in Kuwait City restaurants. BMC Research Notes 2: 108.

Uhlemann, A.-C., J. Knox, M. Miller, C. Hafer, G. Vasquez, M. Ryan, P. Vavagiakis, Q. Shi and F.D. Lowy. 2011. The environment as an unrecognized reservoir for community-associated methicillin-resistant *Staphylococcus aureus* USA300: a case-control study. PLoS One 6: e22407.

Valle, J., A. Toledo-Arana, C. Berasain, J.M. Ghigo, B. Amorena, J.R. Penadés and I. Lasa. 2003. SarA and not σB is essential for biofilm development by *Staphylococcus aureus*. Molecular Microbiology 48: 1075–1087.

Van Duijkeren, E., A. Box, J. Mulder, W. Wannet, A. Fluit and D. Houwers. 2003. Methicillin resistant *Staphylococcus aureus* (MRSA) infection in a dog in the Netherlands. Tijdschrift voor diergeneeskunde 128: 314–315.

Van Duijkeren, E., A. Box, M. Heck, W. Wannet and A. Fluit. 2004. Methicillin-resistant staphylococci isolated from animals. Veterinary Microbiology 103: 91–97.

Van Rijen, M., P. van Keulen and J. Kluytmans. 2007. P1591 Increase of pig-and calf-related MRSA in a Dutch hospital. International Journal of Antimicrobial Agents, S446–S447.

Vandenbroucke-Grauls, C. and D. Beaujean. 2006. Methicillin-resistant *Staphylococcus aureus* in pig breeders and cattle breeders. Nederlands tijdschrift voor geneeskunde 150: 1710–1712.

Vannuffel, P., J. Gigi, H. Ezzedine, B. Vandercam, M. Delmée, G. Wauters and J.-L. Gala. 1995. Specific detection of methicillin-resistant *Staphylococcus* species by multiplex PCR. Journal of Clinical Microbiology 33: 2864–2867.

Voss, A., F. Loeffen, J. Bakker, C. Klaassen and M. Wulf. 2005. Methicillin-resistant *Staphylococcus aureus* in pig farming. Emerging Infectious Diseases 11: 1965.

Wagenvoort, J., W. Sluijsmans and R. Penders. 2000. Better environmental survival of outbreak vs. sporadic MRSA isolates. Journal of Hospital Infection 45: 231–234.

Waknine, Y. 2006. Highlights From MMWR: Outbreaks of MRSA Infection in Newborns and More. The United States Centers for Disease Control and Prevention.

Wan, M.T., T.L. Lauderdale and C.C. Chou. 2013. Characteristics and virulence factors of livestock associated ST9 methicillin-resistant *Staphylococcus aureus* with a novel recombinant staphylocoagulase type. Veterinary Microbiology 162: 779–784.

Wan, M.T. and C.C. Chou. 2014. Spreading of β-lactam resistance gene (mecA) and methicillin-resistant *Staphylococcus aureus* through municipal and swine slaughterhouse wastewaters. Water Research 64: 288–295.

Wang, Y.-T., Y.-T. Lin, T.-W. Wan, D.-Y. Wang, H.-Y. Lin, C.-Y. Lin, Y.-C. Chen and L.-J. Teng. 2019. Distribution of antibiotic resistance genes among *Staphylococcus* species isolated from ready-to-eat foods. Journal of Food and Drug Analysis 27: 841–848.

Watkins, R.R., M.Z. David and R.A. Salata. 2012. Current concepts on the virulence mechanisms of meticillin-resistant *Staphylococcus aureus*. Journal of Medical Microbiology 61: 1179–1193.

Wildeman, P., S. Tevell, C. Eriksson, A.C. Lagos, B. Söderquist and B. Stenmark. 2020. Genomic characterization and outcome of prosthetic joint infections caused by *Staphylococcus aureus*. Scientific Reports 10: 1–14.

Wolny-Koładka, K. 2018. Resistance to antibiotics and the occurrence of genes responsible for the development of methicillin resistance in *Staphylococcus* bacteria isolated from the environment of horse riding centers. Journal of Equine Veterinary Science 61: 65–71.

Wu, M., X. Tong, S. Liu, D. Wang, L. Wang and H. Fan. 2019a. Prevalence of methicillin-resistant *Staphylococcus aureus* in healthy Chinese population: A system review and meta-analysis. PLoS One 14: e0223599.

Wu, S., J. Huang, F. Zhang, Q. Wu, J. Zhang, R. Pang, H. Zeng, X. Yang, M. Chen and J. Wang. 2019b. Prevalence and characterization of food-related methicillin-resistant *Staphylococcus aureus* (MRSA) in China. Frontiers in Microbiology 10: 304.

Xu, X., D. Lin, G. Yan, X. Ye, S. Wu, Y. Guo, D. Zhu, F. Hu, Y. Zhang, F. Wang, G.A. Jacoby and M. Wang. 2010. *vanM*, a new glycopeptide resistance gene cluster found in *Enterococcus faecium*. Antimicrobial Agents and Chemotherapy 54: 4643–4647.

Yoong, P. and V.J. Torres. 2013. The effects of *Staphylococcus aureus* leukotoxins on the host: cell lysis and beyond. Current Opinion in Microbiology 16: 63–69.

Zheng, Y., L. He, T.K. Asiamah and M. Otto. 2018. Colonization of medical devices by staphylococci. Environmental Microbiology 20: 3141–3153.

Zhu, W., K. Sieradzki, V. Albrecht, S. McAllister, W. Lin, O. Stuchlik, B. Limbago, J. Pohl and J.K. Rasheed. 2015. Evaluation of the Biotyper MALDI-TOF MS system for identification of *Staphylococcus* species. Journal of Microbiological Methods 117: 14–17.

Part III
Diagnosis of Human-Pathogenic Virus

Chapter 5

Advances and Opportunities in Hepatitis B Virus and Hepatitis C Virus Diagnostics[#]

Matthew D. Pauly,* Sabrina Weis-Torres and *Saleem Kamili*

Introduction

Viral hepatitis is a major global health burden caused by five diverse hepatotropic viruses. Hepatitis A, hepatitis B, hepatitis C, hepatitis D, and hepatitis E viruses are unrelated viruses with distinct replication mechanisms that are named in the order of their discovery (Table 1). Clinical signs and symptoms are similar among infections caused by these viruses and include jaundice, fever, fatigue, abdominal pain, dark urine, and clay-colored stool. Hepatitis A virus (HAV) and hepatitis E virus (HEV) are both small, non-enveloped, RNA viruses transmitted via the fecal-oral route, usually through the ingestion of contaminated food or water. Disease caused by HAV or HEV infection is acute and typically resolves in weeks to months in most individuals. HAV is vaccine-preventable. Hepatitis B virus (HBV) and hepatitis C virus (HCV) are unique bloodborne viruses that can establish long-term chronic infections. HBV is a small, enveloped DNA virus that is primarily transmitted perinatally or via unsafe sexual and injection practices. HCV is an enveloped RNA virus that is mostly associated with transmission via unsafe injection practices or other percutaneous exposure to contaminated blood. Recent therapeutic advancements have led to a reliable cure for HCV infection. Hepatitis D virus (HDV)

Centers for Disease Control and Prevention, Division of Viral Hepatitis, 1600 Clifton Rd. NE Atlanta, GA 30329, USA.
* Corresponding author: omx4@cdc.gov
# Disclaimer: The findings and conclusions in this report are those of the authors and do not necessarily represent the official position of the Centers for Disease Control and Prevention. Use of trade names and commercial sources is for identification only and does not constitute an endorsement by the U.S. Department of Health and Human Services or the U.S. Centers for Disease Control and Prevention.

Table 1. Hepatitis virus overview.

	Hepatitis A virus (HAV)	Hepatitis B virus (HBV)	Hepatitis C virus (HCV)	Hepatitis D virus (HDV)	Hepatitis E virus (HEV)
Virus family	*Picornaviridae*	*Hepadnaviridae*	*Flaviviridae*	*Kolmioviridae*	*Hepeviridae*
Genome (size in kb)	RNA (7.5)	DNA (3.2)	RNA (9.6)	RNA (1.7)	RNA (7.2)
Transmission route	Fecal-oral	Percutaneous/permucosal	Percutaneous/permucosal	Percutaneous/permucosal	Fecal-oral
Vaccine preventable	Yes	Yes	No	Yes (hepatitis B vaccine)	Yes (approved in few countries)
Progression to chronic infection	No	> 90% in neonates; < 10% in adults	> 50%	Only with chronic HBV infection	Only in immunocompromised people
Cure for chronic infection	N/A	No	Yes	No	Yes

is a bloodborne virus that can exacerbate the clinical manifestation of HBV infection. HDV is a satellite virus that is dependent on HBV for its replication. Since infection with each hepatitis virus can present similar signs and symptoms, accurate diagnoses rely on the detection of virus-specific and host immune markers. This chapter will focus on HBV and HCV infection diagnostics, particularly current advances and opportunities for further development and improvements in diagnostic assays.

HBV and HCV are major worldwide health problems, infecting an estimated 296 million and 58 million people worldwide in 2019, respectively (WHO 2021a). Both viruses can cause chronic infections that lead to long-term liver damage and cancer. Despite the availability of an efficacious hepatitis B vaccine and the recent development of highly effective drugs against HCV infection, there continue to be over one million deaths attributed to these viral infections annually. The World Health Organization (WHO) has set goals for the elimination of HBV and HCV infections, which include a 90% reduction in the number of new cases and a 65% reduction in deaths by the year 2030 (Figure 1; WHO 2016). To realize these goals, it will be necessary to make improvements to the entire cure cascade, including access to affordable and reliable diagnostic testing, and expeditious referral to curative treatment after diagnosis. There continues to be a high rate of new cases in certain populations and only 10% of chronic HBV and 21% of chronic HCV infections have been diagnosed (WHO 2021a). The recent global SARS-CoV-2 pandemic has hindered testing and linkage to care efforts for HBV and HCV infections, but it has

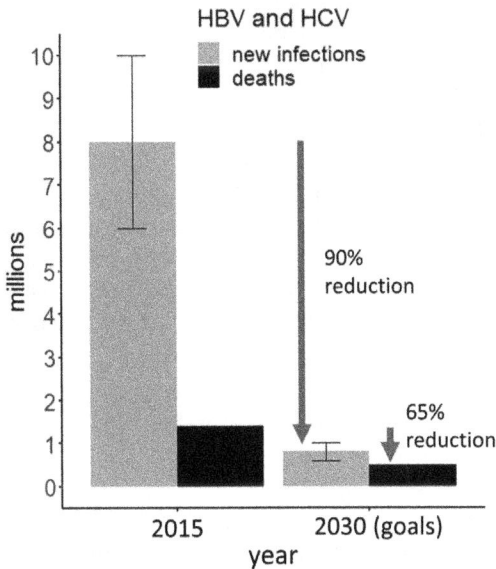

Figure 1. World Health Organization viral hepatitis elimination goals. The WHO has set worldwide goals for reducing the number of annual new infections and deaths caused by HBV and HCV by 2030 (WHO 2016). In 2015 there were estimated to be between six and ten million new HBV and HCV infections and 1.5 million deaths attributed to these viral infections annually. Current goals for the year 2030 are to reduce new infections by 90% and deaths by 65%.

also been the impetus for the development of stronger healthcare infrastructures, and novel diagnostic platforms and technologies that may be leveraged for infections by other pathogens including hepatitis viruses (Laury et al. 2021).

Virus-specific antibodies, viral antigens, and viral nucleic acids, alone or in various combinations, can be used to diagnose or determine the stage of infection with any of the hepatitis viruses. Many of the current methods for detecting diagnostic markers of HBV and HCV infection need to be performed in laboratory settings and require trained personnel for testing. Most tests use serum and/or plasma samples, which require phlebotomists to collect blood and cold storage conditions during transportation. These requirements make access to HBV and HCV infection testing challenging for some populations with high burdens of disease, such as persons who inject drugs (PWID), men who have sex with men (MSM), sex workers, incarcerated persons, and persons infected with human immunodeficiency virus (HIV), especially in low- and middle-income countries (LMICs). Accurate and timely diagnosis of these viral infections is important for preventing further transmission and initiating antiviral therapy. There are several areas where additional diagnostic markers, novel technologies, or expansion of existing technologies could improve options for diagnosing and monitoring treatment against viral hepatitis (Figure 2). Recent technological advances in testing methods that are rapid or can be performed at the site of sample collection have the potential to increase access to accurate diagnostics. Additional opportunities to increase access to testing can be achieved by using alternative sample types, such as dried blood spots (DBSs) and oral fluids, that are simpler to collect and transport than the commonly used plasma or serum. Advances in laboratory techniques have opened possibilities for testing multiple diagnostic markers in a multiplex format with the potential to streamline testing, reduce cost, and decrease the time to diagnosis. Since HBV and HCV infections continue to be common in some populations, accurate incidence measurements are needed to monitor progress toward disease control and prevention. Methods for differentiating recent HCV infections from long-term chronic infections that could assist with incidence measurements and monitoring progress toward elimination have recently regained attention. Finally, with the lack of an effective cure for HBV infection, monitoring the effectiveness of antiviral drug therapy and predicting when it is advisable to cease therapy has renewed importance for achieving elimination goals.

Hepatitis Virus Infection and Diagnostics

Hepatitis B Virus

It was known by the 1940s that a blood-transmitted form of hepatitis was distinct from other forms of infectious hepatitis. Seminal work characterizing the agent that would come to be known as HBV and the discovery of the Australia antigen, now known as HBV surface antigen (HBsAg), by Baruch S. Blumberg earned him the 1976 Nobel Prize in Physiology and Medicine. HBV contains a small, partially double-stranded DNA genome approximately 3.2 kb in length and utilizes a replication mechanism that involves the reverse transcription of an RNA intermediate

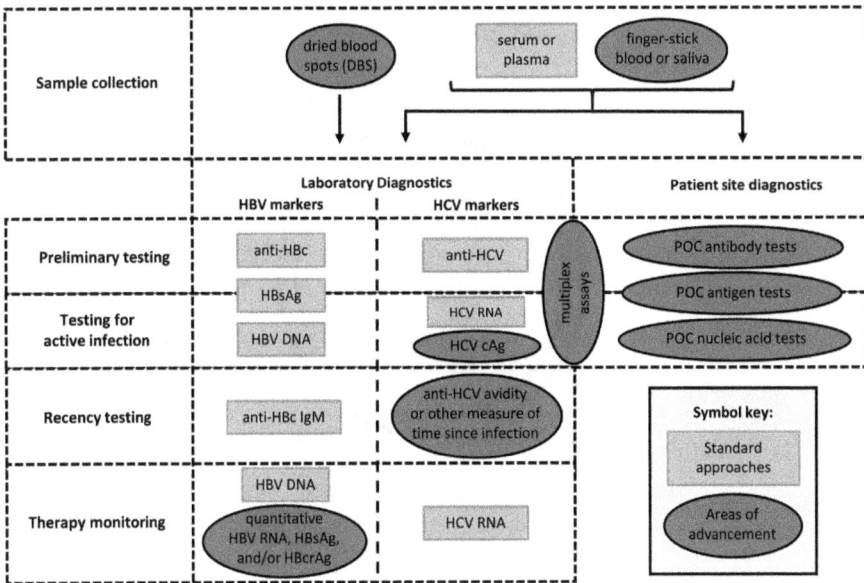

Figure 2. Opportunities for advancement in HBV and HCV infection diagnostics. Areas, where testing-related advancements can improve sampling, diagnostics, and therapy monitoring, are shown in dark gray circles. Dried blood spots (DBSs) and other easy-to-collect bodily fluids, such as saliva, show potential for improving access to testing in resource-limited settings. Point-of-care (POC) diagnostic tests for the detection of virus-specific antibodies, antigens, and nucleic acids have been developed or are in development for diagnosis of HBV and HCV infections. These tests allow diagnosis at the patient site and reduce the need for centralized laboratory testing. New multiplex assays that allow for the simultaneous detection of multiple viral markers could lead to easier and more efficient sample testing, both in the laboratory and at the patient site. Diagnosis of current HCV infection is done primarily by the detection of HCV RNA. Expansion of the use of HCV core antigen (HCV cAg) could promote simpler or less-expensive diagnoses of HCV infection. Recent work suggests that the avidity of antibodies targeting HCV (anti-HCV) is a reliable marker for detecting recent HCV infections and would aid in incidence estimations. Quantitative levels of HBV surface antigen (HBsAg), HBV RNA, and HBV core-related antigen (HBcrAg) show promise for monitoring and more accurately predicting HBV drug therapy outcomes. Anti-HBc = Antibodies Targeting HBV Core Antigen.

(Figure 3). HBV genetic sequences can be divided into at least nine genotypes (A–I) that differ by more than 7.5% of nucleotide positions (Velkov et al. 2018). These genotypes are geographically distributed around the world and may differ in their disease progression and clinical outcomes. Genetic populations of HBV are diverse, leading to rapid intra-host evolution (Revill et al. 2020). Viral particles are composed of the DNA genome, a capsid made of HBV core antigen (HBcAg), and a lipid envelope containing HBsAg (Seeger and Mason 2015). Although an effective vaccine has been available since the late 1980s, HBV is estimated to infect 296 million people worldwide with 1.5 million new infections each year, most of them undiagnosed (WHO 2021a). HBV replicates in the liver and is transmitted by blood, primarily perinatally and through unsafe injection and sexual practices (CDC 2021a). Upon infection, the first diagnostic marker to appear is HBV DNA followed by HBsAg. HBsAg is secreted from infected cells into the blood in viral particles and

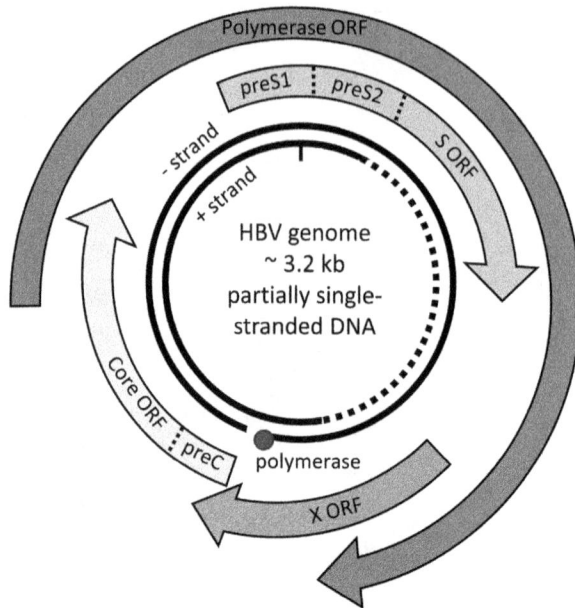

Figure 3. HBV genome organization. The HBV genome is a partially double-stranded circular DNA molecule (relaxed circular DNA; rcDNA) approximately 3.2 kilobases in length. The negative-sense strand is full-length, not covalently closed, and bound by the viral polymerase. The positive-sense strand is variable in length. Upon infection, the genome traffics to the nucleus where it is repaired into a covalently closed circular DNA (cccDNA) containing two full-length strands. The cccDNA is the transcriptional template for four mRNAs (not shown) that encode open reading frames (ORFs, indicated by arrows) for seven proteins. The core ORF encodes the HBV core antigen protein (HBcAg) that makes the virus nucleocapsid while the longer preC ORF encodes the HBV e antigen protein (HBeAg) that is secreted from infected cells and may modulate the infected host's immune system. The polymerase ORF encodes the HBV polymerase that carries out a reverse transcription of the HBV DNA genome from pre-genomic RNA. The S, preS2, and preS1 ORFs have the same 3' ends and encode the small, medium, and large HBV surface antigen proteins (HBsAg) that are found in the envelope of HBV virions. The X ORF encodes the X protein, which modulates cellular functions and viral replication.

subviral particles that lack HBcAg and viral nucleic acid. The virus also expresses and secretes high concentrations of HBV e antigen (HBeAg) into the blood. The function of HBeAg is poorly understood, but it is thought to modulate host immunity (Kramvis et al. 2018). A few weeks after the appearance of HBsAg, antibodies targeting HBcAg (anti-HBc) begin to be produced and IgG antibodies persist for life. Anti-HBc IgM antibodies appear first and remain detectable for approximately six months (Mast et al. 2006). Many commonly used diagnostic tests measure total anti-HBc, which includes both IgG and IgM. These antibodies do not provide immune protection against HBV infection. Greater than 95% of adults resolve HBV infection and become negative for HBsAg and HBV DNA by six months after exposure (Figure 4A; Terrault et al. 2018). Antibodies targeting HBsAg (anti-HBs) appear in patients with resolved HBV infection and are a marker of immunity to the virus. On the other hand, infants and young children have a high probability of

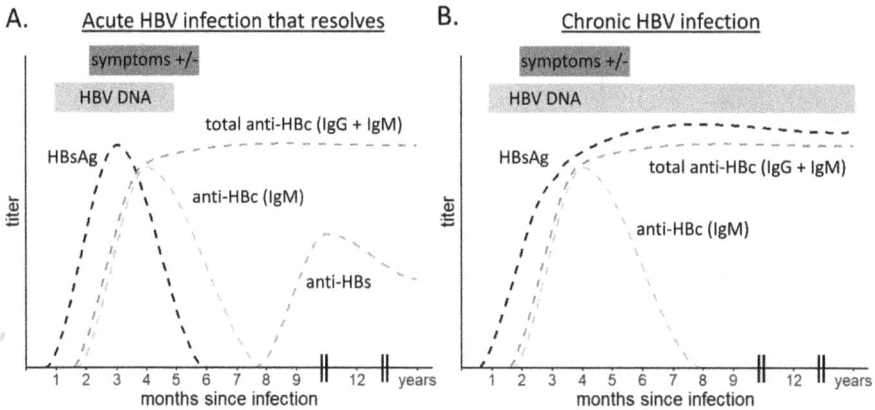

Figure 4. Timing of HBV markers in acute and chronic infections. (A) During an acute HBV infection that resolves, HBV DNA (light gray rectangle) and HBsAg (black dashed line) are detectable in the blood several weeks after infection. Antibodies targeting HBcAg (anti-HBc) appear after HBsAg, but they do not provide immunity to HBV infection. Anti-HBc IgM appears earlier than IgG but is short-lived, typically declining to undetectable levels six months after its first appearance. Anti-HBc IgG is long-lived and indicates exposure to HBV. Most commercial assays detect total anti-HBc which includes both IgG and IgM. Upon resolution of HBV infection, generally within six months, HBV DNA and HBsAg become undetectable. Antibodies targeting HBsAg (anti-HBs) begin to appear during or shortly after resolution and indicate immunity to HBV infection. Symptoms are rare and can range from two to six months after infection (dark gray rectangle). (B) When an acute infection does not resolve within six months it is classified as a chronic infection. During chronic infection, HBV DNA and HBsAg persist and anti-HBs antibodies do not appear.

developing chronic HBV infections that can increase the risk for liver cirrhosis and hepatocellular carcinoma (Figure 4B; Fattovich et al. 2008). Most acute infections are asymptomatic, but when signs and symptoms such as jaundice, fever, fatigue, and abdominal pain are present, they usually occur approximately two months after exposure. It is estimated that only 10% of HBV infections are diagnosed because of low testing rates and the asymptomatic nature of most infections (WHO 2021a).

The diagnostic landscape of hepatitis B provides a rich array of markers that can be detected using various serological assays. These markers allow for the determination of a patient's HBV infection status, the stage of chronic infection, and whether immunity in a patient is from vaccination or natural infection. The presence of total anti-HBc indicates that a person has been exposed to HBV during their lifetime. Whether someone is currently infected is determined by the presence of HBsAg and/or HBV DNA. Immunity is identified by the presence of anti-HBs. In the absence of other HBV markers, anti-HBs suggest immunity due to vaccination. When total anti-HBc exists, anti-HBs indicate immunity due to a resolved infection (Table 2; Mast et al. 2006). Acute HBV infection can be differentiated from chronic infection by the presence of IgM antibodies targeting HBcAg (Gerlich et al. 1986).

Chronic HBV infections can be divided into stages that are differentiated by quantifying HBV DNA, determining HBeAg and anti-HBe status, and assessing liver damage using the level of liver enzymes, such as alanine aminotransferase (ALT), in serum. Multiple classification and naming schemes for these stages exist, but they

Table 2. Interpretation of HBV serological markers.

Interpretation	HBV serological marker			
	HBsAg	Total anti-HBc	Anti-HBc IgM	Anti-HBs
Susceptible	-	-	-	-
Immune (natural infection)	-	+	-	+
Immune (vaccination)	-	-	-	+
Infected (acute)	+	+	+	-
Infected (chronic)	+	+	-	-

HBsAg = HBV surface antigen, anti-HBc = antibodies targeting HBV core antigen, IgM = immunoglobulin M, anti-HBs = antibodies targeting HBsAg.

are generally similar (Villeneuve 2005; Gish et al. 2015; EASL 2017; Terrault et al. 2018). The first stage (immune tolerance, HBeAg-positive chronic HBV infection, or high replicative and low inflammatory) is typically short in adults but can last decades in people infected at birth. It is characterized by high HBV DNA levels, HBeAg expression, and normal ALT levels. During the second stage (immune clearance, immune-active, or HBeAg-positive chronic hepatitis B), the immune system targets the HBV infection leading to liver necrosis, inflammation, and elevated ALT levels. High viral titers persist during this phase and the virus continues to express HBeAg. This stage generally lasts until HBeAg seroconversion (presence of anti-HBe), which occurs in approximately 8–12% of chronically infected adults annually (Lok and McMahon 2007).

Once HBeAg seroconversion is achieved, patients enter the third stage (inactive carrier, non-replicative, or HBeAg-negative chronic HBV infection), which is characterized by immune control of virus replication, low viral titers, undetectable HBeAg, normal ALT levels, and minimal liver inflammation. Most patients in the inactive carrier state will have prolonged control of viral replication with a small percentage resolving their HBV infection indicated by HBsAg loss and anti-HBs seroconversion. The likelihood of resolution increases with drug therapy. The probability of transmission, especially during childbirth, is reduced during the inactive carrier stage of chronic infection (Cryer and Imperial 2019). The partial immune control of anti-HBe can select for mutations in the HBV genome that prevent or downregulate HBeAg expression (Hadziyannis and Papatheodoridis 2006). If these mutations are selected, patients can experience reactivation of the virus. This reactivation stage of chronic infection (HBeAg-negative chronic hepatitis B) is characterized by elevated serum HBV DNA and ALT levels and the recurrence of liver necrosis and inflammation that can lead to serious health complications.

The final stage of chronic HBV infection (HBsAg-negative, or occult) is classified by the loss of detectable HBsAg while maintaining replication-competent HBV DNA within the liver. During this stage, HBV DNA can sometimes be detected at low levels and ALT levels are normal. Some, but not all, patients at this stage have anti-HBs antibodies. If this stage is reached prior to the development of cirrhosis or hepatocellular carcinoma (HCC) there is a low risk of new liver damage.

Several drugs have been approved for the treatment of HBV infection (Terrault et al. 2016; 2018). These drugs, including Lamivudine, Telbivudine, Entecavir, Adefovir, and Tenofovir, are nucleos(t)ide analogues (NUCs) that inhibit the viral polymerase to prevent viral replication. They can be used alone or in combination with interferon treatment. These therapies are not a cure but suppress viral replication. As a result, HBV-infected individuals are often on therapy for long periods.

Hepatitis C Virus

Even after the discovery of HBV, many cases of bloodborne hepatitis were unexplainable. During the 1970s, 80s, and 90s, groundbreaking work to identify and characterize this "non-A, non-B" form of infectious hepatitis led to the discovery of HCV. The importance of this monumental effort was recently recognized with the 2020 Nobel Prize in Physiology and Medicine awarded to Harvey J. Alter, Michael Houghton, and Charles M. Rice. HCV has an approximately 9.6 kb single-stranded RNA genome that encodes 10 proteins (Figure 5). HCV is classified into at least eight genetically distinct genotypes (1–8) that have different geographical distributions (Gower et al. 2014; Smith et al. 2014). As an RNA virus, HCV has high levels of intra- and interhost genetic diversity that can promote rapid evolution. Most HCV proteins play non-structural roles in the replication cycle and are not found as part of the viral particles. Viral particles are composed of the RNA genome, a capsid made of core antigen protein, and a lipid envelope containing two envelope proteins (E1 and E2) (Dubuisson and Cosset 2014). HCV infects an estimated 58 million people worldwide and causes 1.5 million new infections each year, although most cases are undiagnosed (WHO 2021a). Major risk factors for infection include unsafe

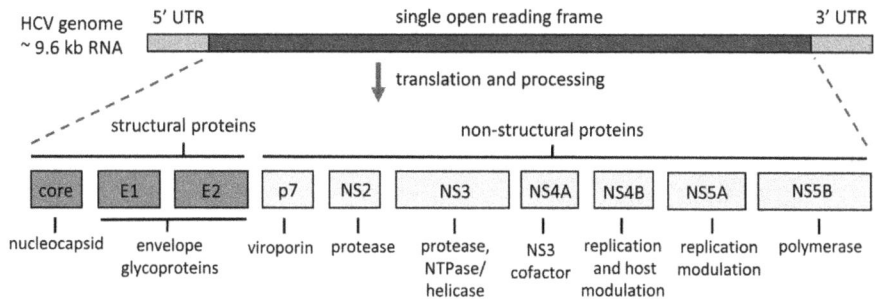

Figure 5. HCV genome organization and protein functions. The HCV genome is a positive-sense RNA approximately 9.6 kilobases in length. It encodes a single open reading frame (ORF) flanked by the 5' and 3' untranslated regions (UTRs). A single polyprotein is translated from the ORF and subsequently processed into 10 proteins by cellular and virally encoded proteases. The three structural proteins that are found in mature HCV virions are encoded at the N-terminus of the polyprotein. Core protein (HCV cAg) makes the viral nucleocapsid while E1 and E2 are glycoproteins found in the viral envelope that allow for virus attachment and entry into cells. The p7 protein (sometimes called NS1) creates small pores in membranes and is involved in virus assembly and release. NS2, NS3, and NS4a are involved in polyprotein processing and viral replication. NS4B and NS5A modulate cellular functions and are involved in the formation and activity of viral replication complexes. NS5B is the RNA-dependent RNA polymerase that replicates the viral genome.

infection practices and having a blood transfusion prior to routine screening of donated blood in the early 1990s (CDC 2021b). The first diagnostic markers that appear are HCV RNA and HCV cAg, which can be detected as early as one-to-two weeks after infection (Figure 6A). Only 15–30% of HCV infections are symptomatic with jaundice, fever, joint pain, nausea, and abdominal pain. If symptoms do occur, they are typically first observed six to seven weeks after infection. Antibodies targeting HCV (anti-HCV IgG) can be detected in most people using immunoassays by 4–12 weeks after HCV exposure (Colin et al. 2001). Antibodies target various structural and non-structural proteins, primarily core, NS3, NS4a, NS4b, and NS5a. The HCV targets of the antibody response vary from person to person and do not typically provide protective immunity. Neutralizing antibodies are thought to target the E1 and E2 proteins, but these antigens are highly variable and can evolve quickly in response to immune pressures (Law 2021). Some HCV infections resolve in six months, but more than half of infections persist and become chronic (Figure 6B; EASL 2015; Seo et al. 2020). Chronic HCV infections can last for the life of the patient, increasing the risk of liver cirrhosis and cancer (Westbrook and Dusheiko 2014). Rapid advancements in the treatment of HCV infection have occurred in the past decade. Current-generation HCV direct-acting antiviral (DAA) drug therapies are well tolerated, effective against all HCV genotypes, and have cure rates > 95% after 12–24 weeks of therapy (Cuypers et al. 2016; Falade-Nwulia et al. 2017; EASL 2018). Unfortunately, most chronic HCV infections remain undiagnosed. With the increasing availability of highly effective DAAs, expanding access to accurate testing is needed to effectively link infected persons to treatment.

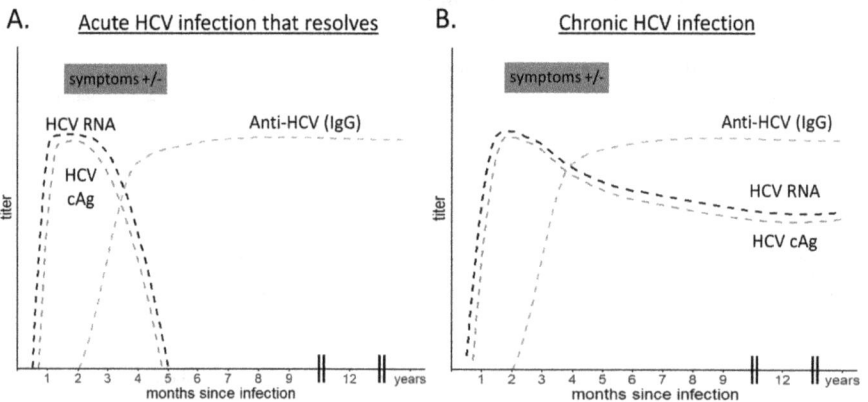

Figure 6. Timing of HCV markers in acute and chronic infections. (A) During an acute HCV infection that resolves, HCV RNA and HCV core antigen (HCV cAg), which indicate current infection, are detectable in the blood several weeks after exposure. Antibodies targeting HCV (anti-HCV IgG) begin to be detectable approximately one to three months after HCV infection and gradually increase. These antibodies indicate exposure to HCV and do not indicate immunity to HCV infection. HCV RNA and HCV cAg levels begin to decrease around the time that anti-HCV titers begin to increase. In resolved acute infections, HCV RNA and HCV cAg become undetectable by six months after HCV infection. Symptoms are rare but generally begin to appear one to two months after infection (dark gray rectangle). (B) When an acute HCV infection does not resolve within six months it is classified as a chronic HCV infection. During chronic infection, HCV RNA and HCV cAg often decline slightly after the appearance of anti-HCV but then persist.

Initial screening for HCV infection relies upon the detection of anti-HCV, which indicates exposure to HCV. Most anti-HCV tests detect a combination of antibodies targeting antigens in the core, NS3, NS4, and NS5 proteins, which increases sensitivity and allows for detection earlier after infection (Colin et al. 2001; Kamili et al. 2012). Reflex or follow-up testing for HCV RNA or core antigen is needed to determine if an anti-HCV positive individual is currently infected with HCV (CDC 2013). While assays that detect HCV RNA are more commonly used for this purpose, testing for HCV cAg may be a promising and cost-saving alternative in most settings (Mixson-Hayden et al. 2015; Freiman et al. 2016; Pollock et al. 2020).

Laboratory Diagnostics

Quality assurance is an important aspect of all *in vitro* diagnostic testing. It is important to have assays that are verified to be robust, accurate, reliable, and clinically valid. The regulatory bodies that grant quality certifications to commercially available assays vary by country. In the United States, the Food and Drug Administration (FDA) grants approvals to *in vitro* diagnostics (FDA 2021). In the European Union, the Directorate General for Internal Market, Industry, Entrepreneurship, and SMEs grants a Conformitè Europëenne (CE) Mark to approve diagnostic tests. The WHO grants prequalification as an indication of diagnostic test quality to promote faster regulatory approval by individual member states (WHO 2021c). These quality certifications indicate to a user that a test will perform as advertised if the manufacturer's instructions are followed. Therefore, tests with quality certifications are much easier than laboratory-developed or non-quality-certified tests to implement broadly for patient diagnosis. Throughout this chapter, we will highlight test types that currently have FDA approval, CE Mark, or WHO prequalification. Currently available tests with quality certifications are summarized in Table 3.

Testing for HBV and HCV infection diagnostic markers is generally performed in centralized laboratories. A variety of immunoassay techniques and formats are utilized for the qualitative and quantitative detection of HBsAg, HBeAg, total anti-HBc, anti-HBe, anti-HBs, anti-HCV, and HCV cAg. These include laboratory-developed and commercially available tests in a range of manual and automated formats. The most used are enzyme immunoassays (EIAs) and chemiluminescence immunoassays (CIAs). When used with serum or plasma samples, these HBV and HCV laboratory immunoassay methods generally have very high sensitivities and specificities. Many of the laboratory assays used for the detection of HBV and HCV serological markers have quality certifications. Currently, there are a broad range of commercially available tests and testing platforms that have received FDA approval or are CE-marked for the detection of HBsAg and anti-HCV.

HBV DNA and HCV RNA are most frequently measured using quantitative PCR (qPCR) and quantitative reverse transcription PCR (qRT-PCR), respectively. Depending on the laboratory, these measurements may be performed using laboratory-developed tests, commercial assays, or automated sample-to-answer platforms. Nucleic acid tests (NATs) performed in the laboratory have very high sensitivity and specificity with limits of detection often on the order of 10–1,000 IU/mL when used with serum or plasma samples. There are many options

Table 3. Quality certifications granted to HBV and HCV diagnostic tests.

Analyte	Test type	FDA approved[1]	CE marked[2]	WHO prequalified[3]
HBsAg	Laboratory	> 5	> 5	3
	RDT/POC	0	3	3
HBeAg	Laboratory	> 5	> 5	0
	RDT/POC	0	0	0
HBV DNA	Laboratory	> 5	> 5	0
	RDT/POC	0	1	0
Anti-HCV	Laboratory	> 5	> 5	5
	RDT/POC	1	4	4
HCV cAg	Laboratory	0	1	1
	RDT/POC	0	0	0
HCV RNA	Laboratory	5	> 5	4
	RDT/POC	0	2	2

The numbers indicate how many commercially available tests have been granted quality certifications by each regulatory body.

RDT = rapid diagnostic test; POC = point-of-care.

[1] Approval granted by the United States Food and Drug Administration (FDA 2021).

[2] European Union market approval granted by the Directorate General for Internal Market, Industry, Entrepreneurship, and SMEs.

[3] Prequalification approval granted by the World Health Organization (WHO 2021c).

for quality-certified laboratory-based HBV DNA and HCV RNA tests on the market. Testing for HBV and HCV nucleic acids is typically more expensive than testing for antibody or antigenic markers.

Both HBV and HCV are divided into several genetically distinct genotypes. The genotype of the infecting virus can be important for epidemiological purposes and clinical decision-making (Lin and Kao 2011; Cuypers et al. 2016). Genotyping can be performed for both HBV and HCV using genetic sequencing, genotype-specific PCR primers, or hybridization probes that differentiate among genotypes (Guirgis et al. 2010; Yusrina et al. 2018). For HCV infections, routine genotyping may become less important due to the pan-genotypic activity of many current DAA therapies.

Point-of-Care Testing

Diagnosing infections caused by HBV and HCV is the first step toward decreasing transmission and linking people to therapy. Infections caused by both viruses remain underdiagnosed, with recent estimates suggesting that only 10% of HBV-infected and 21% of HCV-infected people are aware of their infection status. In LMICs or at-risk populations with low access to testing and high infection rates, awareness of HBV or HCV infection status may be below 5% (WHO 2021a). People who are unaware that they are infected may be more likely to transmit the virus to others, particularly through risky behaviors. Therefore, to meet the WHO's 2030 elimination goals, increased access to testing and accurate diagnosis of HBV and HCV infections among all the world's populations are urgently needed. Diagnostic tests that can be performed rapidly or at the patient site without the need for highly trained personnel represent a promising approach to achieving this goal. These tests include rapid diagnostic tests (RDTs), which provide results quickly, often within 30 minutes, and point-of-care (POC) tests that can be performed outside of a laboratory by non-laboratory staff, generally at or near the patient or sample collection site. Most POC tests are RDTs, but RDTs are not necessarily compatible with POC use.

Laboratory-based techniques, including quantitative PCR and immunoassays, such as EIAs and CIAs, remain the most used methods to diagnose HBV and HCV infections. However, these techniques are labor-intensive, require highly trained staff, and depend on expensive instruments and reagents. POC testing options would eliminate or reduce these challenges and facilitate the access by more people (Table 4). The WHO endorses the development of tests that are appropriate for use in resource-constrained settings. They have developed the ASSURED criteria as a point of reference to determine if an assay addresses the following needs: Affordable, Sensitive, Specific, User-friendly, Rapid and robust, Equipment-free, and Deliverable to end-users (Peeling et al. 2006). Also, the WHO recently released guidance on HCV self-testing to support the efforts of national programs to reach people who may not otherwise be tested (WHO 2021b).

POC lateral-flow immunochromatographic RDTs can detect antibodies or antigens. These tests are simple and rapid, use easy-to-collect finger-stick blood or saliva, and do not require the transport of samples to a laboratory. Lateral-flow tests provide qualitative results that are observable by the eye in 15–30 minutes. In these tests, the sample is added to a nitrocellulose membrane and flows laterally over test and control detection lines to create visible bands if the analyte of interest is present, and the assay is performing properly. The test line contains immobilized antigen-specific antibodies for antigen detection or immobilized antigens for antibody detection. The observable bands are generated by colloidal gold-conjugated antibodies that bind to the analyte of interest. The control line contains immobilized antibodies that bind to colloidal-gold-conjugated detection antibodies that do not bind to the analyte at the test line.

Lateral-flow tests for HBsAg are affordable and generally have sensitivities above 90% with near-perfect specificity (Amini et al. 2017; Sullivan et al. 2019; Dembele et al. 2020; Kabamba et al. 2020; Xiao et al. 2020). These tests can detect common HBsAg mutants (Hirzel et al. 2015). Lower sensitivities for HBsAg detection have been reported in HIV-infected populations (Nyirenda et al. 2008; Geretti et al. 2010). The use of HBsAg RDTs to screen for HBV infections can result in less-expensive diagnosis and greater patient linkage to care compared with laboratory serology testing (Ho et al. 2020). Testing for HBeAg can be used to determine the level of HBV replication to evaluate the risk of mother-to-child-transmission but POC tests for this analyte have low sensitivities (30–82%) or poor specificity (Seck et al. 2018; Leathers et al. 2019). Likewise, POC tests for anti-HBs and anti-HBe exist, but they generally have poor sensitivity (20–80%), making them impractical for assessing HBV immunity or vaccine uptake (Bottero et al. 2013; Cruz et al. 2017; Poiteau et al. 2017). There are currently a few RDTs for the detection of HBsAg that have

Table 4. Comparison of sampling and testing approaches for detecting HBV and HCV infection.

Objective	Approach	Advantages	Disadvantages
Sample Collection	Plasma/serum	Gold-standard sample material	Requires phlebotomist, cold storage, and cold transport
	Dried blood spots (DBSs)	Non-invasive collection, room temperature transport, low cost	Lower sensitivity, lengthier processing time in the laboratory
	Saliva/oral fluids	Non-invasive collection, low cost	Lower sensitivity; requires cold storage and transport
Serology Testing	Laboratory (EIA, CIA)	High sensitivity and specificity	Requires trained personnel, expensive instrumentation, and transport of specimens to a laboratory
	POC rapid diagnostic tests (RDTs)	Portable, rapid, low cost, POC use, no need for trained personnel	Lower sensitivity
Nucleic Acid Testing	Quantitative PCR	High sensitivity and specificity, low limits of detection, quantitative results	Requires expensive instrumentation, lengthy procedure
	Isothermal amplification	Rapid, single temperature (simpler instrumentation), compatible with visual detection methods	Lower sensitivity and specificity; only qualitative results possible
	POC platforms	Rapid result to patients, high sensitivity and specificity	May require chemical waste disposal and expensive instrumentation. May require serum/plasma samples, Not yet widely available

EIA = enzyme immunoassay, CIA = chemiluminescence immunoassay.

been CE-marked or WHO-prequalified but no POC tests for HBeAg or HBV-specific antibodies have yet received quality certifications (WHO 2021c).

For HCV, many POC lateral-flow tests for the detection of anti-HCV are accurate, with most studies reporting sensitivities and specificities above 95% (Tang et al. 2017; Chevaliez et al. 2021). Sensitivities are generally lower (76–95%) when these rapid tests are used with saliva or oral fluid compared with finger-stick blood or plasma (Scalioni et al. 2014; Tang et al. 2017; Pallarés et al. 2018; Uuskūla et al. 2021). Several options exist for CE-marked or WHO-prequalified POC anti-HCV tests but only the OraQuick HCV Rapid Antibody Test is currently FDA-approved in the United States. Rapid lateral-flow assays for the detection of HCV cAg have been developed, but their performance needs to be further evaluated using clinically relevant samples (Mikawa et al. 2009; Patel and Sharma 2020).

While HBsAg can indicate current HBV infection, testing for anti-HCV does not identify whether someone is currently infected with HCV. Testing for HCV RNA or HCV cAg is still needed to identify the current infection. Positive test results from

HBsAg or anti-HCV RDTs should be followed by the quantification of viral nucleic acid to identify viremia and to inform decision-making for treatment and monitoring. However, this typically requires samples to be sent to a laboratory or multiple patient visits. Recent advances may allow for POC detection of these analytes.

Adapting nucleic acid detection to a POC format has proven to be more challenging than for antibodies or antigens. The challenges associated with this process are the need for simplified and robust extraction and/or purification of nucleic acids from viral particles and inexpensive, portable, and easy-to-use methods for amplifying and detecting those nucleic acids. Laboratory methods for extracting and purifying nucleic acids require multiple steps that include precision pipetting and, in some cases, high-speed centrifugation. Simplification of these methods for POC use is especially difficult for blood-based samples which contain components that can interfere with downstream amplification. Various protocols including viral lysis buffers, high-temperature incubations, and filter or bead-based purifications have been developed for the extraction of viruses from blood or other bodily fluids, but these may still require a few precision pipetting steps (McFall et al. 2015; Benzine et al. 2016; Joung et al. 2020; Paul et al. 2020). Portable microfluidics devices have been developed to simplify nucleic acid extraction and purification, but most of these technologies have not been advanced beyond proof-of-concept studies (Dong et al. 2019; Zhang et al. 2019; Paul et al. 2020; Sharma et al. 2020). PCR is the most common way that nucleic acids are amplified in the laboratory, but this process requires instrumentation capable of precise temperature changes. Methods that can be performed at a single temperature (isothermal) in a simple heating block or water bath are promising alternatives. There are a wide variety of isothermal nucleic acid amplification techniques that have been developed to detect viral nucleic acids and novel methods are constantly being developed (Ding et al. 2019; Carter et al. 2021). Two isothermal techniques that have been investigated for the detection of HBV and HCV nucleic acids are loop-mediated isothermal amplification (LAMP) and recombinase polymerase amplification (RPA).

LAMP uses four or six primers and a DNA polymerase with high strand displacement activity to amplify a target DNA molecule by auto-cycling strand displacement nucleic acid synthesis at a constant 60–65°C (Notomi et al. 2000). Incorporation of a reverse transcriptase allows LAMP amplification of an RNA analyte. RPA uses two opposing primers with a recombinase, a single-strand binding protein, and a strand displacing DNA polymerase to rapidly amplify DNA at 37–42°C (Piepenburg et al. 2006). These isothermal methods are fast, allowing for the sensitive detection of viral nucleic acids in 30 minutes or less. Strict limits on the timing of the amplification are needed to maintain high specificity as amplification of non-target nucleic acids can occur. LAMP and RPA have been applied to the detection of HBV or HCV nucleic acids. When laboratory-based nucleic acid extraction methods are used with patient serum samples, isothermal methods can achieve high sensitivities (84–100%) and specificities (> 95%) for HBV or HCV nucleic acid detection compared to quantitative PCR methods (Nyan and Swinson 2016; Zhao et al. 2017; Shen et al. 2019; Chen et al. 2020; Yi et al. 2020; Hongjaisee et al. 2021). However, when simple extraction methods like sample heating are used, limits of detection and sensitivities decline (Nyan et al. 2014; Shen et al. 2019; Vanhomwegen et al. 2021). Therefore, for these methods to become more attractive for POC use, alternative rapid and simple nucleic acid extraction methods should be investigated.

For POC nucleic acid detection, several methods are compatible with isothermal amplification, including fluorescent, colorimetric, or lateral-flow techniques that can be visualized by the eye (Goto et al. 2009; Zhao et al. 2017; Daskou et al. 2019; Yi et al. 2020). Some of these easily visualized detection techniques are generic in nature, detecting either a pH change, a reduction in magnesium ion concentration, or the presence of double-stranded DNA, while others specifically recognize the amplified nucleic acid of interest. One recently developed approach that has gained attention for its ability to both increase analytical sensitivity and specificity of detecting isothermally amplified nucleic acids involves Cas12- and Cas13-based techniques.

Clustered regularly interspersed short palindromic repeat (CRISPR)-associated (Cas) proteins come from the adaptive immune systems of prokaryotes. These systems adapt to recognize and destroy foreign nucleic acids found within bacterial or archaeal cells. Trans-activating CRISPR RNA (crRNA) direct their associated Cas proteins to specifically target complementary DNA or RNA molecules. The recently characterized Cas12 and Cas13 proteins have enzymatic properties that make them useful for nucleic acid detection. When Cas12 encounters its target DNA it indiscriminately cleaves single-stranded DNA molecules (Chen et al. 2018). Similarly, Cas13 indiscriminately cleaves RNA when it encounters RNA complementary to its crRNA (Abudayyeh et al. 2016). These discoveries have led to the development of the DNA endonuclease-targeted CRISPR trans-reporter (DETECTR) assay that uses Cas12 and the specific high-sensitivity enzymatic reporter unlocking (SHERLOCK) assay that uses Cas13 (Gootenberg et al. 2017; Chen et al. 2018). In these approaches, short single-stranded DNA or RNA probes containing fluorophore/quencher or biotin/fluorescein modifications are cleaved by Cas12 or Cas13 when they encounter their targeted nucleic acids. Detection can be fluorescence-based or use lateral-flow paper strips, making these approaches amenable to POC applications. The DETECTR and SHERLOCK approaches have been demonstrated to work for the detection of HBV DNA and HCV RNA (Ackerman et al. 2020; Ding et al. 2021).

Integrating nucleic acid extraction, purification, amplification, and detection using microfluidics-based chips or cartridges may be a promising approach for POC testing that requires minimal user intervention, but many of these approaches have not advanced beyond use in laboratory research studies (Song et al. 2016; Sharma et al. 2020). However, some portable, easy-to-use, and commercially available platforms, including the Cepheid GeneXpert Xpress and the Genedrive System, have recently been developed and received WHO prequalification and are CE-marked. These platforms integrate extraction, purification, and quantitative PCR into portable platforms that can detect HBV DNA or HCV RNA in 60–120 minutes from finger-stick blood and/or serum and plasma (Lamoury et al. 2018; Llibre et al. 2018; Poiteau et al. 2020). These assays have comparable sensitivities and limits of detection to traditional laboratory quantitative PCR assays. However, drawbacks include that these tests may still need trained personnel, require an initial investment in equipment, involve precision pipetting steps, or require access to appropriate waste disposal for hazardous chemicals and cartridges.

As low-cost POC diagnostic tools for the detection of HBV and HCV markers continue to be improved and validated across diverse patient populations and settings, they will gain more widespread use and reduce the need for follow-up or confirmatory testing in a laboratory. Testing, from screening through diagnosis and long-term monitoring, may occur at the patient site with quick and reliable results that can inform immediate decisions regarding care. As we make progress toward the WHO's 2030 viral hepatitis elimination goals, the expansion of existing technologies and continued research into new POC technologies will be an important component of ensuring that every person and population that needs testing has access to it.

Alternative Sample Types for Diagnostic Assays

Laboratory diagnostics for HBV and HCV infections commonly use patient serum or plasma as the testing matrix. These bodily fluids reliably contain the diagnostic markers of infection but require skilled personnel to perform venipuncture and cold storage until they can be tested in a laboratory. These requirements can prohibit sample collection for diagnostic testing from certain populations and in low-resource settings or sites that are distant from a central laboratory. Sample materials that are easier to collect, store, or transport to the testing laboratory have been investigated for their utility in hepatitis virus diagnostics (Table 4). Here we describe the potential use of DBSs and saliva as sample types for HBV and HCV diagnostics.

DBS specimens are collected by blotting blood from a finger-stick or heel-stick onto a filter paper card. Whatman 903 paper cards are the most used and evaluated, although other products, such as FTA cards that preserve nucleic acids, are also available. Once dried, the paper card can be stored at ambient temperature, refrigerated, or frozen until tested for markers of viral infection (Grüner et al. 2015). Tolerance of ambient temperature storage simplifies the transportation of samples from the site of collection to the testing laboratory. Blood and viral analytes can be easily eluted or extracted from the filter paper in a buffered solution and used with various diagnostic assays. DBS specimens are amenable to at-home self-collection, which could lead to increased testing of at-risk populations (van Loo et al. 2017). DBSs have been extensively evaluated for the detection of several HBV and HCV diagnostic markers with varying levels of success. The WHO's current hepatitis B testing guidelines recommend DBS use in locations where testing capacity or access is limited (WHO 2017).

Detection of the two markers of current HBV infection, HBsAg and HBV DNA, from DBSs, have shown high sensitivity and specificity when compared to paired serum or plasma samples. Sensitivities for the detection of HBsAg are generally above 96% and specificities are above 97% (Mössner et al. 2016; Lange et al. 2017a; Villar et al. 2019). Some studies, however, report lower sensitivities (62–85%) in certain populations including people co-infected with HIV (Flores et al. 2017a; Cruz et al. 2021; Flores et al. 2021). Reported sensitivities for the detection of HBV DNA from DBSs are between 86% and 100% and specificities are above 99% (Lange et al. 2017b; Villar et al. 2019). False negatives in these studies have generally been from samples with low HBsAg or HBV DNA levels. This loss of sensitivity can be attributable to the small volume of the spotted blood (~ 50 µL) and the inherent sample dilution that occurs during elution from the filter paper. HBsAg assay cutoff limits may be different for serum and DBS samples and should be determined empirically for the assay being used (Lange et al. 2017a). Limits of detection for HBV DNA from DBSs are often 10–100-fold higher than from serum, but most assay protocols can still reliably detect HBV DNA above 2,000 IU/mL (Lange et al. 2017b; Villar et al. 2019). Despite this sensitivity loss, most chronic infections would be detected at this level. Current studies suggest that there is no effect of DBS storage condition on HBsAg or HBV DNA detection. The detection of the HBV antibody markers, anti-HBc, and anti-HBs, have shown more variable results than HBsAg and HBV DNA. Some studies report high sensitivities above 90% for these antibody markers from DBSs, while others report much lower sensitivities or poor specificity, particularly for anti-HBs (Mössner et al. 2016; Flores et al. 2017a; Villar et al. 2019; Cruz et al. 2021; Flores et al. 2021). Therefore, the utility of HBV antibody testing from DBSs may depend upon the testing method being used and the user's acceptable levels of sensitivity and specificity.

DBSs are a promising sample material for the detection of anti-HCV, HCV RNA, and HCV cAg. Most studies evaluating the detection of anti-HCV from DBSs report sensitivities above 90% and specificities above 95% compared to serum samples, although a few found lower sensitivity (Mössner et al. 2016; Soulier et al. 2016; Lange et al. 2017a; Flores et al. 2017b; Vázquez-Morón et al. 2018; 2019; Carty et al. 2021; Flores et al. 2021). Published protocols for the detection of HCV RNA from DBSs generally have sensitivities above 90% and specificities above 95% when compared to serum samples (Soulier et al. 2016; Lange et al. 2017b; Vázquez-Morón et al. 2018; 2019; Saludes et al. 2019; Carty et al. 2021). As with all testing from DBSs, there is a decreased ability to detect low-titer analytes compared to testing from serum due to the volume of the DBS and the elution process. Limits of detection for both anti-HCV and HCV RNA from DBSs are generally at least 10-fold higher than from serum, but still adequate to detect most chronic infections (Lange et al. 2017b; Weber et al. 2019). Storing DBSs at ambient temperature may reduce the quantitation of HCV RNA compared with refrigerated or frozen storage, but qualitative detection of samples is unlikely to be affected unless the sample is near the assay limit of detection (Abe and Konomi 1998; Tuaillon et al. 2010; Tejada-Strop et al. 2015). Recently, plasma separation cards that separate and preserve plasma from spotted blood have been developed. These cards may have higher sensitivity for detecting HIV nucleic acids than traditional DBSs and should be evaluated for HCV RNA testing (Carmona et al. 2019). Detection of HCV cAg from DBSs has been found to have sensitivities in the 64–94% range with near 100% specificity when compared to HCV RNA from serum (Soulier et al. 2016; Nguyen et al. 2018; Biondi et al. 2019; Catlett et al. 2019; Carty et al. 2021). These studies also suggest that the detection of HCV cAg from DBSs is less sensitive than HCV RNA detection from DBSs. As with other markers, HCV cAg detection signal from DBSs is systematically lower than from serum. Storage temperature of the DBS does not appear to have a major impact on HCV cAg detection. DBSs are also compatible with immunoassays that simultaneously detect both anti-HCV and HCV cAg, maintaining high sensitivity and specificity compared to serum samples (Brandão et al. 2013).

Saliva or oral fluids have also been proposed as an alternative bodily fluid to test for HBV and HCV infection markers. Saliva can be collected and transported using commercial devices and then tested in a laboratory. The collection is easy and non-invasive making it a useful tool for sampling children, the elderly, incarcerated persons, people who inject drugs, and other individuals with difficult venous access. Using saliva greatly simplifies sample collection but is generally less sensitive than serum due to its much lower concentrations of antibodies and bloodborne viruses or viral markers. Additionally, unlike DBSs, saliva samples need to be stored and transported to the testing site while frozen. Several studies have evaluated the use of saliva for testing HBsAg and found sensitivities greater than 75% but with sometimes poor specificity (Flores et al. 2017a; Villar et al. 2019). Few studies have assessed saliva for anti-HBc and anti-HBs detection, but these have identified wide-ranging sensitivities (13–85%) that are generally lower than for HBsAg (Villar et al. 2019). Similarly, HBV DNA detection from saliva appears to be impractical due to low sensitivity, with most studies reporting less than 50% of serum-positive patients having detectable HBV DNA in saliva (Villar et al. 2019).

Detection of anti-HCV from saliva may be a promising screening approach for HCV infections. Sensitivities are generally in the 80–95% range, and the false negatives may be primarily from HCV RNA-negative individuals that have resolved their infections and have low antibody titers (Elsana et al. 1998; Cameron et al. 1999; De Cock et al. 2004; Elsana et al. 1998; Flores et al. 2017b). Immunoassay cutoffs are often adjusted for use with saliva samples to preserve sensitivity. Saliva can be used as a sample type in some anti-HCV RDTs that are marketed commercially (Kimble et al. 2019). On the other hand, detection of active HCV infection using saliva would be impractical. While HCV RNA can be detected in the saliva of some infected patients, it has poor sensitivity when compared to serum for patients with low viral titers (Hermida et al. 2002; Gonçalves et al. 2005).

Though DBSs have been extensively investigated for their use in HBV and HCV infection diagnostics, quality certification of commercial assays for use with this sample type is uncommon. Future work should be directed toward standardizing best practices for DBS collection, transport, elution, and testing to ensure that results from this sample type are reliable and of high quality (Tuaillon et al. 2020). Additionally, DBSs may be poorly amenable to high throughput testing due to the need for lengthy elution and pre-testing preparation procedures. Saliva testing for HBsAg or anti-HCV may be appropriate for certain uses or populations, but a general lack of sensitivity for detecting viral nucleic acids and HBV-specific antibodies in saliva limits broad applicability for these markers. Despite these shortcomings, DBSs and oral fluid samples for testing HBV and HCV infection markers may still represent important opportunities for expanding the reach of centralized laboratory testing to low-resource or difficult-to-access populations, many of whom have high HBV and HCV infection burdens.

Multiplex Diagnostic Assays

Most assays used for the diagnosis or characterization of HBV and HCV infections are singleplex, meaning that they measure or detect a single marker at a time. While this approach is established and works well for cases with straightforward diagnoses, it may not be the most efficient approach for population screening purposes or syndromic surveillance where patient symptoms could be caused by a variety of agents. In these situations, having the ability to multiplex, or assess multiple markers simultaneously using a single sample input, could streamline the laboratory testing process. For example, screening populations for antibody markers of HBV and HCV infection, or vaccination against vaccine-preventable diseases could be streamlined if these markers were tested simultaneously. Likewise, multiplexing the detection of diagnostic markers for all hepatitis viruses would simplify diagnosis because acute infections by these viruses have similar clinical signs and symptoms that are difficult to distinguish in the absence of known risk factors or exposure. Including markers for HIV infection in multiplex assays could be advantageous because co-infections of HBV or HCV with HIV are common in some populations. Many assays for the multiplexed detection of hepatitis virus markers, as well as other infectious disease markers have been developed. These assays include tests for the detection of antigens, antibodies, and nucleic acids and range from measuring two markers to hundreds of markers simultaneously. Here, we briefly describe the types of assays that have been developed by research laboratories or commercial test manufacturers that allow for multiplexed detection of analytes and include markers specific to HBV and HCV infections.

Viral NATs are often the gold standard for the detection of current viral infections. Several methods for the multiplexed detection of nucleic acids from different viruses or different viral sequences from a single sample input have been developed. These methods all involve a means of amplifying multiple targeted nucleic acids and differentially detecting which targets amplify successfully. The simplest of these approaches is a quantitative PCR using multiple primer and fluorescent probe sets, with one set specific for each target. Alternatively, a general marker for nucleic acid amplification like SYBR green can be used with melting curve analysis to distinguish which analyte amplified (De Crignis et al. 2010; Zhang et al. 2020). Since nucleic acid amplification and detection occur in a single well, assay conditions need to be identical for all analytes. Fluorescent probe detection is limited to three or four targets due to spectral compatibility, and multiplex melt curve analysis is generally limited by the inability to reliably discriminate more than two or three DNA products by their melting temperatures. Multiplex quantitative PCR assays have been used for genotyping, subtyping, identifying genetic variants, and the simultaneous detection of hepatitis and other viruses (Irshad et al. 2016; Singh et al. 2017). Automated commercial assays to simultaneously detect HBV, HCV, and HIV for blood screening purposes have been developed (Margaritis et al. 2007). Multiplex quantitative PCR assays often have similar performance to their singleplex counterparts. Isothermal methods that could be amenable to POC settings can also be multiplexed (Xie et al. 2021).

Alternative detection techniques have been applied to increase the number of amplified nucleic acid targets that can be detected or the genotypes/subtypes that can be differentiated in a single assay. These include various capture-probe, line-probe, and oligonucleotide microarray assays (Song et al. 2006; Bouchardeau et al. 2007; Khodakov et al. 2008; Yang et al. 2015; Warkad et al. 2020). Microfluidics approaches have also been leveraged to simplify reagent mixing or increase the number of nucleic acid targets that can be detected in a single assay. TaqMan array cards, which can test a single sample in 48 individual miniature qPCR reactions, have been successfully applied to the detection of the five hepatitis viruses: HAV, HBV, HCV, HDV, and HEV, as well as HIV-1 and HIV-2 (Kodani et al. 2014; Granade et al. 2018). This approach maintains high specificity but loses some sensitivity for detecting low-titer samples due to the small volumes of the qPCR reactions. A recently reported multiplex assay format that includes the detection of HBV, HCV, and other hepatitis viruses is the combinatorial assayed reactions for multiplexed evaluation of nucleic acids (CARMEN). This method combines microfluidics and nucleic acid detection by Cas13 to allow for greater than one-hundred nucleic acid targets to be detected simultaneously in a single sample (Ackerman et al. 2020). The CARMEN method has good sensitivity with reported limits of detection of one nucleic acid copy per microliter and very high specificity due to its reliance on Cas13 for detection (see POC Testing section for details).

Multiplex assay formats have also been adapted to the detection of virus-specific antibodies and antigens. These include the various immunochromatography or recombinant immunoblot lateral-flow assays that have been used as supplemental or confirmatory anti-HCV tests (Martin et al. 1998; Maertens et al. 1999). In these tests, antibodies targeting multiple HCV protein antigens can be detected as distinct bands on the detection strip. Detection of multiple different HCV targeting antibodies accounts for the diverse immune responses by HCV-infected people and allows for more accurate results. Rapid, POC lateral-flow immunochromatographic assays have also been developed to detect HBsAg, anti-HCV, and anti-HIV in the same test (Robin et al. 2018). Lateral-flow assays are quick to perform, but results are qualitative or semi-quantitative and the number of analytes that can be multiplexed is limited.

Fluorescent-microsphere-based serological assays allow for quantitative multiplexing of antibody or antigen detection of dozens of analytes. These assays are analogous to ELISAs except that each capture antigen or antibody is covalently linked to magnetic microspheres with a characteristic fluorescence signal that allows for the multiplexed detection of different antibodies or antigens. This approach has been applied to the development of assays for the simultaneous detection of antibodies targeting various HCV antigens and assays for anti-HBc, anti-HBe, anti-HCV, and antibodies targeting other pathogenic microorganisms (Fonseca et al. 2011; Araujo et al. 2011; Brenner et al. 2019). Antigenic characterization of HBV is also possible using this technology (Ijaz et al. 2012). With this approach, the detection of antigens and antibodies on the same sample in the same well is possible, but only if the antibodies of interest do not target the antigens of interest. For example, anti-HBs and HBsAg could not be measured simultaneously because the antibodies used to measure HBsAg would bind to the antigens used to detect anti-HBs.

Enzyme immunoassays that detect cumulative antigens and antibodies have also been developed. These include assays that measure the sum of anti-HCV and HCV cAg, which maintain the sensitivity of antibody tests while potentially allowing for earlier detection of infection using core antigen, which appears sooner after infection than antibodies (Laperche et al. 2005). Other approaches that have been applied to the multiplexed detection of hepatitis-specific antibodies, antigens, and nucleic acids are protein microarray, array-in-well, and electrochemical sensor assays that utilize antigens or antibodies coating specific regions of a detection chip or well (Perrin et al. 2003; Xu et al. 2007; Tang et al. 2010; Talha et al. 2016). The location of the signal in these assay formats identifies the type of antibody or antigen being detected. It is worth noting that many of the multiplex assays that have been reported show good sensitivity and specificity when compared to standard singleplex assays.

There are many promising technologies that allow for the multiplexed detection of HBV and HCV nucleic acid and antibody markers. Most of these technologies have been developed by individual diagnostic or research laboratories, but some commercial multiplex assays are also available, including a few multiplex nucleic acids and rapid serological assays that have been CE-marked. Further evaluation and development are warranted to make these methods more widely available and used. Screening and surveillance activities could be streamlined as multiple markers would be detected simultaneously. These assays would save not only time but also financial resources and patient sample volume as they eliminate the need to run multiple tests in parallel or sequentially.

Identifying Recently Acquired Infections

HBV and HCV can both establish chronic infections that can persist for the lifetime of the patient. It is estimated that hundreds of millions of people are living with HBV and HCV infection, yet only 10–20% of them have been diagnosed (WHO 2021a). From an epidemiological perspective, it is important to be able to identify which newly diagnosed infections are acute and which are chronic. There are several uses for the identification of acute cases including estimating the incidence of HBV and HCV infection in a population and identifying cases associated with an outbreak. Incidence measurements are important for informing public health policy because they can be used to identify populations with high levels of transmission and evaluate the effectiveness of interventions for preventing new infections.

Acute and chronic are clinical states of disease progression typically defined by the timing of symptoms or infection lasting longer than six months. Infections by HBV and HCV are typically asymptomatic, preventing clinical criteria, such as jaundice and elevated levels of ALT and bilirubin in serum, from being reliable markers for acute infection. Additionally, clinical symptoms can be observed in some patients with infections in the chronic phase. Serially collecting blood from healthy individuals and looking for the initial appearance of markers of HBV or HCV infections allow for precise classification of acute and chronic phases of infection, but this is not a practical approach for most populations and settings. Therefore, other laboratory markers inherent to the sample have been investigated to distinguish acute infections from chronic infections. Since these laboratory markers may not precisely coincide with the clinical definitions of acute and chronic, but instead vary with the length of time since infection, we will use the terms recent infection and long-term infection in this text. This classification dichotomy is more descriptive of its purpose which is to differentiate an infection that was recently acquired from one that has existed for a long-term period. The length of time that is classified as recent depends on the specific method and assay being used and can range from a few months to a year. Methods for this purpose are well established for HBV infections but are still in the developmental stage for HCV infections. In this section, we discuss current and promising approaches for identifying recent infections.

Recent HBV infections can be differentiated from long-term HBV infections through the presence of anti-HBc IgM. Immunoglobulin M is typically the first antibody isotype produced in response to a foreign antigen. This is true for HBV infections where anti-HBc IgM often appears one to two months after infection around the time of symptom onset in symptomatic patients. It quickly rises to high titer levels that typically decline after 6 months (Mast et al. 2006). Since low-titer anti-HBc IgM can occasionally be detected years after infection or during exacerbations of chronic infections, appropriate assay cutoffs can be used to reliably differentiate recent from long-term HBV infections (Gerlich et al. 1986; Park et al. 2015). Use of this method requires confirmation of active infection by the presence of HBsAg or HBV DNA and that the patient has generated an antibody response by the presence of total anti-HBc. This approach is well established and routinely used in hepatitis diagnostic laboratories.

Differentiation of recent from long-term HCV infections is not as straightforward as for HBV infections and there are currently no standardized methods for this purpose. Anti-HCV IgM is not a reliable marker for recent infections as it is not observed in all cases and its longevity varies among infected people (Chau et al. 1991; Sagnelli et al. 2005). Additionally, it has also been detected in some chronic cases, especially those that have exacerbations of symptomatic infection (Chen et al. 1992; Yamaguchi et al. 2000). Therefore, several alternative markers or measurements that can be used to differentiate recent from long-term HCV infection have been investigated. These include the diversity and titer of the antibody response, the avidity of the antibodies produced by an infected person, and the evolutionary dynamics of intra-host HCV populations (Figure 7).

Antibody titers generally increase with the time of infection. This is the result of B-cell activation, proliferation, and differentiation leading to increasing numbers of antibody-secreting plasma cells. The level of anti-HCV antibodies in the blood starts low and generally increases through the first several months of infection before reaching a plateau. In long-term infections, where the immune system is continually exposed to HCV antigens, anti-HCV titers remain high. The nature of the anti-HCV antibody response is variable among infected people. Most infected people have antibodies that recognize antigens from the core and NS3 proteins, while antibodies targeting the E1, E2, NS4, and NS5 proteins are more variable.

Time since virus infection

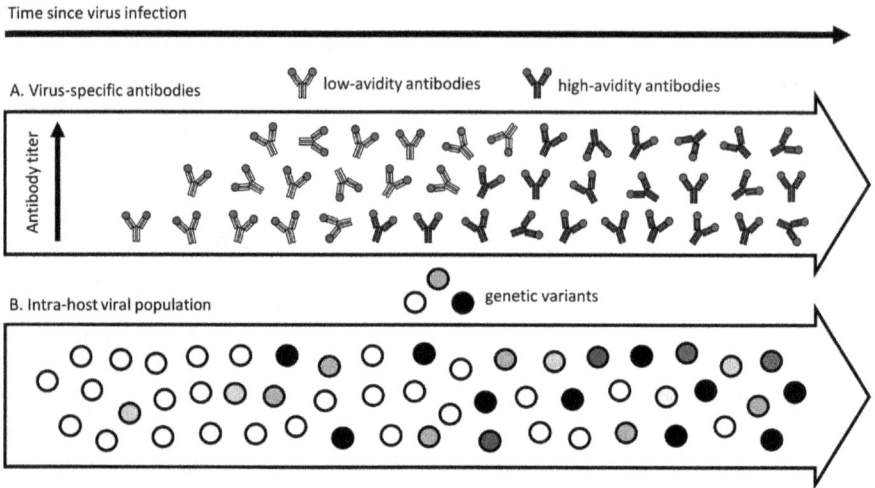

Figure 7. Differentiating recent from long-term HCV infections. (A) The titer and avidity of anti-HCV antibodies increase with the time since infection (left to right inside of arrows). Avidity describes the strength of the antibody binding to its target. Initially, antibodies have low avidity. Affinity maturation leads to an increase in the avidity of antibodies over time (white = low-avidity antibodies, dark gray = high-avidity antibodies). (B) Intra-host HCV populations evolve in characteristic ways. Genetic diversity increases over the course of the infection and particular genetic signatures are selected for as the infection progresses. Circles represent a simplified viral population with colors indicating genetic variants. Early in infection (left), there is low genetic diversity, while later in infection (right) there is higher genetic diversity and selection of characteristic mutations, genetic signatures, or genetic structure.

Antibodies targeting Core and NS3 often rise to detectable levels earliest in infection (Chen et al. 1999; Netski et al. 2005). Measuring antibody titers targeting different HCV antigens has shown promise for differentiating recent from long-term HCV infections, particularly when multivariate predictive models are used (Nikolaeva et al. 2002; Araujo et al. 2011). However, active infection needs to be confirmed due to the waning of antibody titers upon resolution of infection which could lead to samples from uninfected persons being misclassified as recent. While this approach has demonstrated promising results, it has not garnered as much attention as other approaches, perhaps due to the inefficiency of making multiple EIA measurements, the need for multiplexed immunoassay capabilities, or the relative simplicity of avidity-based measurements.

In addition to increasing in titer over time, antibodies also increase in avidity, or antigen-binding strength. This occurs through the process of affinity maturation, during which somatic hypermutation diversifies the variable region of immunoglobulin genes in the B-cells that proliferate after encountering a foreign antigen (Bannard and Cyster 2017). Mutant B-cell receptors that better recognize and bind to their target antigens are selected for as the humoral immune response adapts in response to re-exposure or continued exposure to a foreign antigen. This process leads to an increase in the avidity of an antibody for its antigen. Measurement of antibody avidity to HCV antigens may be the most studied method for the differentiation of recent and long-term HCV infections (Ward et al. 1994; Klimashevskaya et al. 2007;

Gaudy-Graffin et al. 2010). Antibody avidity is measured by an immunoassay in which either the patient sample or the antigen-antibody complex is treated with a chaotropic agent such as urea or diethylamine. The chaotropic agent interferes with hydrogen bonding and hydrophobic interactions, effectively weakening interactions between antibodies and antigens. Calculating the ratio of immunoassay signal in the presence and absence of a chaotropic agent determines the avidity index or proportion of antibodies that have a high avidity.

Reported methods for anti-HCV avidity index measurement utilize either laboratory-developed or modified commercial immunoassays, some of which have quality certifications, but are used "off-label." These assays have been applied to serum and plasma samples with good performance. In general, they have low probabilities of misclassifying a long-term sample as recent (false recency rates (FRR) of 0.4–4.0%) and long mean durations from the date of seroconversion that patient would be classified as recent (mean duration of recent infections (MDRI) of 114–147 days) (Patel et al. 2016; Shepherd et al. 2018; Boon et al. 2020). The FRR and MDRI are important parameters to consider when using a recency assay (UNAIDS/WHO 2011). Characteristics of the population being tested, such as the expected incidence and infection levels, will determine whether an assay's FRR and MDRI are good enough to accurately estimate incidence (Patel et al. 2016; Boon et al. 2020). It may be possible to tune the FRR and MDRI performance of an avidity immunoassay by selecting certain combinations of antigens that either lengthen or shorten the period since infection over which the assay classifies a sample as recent. Similarly, the antigens used in an avidity assay may affect its performance with different HCV genotypes. Some avidity assays have shorter MDRI values for genotype 1 samples than for others (Shepherd et al. 2018; Eshetu et al. 2020). This may be due to the capture antigens used in the assays coming from genotype 1 HCV and being recognized with higher avidity by antibodies from genotype 1 infected patients than from patients infected with other genotypes. Anti-HCV avidity immunoassays have been reported to perform less well when used with dried blood or serum/plasma spot specimens, which could decrease their utility for incidence estimation in certain populations where this is the most practical sample type (Shepherd et al. 2013; Eshetu et al. 2020). Similarly, populations with high rates of HCV and HIV coinfection may not be ideal for the use of antibody avidity assays due to the immunocompromised status of some long-term patients causing them to be misclassified as recent (Patel et al. 2016; Boon et al. 2020). It Is worth noting that antibody avidity is a potential alternative measurement for the identification of recent HBV infections. Antibody avidity measurements have shown similar performance to the anti-HBc IgM assays that have been widely adopted for this purpose (Rodella et al. 2006). If avidity immunoassays are to gain widespread use, establishing more standardized methods would allow for easier comparison and validation of results among studies and populations.

A final potential marker for differentiating recent from a long-term infection that has gained attention is the structure of the intra-host HCV population. Viral populations may change in predictable and reproducible ways within a host over time (Farci et al. 2000). Measuring the genetic diversity, genetic structure, and the physicochemical properties in short regions of the HCV E1, E2, or NS5b genes has shown promise for differentiating recent from long-term infections (Astrakhantseva et al. 2011; Montoya et al. 2015; Lara et al. 2017; Baykal et al. 2020). Using these population features by themselves or in multivariate predictive models can differentiate recent from long-term with high predictive power. This approach can be used with DBS specimens in addition to serum and plasma specimens (Antuori et al. 2021). The length of time that differentiates a recent infection from a long-term infection has not been as precisely defined with these methods as they have with avidity assays. Drawbacks to this approach are the time and cost of population sequencing methods, the expertise required for analysis, and the inability to test samples that have low HCV RNA titers. As population-based sequencing approaches become more affordable and widely used for routine medical diagnostics purposes, this approach will become more practical for identifying recent HCV cases or measuring incidence within a population.

The differentiation of recent from long-term HBV and HCV infections is important for estimating population incidence and identifying outbreaks. While the presence of anti-HBc IgM is a reliable method for identifying recent HBV infection, there is no standard method for identifying recent HCV cases. The avidity, titer, and diversity of anti-HCV antibodies and the genetic structure of intra-host HCV populations show promise for this purpose, but each approach has drawbacks. Further studies are needed to establish a reliable and standardized method for identifying recently acquired cases of HCV infection.

Monitoring and Predicting Therapy Outcomes

HBV and HCV are both capable of causing lifelong chronic infections that can lead to liver cirrhosis, HCC, and other serious liver problems. Fortunately, therapeutic drugs are available for both, although current HCV drugs are more effective and require shorter treatment regimens than those used for HBV infection. Monitoring the outcome of drug therapy and predicting when it is advisable to discontinue therapy are important for ensuring effective and efficient treatment. Here, we will discuss how current HCV therapies have led to the simplification of these processes and that there is still no widely agreed-upon approach for monitoring HBV activity in patients who are on treatment.

The latest generation of DAAs that target the NS3/NS4a protease, NS5a, or the NS5b polymerase of HCV are very effective at treating HCV infections especially when taken in combination. Recent studies indicate cure rates of > 95% after 12–24 weeks for many viral genotypes and treatment regimens (Cuypers et al. 2016; Falade-Nwulia et al. 2017). Therapy outcomes are determined by assessing whether a patient no longer has detectable HCV RNA in the blood, which was previously termed a sustained virological response (SVR). Whether HCV RNA is still detectable after therapy can be determined by using highly sensitive NATs. Twenty-four weeks is the typical time point for assessing SVR, but shorter time points may work equally well for evaluating whether a patient was cured (Burgess et al. 2016). Individuals who achieve SVR at 12 weeks after treatment are considered cured with fewer than 0.5%

having subsequent detectable HCV RNA (Sarrazin et al. 2017). The best predictor for successful therapy is completing the full 12–24-week treatment regimen. With the high cure rates of current-generation DAAs, the future may bring a time when monitoring for SVR is no longer required for patients who complete their therapy regimen.

Unlike HCV which has therapeutics that cure almost all patients, current HBV therapies suppress viral replication without sterilizing the host of virally infected cells. HBV is typically treated using nucleos(t)ide analog (NUC) reverse transcriptase inhibitors, interferon, or a combination of the two (Grossi et al. 2017; Tang et al. 2018). Currently approved NUCs include Lamivudine, Telbivudine, Entecavir, Adefovir, and Tenofovir (Terrault et al. 2016; 2018). Of these, Entecavir and Tenofovir may have a higher barrier to evolving resistance. Cure is often defined by the loss of HBsAg and anti-HBs seroconversion. Drug therapy rarely leads to this outcome, so viral suppression (undetectable HBV DNA) and HBeAg seroconversion (HBeAg loss with development of anti-HBeAg) indicating partial immune control are alternative indicators of therapeutic success (Hadziyannis and Hadziyannis 2020). However, in these cases, long-term drug therapy is often required to maintain viral suppression as reactivation of HBV replication is common once treatment is discontinued (Hall et al. 2020; Liem et al. 2020). Several markers have been proposed to predict when patients can safely discontinue antiviral therapy in the absence of HBsAg loss or seroconversion. These include quantitative HBsAg levels, serum HBV-RNA levels, and HBcrAg levels.

Before discussing the utility of HBV markers for monitoring and predicting therapy outcomes, we will briefly discuss the HBV replication cycle and the effects of NUC and interferon therapies (Figure 8). Upon infection of a cell, the relaxed circular partially double-stranded genomic DNA of the virus traffics to the nucleus where it is repaired into a fully double-stranded covalently closed circular DNA (cccDNA) (Seeger and Mason 2015; Tsukuda and Watashi 2020). The cccDNA is a template for the transcription of four different viral mRNAs and the 3.5 kb pre-genomic RNA (pgRNA) that gets assembled into nascent viral particles. During the maturation of the viral particle, the HBV polymerase generates the partially double-stranded genomic DNA while degrading the pgRNA. This maturing viral particle containing the genomic DNA either traffics back to the nucleus to generate more cccDNA or is secreted from the cell as a fully mature viral particle. NUCs used in the treatment of HBV inhibit the reverse transcriptase activity of the viral polymerase, preventing the production of genomic DNA from pgRNA (Fung et al. 2011). Treatment with this class of drugs effectively inhibits the production of new infectious viral particles and the generation of more cccDNA transcription templates. Since cccDNA is long-lived in infected cells, NUC therapy does not cure infected cells or inhibit the transcriptional activity of cccDNA in infected cells. On the other hand, interferon therapy does not directly target HBV but rather stimulates the immune system with the goal of limiting viral replication and destroying cccDNA reservoirs within the liver (Konerman and Lok 2016). Interferon therapy has low rates of success and is poorly tolerated. Due to the difficulty of eliminating all cccDNA reservoirs, many patients are on therapy for extended periods of time. The goal of HBV therapy is

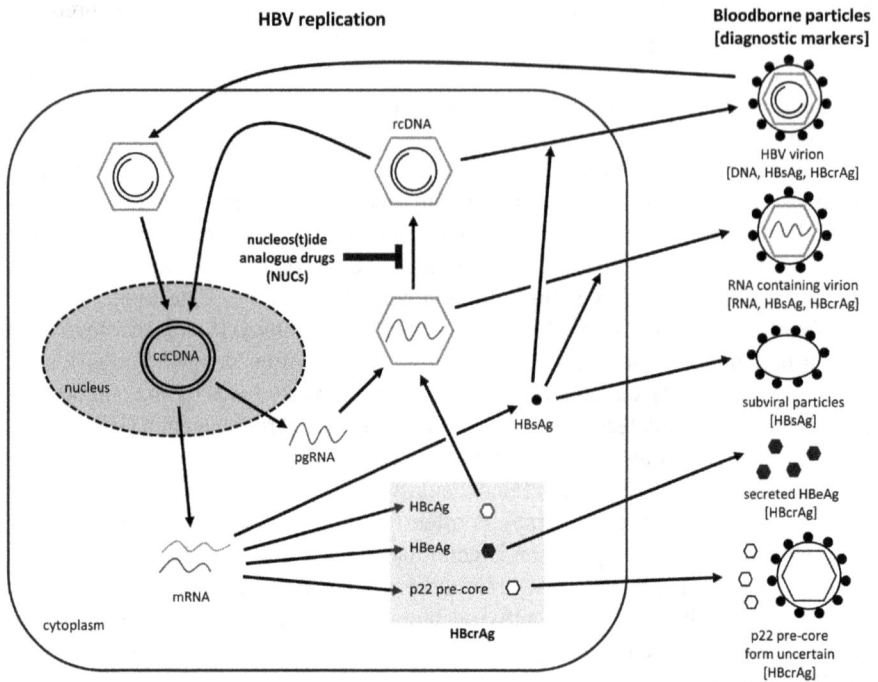

Figure 8. HBV markers and the effect of nucleos(t)ide analog therapy. Covalently closed circular DNA (cccDNA; black double circle) is the transcriptional template for viral mRNAs and pre-genomic RNA (pgRNA). Viral mRNA translation products include HBV surface antigen (HBsAg; small black circles) and the HBV core-related antigens (HBcrAg, gray box), which include HBV core antigen (HBcAg), HBV e antigen (HBeAg), and p22 pre-core. The pgRNA is encapsidated in viral capsids composed of HBcAg. The encapsidated pgRNA serves as a template for the viral polymerase to produce relaxed circular DNA (rcDNA; partial double circles) which is the viral genome. rcDNA can either become part of HBV virions or traffic to the nucleus to become cccDNA. HBV virions and RNA-containing virions form when HBsAg-containing envelopes form around DNA or RNA-containing capsids and are secreted from the cell. Subviral particles lacking capsids and viral nucleic acids are also secreted from the cell. In the blood, secreted HBeAg is found in its dimeric form, while it is uncertain if p22 pre-core is found as individual proteins or as part of higher-order capsid structures in empty viral particles. Nucleos(t)ide analogs (NUCs) inhibit the HBV polymerase and prevent the conversion of pgRNA to rcDNA (bolded line). These drugs prevent the production of HBV virions and *de novo* cccDNA. They do not inhibit the production or secretion of viral proteins or RNA.

ending HBsAg expression and anti-HBs seroconversion which indicates immune control of the virus. Since this is rarely achieved, alternative indicators of success that are sometimes used as therapeutic endpoints are therapy duration of at least one year, having undetectable serum HBV DNA, normalized serum ALT levels, and HBeAg seroconversion indicating partial immune control. Using these therapeutic endpoints poorly predicts sustained virologic response or future HBsAg loss and anti-HBs seroconversion. While on therapy, HBV replication is effectively suppressed, but upon stopping therapy, HBV replication often rebounds (Hall et al. 2020; Liem et al. 2020). For this reason, alternative and reliable predictive markers of successful therapy are urgently needed.

Directly assessing cccDNA in infected patients would be ideal, but it is impractical, requiring invasive liver biopsies. A more practical approach would utilize a readily accessible marker and sample type. Measuring HBV DNA in the serum is typically how virus activity is measured in patients who are not undergoing therapy. However, this approach is inappropriate for monitoring therapy with NUC drugs because they inhibit the production of mature viral particles without eliminating the transcriptional cccDNA reservoir (Gao et al. 2017). HBV DNA levels often drop below detectable levels during therapy but rebound once therapy is stopped (Wang et al. 2016; Hall et al. 2020; Liem et al. 2020). Therefore, alternative surrogates for cccDNA levels or activity are being investigated.

It has been known for several decades that HBV RNA can be found in the serum of infected patients. The nature and biological role of this RNA is controversial, but it is likely that pgRNA found in enveloped viral particles or unenveloped viral capsids is a major source of this RNA (Bai et al. 2018; Anderson et al. 2021). Several methods have been developed to measure HBV RNA, the most common of which are rapid amplification of cDNA ends (RACE)-quantitative PCR, and DNase treatment prior to quantitative reverse transcription PCR (van Bömmel et al. 2015; Huang et al. 2018; Prakash et al. 2018; Liu et al. 2019). In the absence of treatment, serum HBV RNA correlates well with serum HBV DNA and intrahepatic cccDNA activity (Giersch et al. 2017; Li et al. 2018; Wang et al. 2018a; 2018b). Serum HBV-RNA levels are typically 10–100-fold lower than HBV DNA levels. While on NUC therapy, however, serum HBV RNA generally decreases less than serum HBV DNA and can be observed in the absence of serum HBV DNA (Wang et al. 2016; Gao et al. 2017; Butler et al. 2018). Correlation between serum HBV RNA and cccDNA may depend upon a patient's HBeAg status and the nature of the method used to detect HBV RNA (Gao et al. 2017; Li et al. 2018; Prakash et al. 2018). Whether these correlations are maintained during NUC therapy is unclear, but serum HBV-RNA levels may predict whether a patient remains negative for serum HBV DNA upon stopping therapy (Wang et al. 2016; Gao et al. 2017).

The quantitative levels of two antigenic markers, HBsAg and HBcrAg, in patient serum also show promise for predicting HBV therapeutic outcomes (Inoue and Tanaka 2019; Lee et al. 2021). HBsAg is commonly used as a qualitative marker for HBV infection and HBcrAg is an aggregate measurement of three HBV proteins; core antigen (HBcAg), e antigen (HBeAg), and p-22 pre-core, each sharing a 149 amino acid sequence and expressed from the pre-core/core ORF (Kimura et al. 2005; Hong et al. 2021). Both antigenic markers can be measured using quantitative immunoassays. These antigenic markers in the blood have been shown to correlate reasonably well with the amount of cccDNA in the liver and its transcriptional activity in HBeAg-positive and HBeAg-negative chronic HBV patients (Wong et al. 2017; Chen et al. 2019; Testoni et al. 2019). A better correlation of cccDNA with HBcrAg may be because it is typically expressed only from cccDNA, while HBsAg expression can have additional contributions from linear HBV DNA fragments integrated into the host cell genome. Both antigenic markers can be observed in serum HBV DNA-negative patients during NUC therapy because their expression is not directly inhibited by NUCs. While patients with chronic HBV are on NUC therapy, HBsAg and HBcrAg decline less rapidly than serum HBV RNA (Liao et al. 2019). Low quantitative serum HBsAg levels at the end of NUC treatment could predict the sustained viral response (maintenance of undetectable or low serum HBV DNA) and HBsAg seroconversion within 6 years in HBeAg-positive and HBeAg-negative patients with 55% and 79% accuracy, respectively (Chen et al. 2014). Low baseline HBcrAg levels have been shown to predict sustained viral response and HBsAg seroconversion similarly to HBsAg levels with greater than 70% accuracy in HBeAg-negative patients treated with pegylated interferon (Martinot-Peignoux et al. 2016). Changes in the quantitative HBcrAg levels during NUC therapy or absolute levels at the end of therapy may also be correlated with HBeAg seroconversion and sustained viral response in HBeAg-positive patients (van Campenhout et al. 2016; Sonneveld et al. 2019).

Quantitative levels of serum HBV RNA, HBsAg, HBcrAg, at the beginning, middle, or end of therapy can predict sustained virological response, HBeAg seroconversion, HBsAg seroconversion, or the occurrence of hepatocellular carcinoma (Inoue and Tanaka 2019; Liu et al. 2019; Lee et al. 2021). However, it is unclear which marker is superior due to variations in the tested populations and the methods used in currently published studies. It does appear that using HBV RNA, HBsAg, and/or HBcrAg in combination improves predictive ability compared to their use individually (Wang et al. 2018; Fan et al. 2020; Papatheodoridi et al. 2020; Xie et al. 2021). While commercial assays for the detection and quantitative measurement of these three HBV markers exist, none are FDA approved or WHO-prequalified and only quantitative HBsAg assays have been granted CE marking. Continued research on the monitoring of HBV therapy outcomes that compares all available markers, uses standardized methods, and assesses therapeutic outcomes at longer time points after the end of therapy is warranted. As better HBV therapies are developed, a more thorough understanding of predictive markers and what they indicate regarding patient outcomes will greatly improve the efficiency of treatment regimens, improve patient care, and assist in achieving global elimination goals.

Conclusion

As we approach the year 2030, it will be necessary to critically assess all aspects of the continuum from patient testing through diagnosis and therapy for areas where improvements can aid in attaining the WHO's elimination goals for HBV and HCV infection. While the standard diagnostic markers for HBV and HCV infections have been well-established for decades, there remains room for improved patient access and efficiency. Promising technological advances may make routine rapid and POC testing for HBV and HCV infection a reality in the near future. Revisiting the idea of using easy-to-collect and transport sample types like saliva and DBSs to expand access to laboratory-based testing would improve diagnosis and linkage to care in many underserved populations with high disease burdens. Elimination efforts will require improvements in the ways that we measure incidence. Currently, no standardized method is used to identify recently acquired HCV infections, but several promising candidate methods exist that could aid in making more accurate incidence measurements. As both HBV and HCV can cause lifelong chronic infections, linkage to effective antiviral therapy is an important elimination goal. Current-generation HCV antivirals are very effective cures while HBV infection treatments may require prolonged therapy. The current predictors of HBV infection treatment success are imperfect, but there are several markers that may improve therapeutic decision-making regarding when it is advisable to cease therapy. Further research and development of methods that would improve access to testing, diagnostic efficiency, and ensure positive antiviral therapy outcomes are needed if we are to meet our goals of greatly reducing the burden of HBV and HCV infection within the next decade.

Acknowledgments

We thank Dr. Tonya Hayden and Amy Sandul for their critical comments and helpful suggestions.

References

Abe, K. and N. Konomi. 1998. Hepatitis C virus RNA in dried serum spotted onto filter paper is stable at room temperature. J. Clin. Microbiol. 36: 3070–3072.

Abudayyeh, O.O., J.S. Gootenberg, S. Konermann, J. Joung, I.M. Slaymaker, D.B.T. Cox et al. 2016. C2c2 is a single-component programmable RNA-guided RNA-targeting CRISPR effector. Science 353: 5573.

Ackerman, C.M., C. Myhrvold, S.G. Thakku, C.A. Freije, H.C. Metsky, D.K. Yang et al. 2020. Massively multiplexed nucleic acid detection with Cas-13. Nature 582: 277–282.

Amini, A., O. Varsaneux, H. Kelly, W. Tang, W. Chen, D.I. Boeras et al. 2017. Diagnostic accuracy of tests to detect hepatitis B surface antigen: a systematic review of the literature and meta-analysis. BMC Inf. Dis. 17: 19–37.

Anderson, M., J. Gersch, K.C. Luk, G. Dawson, I. Carey, K. Agerwal et al. 2021. Circulating pregenomic hepatitis B virus RNA is primarily full-length in chronic hepatitis B patients undergoing nucleos(t)ide analogue therapy. Clin. Inf. Dis. 72: 2029–2031.

Antuori, A., V. Montoya, D. Pineyro, L. Sumoy, J. Joy, M. Krajden et al. 2021. Characterization of acute HCV infection and transmission networks in people who currently inject drugs in Catalonia: usefulness of dried blood spots. Hepatol. 74: 591–606.

Araujo, A.C., I.V. Astrakhantseva, H.A. Fields and S. Kamili. 2011. Distinguishing acute from chronic hepatitis C virus (HCV) infection based on antibody reactivities to specific HCV structural and nonstructural proteins. J. Clin. Microbiol. 49: 54–57.

Astrakhantseva, I.V., D.S. Campo, A. Araujo, C.G. Teo, Y. Khudyakov and S. Kamili. 2011. Differences in variability of hypervariable region 1 of hepatitis C virus (HCV) between acuta and chronic stages of HCV infection. *In Silico* Biol. 11: 163–173.

Bannard, O. and J.G. Cyster. 2017. Germinal centers: programmed for affinity maturation and antibody diversification. Curr. Opin. Immunol. 45: 21–30.

Bai, L., X. Zhang, M. Kozlowski, W. Li, M. Wu, J. Liu et al. 2018. Extracellular hepatitis B virus RNAs are heterogenous in length and circulate as capsid-antibody complexes in addition to virions in chronic hepatitis B patients. J. Virol. 92: e00798-18.

Baykal, P.B.I., J. Lara, Y. Khudyakov, A. Zelikovsky and P. Skums. 2020. Quantitative differences between intra-host HCV populations from persons with recently established and persistent infections. Virus Evolution 6: veaa103.

Benzine, J.W., K.M. Brown, K.N. Agansa, R. Godiska, C.E. Mire, K. Gowda et al. 2016. Molecular diagnostic field test for point-of-care detection of ebola virus directly from blood. J. Inf. Dis. 214: S234–242.

Biondi, M.J., M. van Tilborg, D. Smookler, G. Heymann, A. Aquino, S. Perusini et al. 2019. Hepatitis C core-antigen testing from dried blood spots. Viruses 11: 830.

Boon, D., V. Bruce, E.U. Patel, J. Quinn, A.K. Srikrishnan, S. Shanmugam et al. 2020. Antibody avidity-based approach to estimate population-level incidence of hepatitis C. J. Hepatol. 73: 294–302.

Bottero, J., A. Boyd, J. Gozlan, M. Lemoine, F. Carrat, A. Collignon et al. 2013. Performance of rapid tests for detection of HBsAg and anti-HBsAb in a large cohort, France. J. Hepatol. 58: 473–478.

Bouchardeau, F., J.F. Cantaloube, S. Chavaliez, C. Portal, A. Razor, J.J. Lefrere et al. 2007. Improvement of hepatitis C virus (HCV) genotype determination with the new version of the INNO-LiPA HCV assay. J. Clin Microbiol. 45: 1140–1145.

Brandão, C.P.U., B.L.C. Marques, V.A. Marques, C.A. Villela-Nogueira, K.M.R. Do Ó, M.T. de Paula et al. 2013. Simultaneous detection of hepatitis C virus antigen and antibodies in dried blood spots. J. Clin. Virol. 57: 98–102.

Brenner, N., A.J. Mentzer, J. Butt, K.L. Braband, A. Michel, K. Jeffrey et al. 2019. Validation of multiplex serology for human hepatitis viruses B and C, human t-lymphotropic virus 1 and *Toxoplasma gondii*. PLoS ONE 14: e0210407.

Burgess, S.V., T. Hussaini and E.M. Yoshida. 2016. Concordance of sustained virologic response at weeks 4, 12 and 24 post-treatment of hepatitis C in the era of new oral direct-acting antivirals: a concise review. Ann. Hepatol. 15: 154–159.

Butler, E.K., J. Gersch, A. McNamara, K.C. Luk, V. Holzmayer, M. de Medina et al. 2018. Hepatitis B virus serum DNA and RNA levels in nucleos(t)ide analog-treated or untreated patients during chronic and acute infection. Hepatol. 68: 2106–2117.

Cameron, S.O., K.S. Wilson, T. Good, J. McMenamin, B. McCarron, A. Pithie et al. 1999. Detection of antibodies against hepatitis C virus in saliva: a marker of viral replication. J. Viral Hepat. 6: 141–144.

Carmona, S., B. Seiverth, D. Magubane, L. Hans and M. Hoppler. 2019. Separation of plasma from while blood using the cobas plasma separation card: a compelling alternative to dried blood spots for the quantification of HIV-1 viral load. J. Clin. Microbiol. 57: e01336.

Carter, J.G., L.O. Iturbe, J.L.H.A. Duprey, I.R. Carter. C.D. Southern. M. Rana et al. 2021. Ultrarapid detection of SARS-CoV-2 RNA using a reverse transcription-free exponential amplification reaction, RTF-EXPAR. Proc. Natl. Acad. Sci. U.S.A. 118: e2100347118.

Carty, P.G., M. McCarthy, S.M. O'Neill, C.F. De Gascun, P. Harrington, M. O'Neill et al. 2021. Laboratory-based testing for hepatitis C infection using dried blood spot samples: a systematic review and meta-analysis of diagnostic accuracy. Rev. Med. Virol. e2320: 1–21.

Catlett, B., F.M.J. Lamoury, S. Bajis, B. Hajarizadeh, D. Martinez, Y. Mowat et al. 2019. Evaluation of a hepatitis C virus core antigen assay from venepuncture and dried blood spot collection samples: a cohort study. J. Viral Hepat. 26: 1423–1430.

Centers for Disease Control (CDC). 2013. Testing for HCV infection: an update of guidance for clinicians and laboratorians. MMWR 62: 1–4.

Centers for Disease Control (CDC). 2021a. HBV information. https://www.cdc.gov/hepatitis/hbv/index.htm, accessed December 2021.

Centers for Disease Control (CDC). 2021b. HCV information. https://www.cdc.gov/hepatitis/hcv/index.htm, accessed December 2021.

Chau, K.H., G.J. Dawson, I.K. Mushahwar, R.A. Gutierrez, R.G. Johnson, R.R. Lesniewski et al. 1991. IgM-antibody response to hepatitis C virus antigens in acute and chronic post-transfusion non-A, non-B hepatitis. J. Virol. Methods 35: 343–352.

Chen, C.H., S.N. Lu, C.H. Hung, J.H. Wang, T.H. Hu, C.S. Changchien et al. 2014. The role of hepatitis B surface antigen quantification in predicting HBsAg loss and HBV relapse after discontinuation of Lamivudine treatment. J. Hepatol. 61: 515–522.

Chen, C.M., S. Ouyang, L.Y. Lin, L.J. Wu, T.A. Xie, J.J. Chen et al. 2020. Diagnostic accuracy of LAMP assay for HBV infection. J. Clin. Lab. Anal. 34: e23281.

Chen, E.Q., M.L. Wang, Y.C. Tao, D.B. Wu, J. Liao, M. He et al. 2019. Serum HBcrAg is better than HBV RNA and HBsAg in reflecting intrahepatic covalently closed circular DNA. J. Viral Hepatol. 26: 586–595.

Chen, J.S., E. Ma, L.B. Harrington, M. Da Costa, X. Tian, J.M. Palefsky et al. 2018. CRISPR-Cas12a target binding unleashes indiscriminate single-stranded DNase activity. Science 360: 436–439.

Chen, M., M. Sallberg, A. Sonnerborg, O. Weiland, L. Mattsson, L. Jin et al. 1999. Limited humoral immunity in hepatitis C virus Infection. Gastroenterol. 116: 135–143.

Chen, P.J., J.T. Wang, L.H. Hwang, Y.H. Yang, C.L. Hsieh, J.H. Kao et al. 1992. Transient immunoglobulin M antibody response to hepatitis C virus capsid antigen in posttransfusion hepatitis C: putative serological marker for acute viral infection. Proc. Natl. Acad. Sci. 89: 5971–5975.

Chevaliez, S., F. Roudot-Thoraval, C. Hezode, J.M. Pawlotsky and R. Njouom. 2021. Performance of rapid diagnostic tests for HCV infection in serum or plasma. Future Microbiol. 16: 713–719.

Colin, C., D. Lanoir, S. Touzet, L. Meyaud-Kraemer, F. Bailly, C. Trepo et al. 2001. Sensitivity and specificity of third-generation hepatitis C virus antibody detection assays: an analysis of the literature. J. Viral Hepat. 8: 87–95.

Cruz, H.M., L.D.P. Scalioni, V.S. de Paula, J.C. Miguel, K.M. Rodriguez, F.A.P. Milagres et al. 2017. Poor sensitivity of rapid tests for the detection of antibodies to the hepatitis B virus: implications for field studies. Mem. Inst. Oswaldo Cruz. 112: 209–213.

Cruz, H.M., V.S. de Paula, J.C.M. Cruz, K.M. Rodriguez, F.A.P. Milagres, F.I. Bastos et al. 2021. Evaluation of accuracy of hepatitis B virus antigen and antibody detection and relationship between epidemiological factors using dried blood spot. J. Virol. Methods 277: 113798.

Cryer, A.M. and J.C. Imperial. 2019. Hepatitis B in pregnant women and their infants. Clin. Liver Dis. 23: 451–462.

Cuypers, L., F. Ceccherini-Silberstein, K. Van Laethem, G. Li, A.M. Vanamme and J.K. Rockstroh. 2016. Impact of HCV genotype on treatment regimens and drug resistance: a snapshot in time. Rev. Med. Virol. 26: 408–434.

Daskou, M., D. Tsakogiannis, T.G. Dimitriou, G.D. Amoutzias, D. Mossialos, C. Kottaridi et al. 2019. Warmstart colorimetric LAMP for the specific and rapid detection of HPV16 and HPV18 DNA. J. Virol. Methods 270: 87–94.

De Cock, l., V. Hutse, E. Verhaegen, S. Quoilin, H. Vandenberghe and R. Vranckx. 2004. Detection of HCV antibodies in oral fluid. J. Virol. Methods 122: 179–183.

De Crignis, E., M.C. Re, L. Cimatti, L. Zecchi and D. Gibellini. 2010. HIV-1 and HCV detection in dried blood spots by SYBR green multiplex real-time RT-PCR. J. Virol. Methods 165: 51–56.

Dembele, B., R. Affi-Aboli, M. Kabran, D. Sevede, V. Goha, A.C. Adiko et al. 2020. Evaluation of four rapid tests for detection of hepatitis B surface antigen in Ivory Coast. J. Immunol. Res. 2020: 1–6.

Ding, R., J. Long, M. Yuan, X. Zheng, Y. Shen, Y. Jin et al. 2021. CRISPR/Cas12-based ultra-sensitive and specific point-of-care detection of HBV. Int. J. Mol. Sci. 22: 4842.

Ding, X., G. Wang and Y. Mu. 2019. Single enzyme-based stem-loop and linear primers co-mediated exponential amplification of short gene sequences. Anal. Chim. Acta 1081: 193–199.

Dong, T., G.A. Wang and F. Li. 2019. Shaping up field-deployable nucleic acid testing using microfluidic paper-based analytical devices. Anal. Bioanal. Chem. 411: 4401–4414.

Dubuisson, J. and F.L. Cosset. 2014. Virology and cell biology of the hepatitis C virus life cycle—an update. J. Hepatology 61: S3–S13.

European Association for the Study of Liver Disease (EASL). 2015. EASL recommendations on treatment of hepatitis C 2015. J. Hepatol. 63: 199–236.

European Association for the Study of Liver Disease (EASL). 2017. EASL 2017 Clinical Practice Guidelines on the management of hepatitis B virus infection. J. Hepatol. 67: 370–398.

European Association for the Study of Liver Disease (EASL). 2018. EASL recommendations on treatment of hepatitis C 2018. J. Hepatol. 69: 461–511.

Elsana, S., E. Sikuler, A. Yaari, Y. Shemer-Avni, M. Abu-Shakra, D. Buskila et al. 1998. HCV antibodies in saliva and urine. J. Med. Virol. 55: 24–27.

Eshetu, A., A. Hauser, M. an der Heiden, D. Schmidt, K. Meixenberger, R. Ehret et al. 2020. Establishment of an anti-hepatitis C virus igg avidity test for dried serum/plasma spots. J. Immunol. Methods 479: 112744.

Falade-Nwulia, O., C. Suarez-Cuervo, D.R. Nelson, M.W. Fried, J.B. Segal and M.S. Sulkowski. 2017. Oral direct-acting agent therapy for hepatitis C virus infection. Ann. Int. Med. 166: 637–648.

Fan, R., J. Peng, Q. Xie, D. Tang, M. Xu, J. Niu et al. 2020. Combining hepatitis B virus RNA and hepatitis B core-related antigen: guidance for safely stopping nucleos(t)ide analogues in hepatitis B e antigen-positive patients with chronic hepatitis B. J. Inf. Dis. 222: 611–618.

Farci, P., A. Shimoda, A. Coiana, G. Diaz, G. Peddis, J.C. Melpolder, A. Strazzera et al. 2000. The outcome of acute hepatitis C predicted by the evolution of the viral quasispecies. Science 288: 339–344.

Fattovich, G., F. Bortolotti and F. Donato. 2008. Natural history of chronic hepatitis B: special emphasis on disease progression and prognostic factors. J. Hepatology 48: 335–352.

Flores, G.L., H.M. Cruz, D.V. Potsch, S.B. May, C.E. Brandao-Mello, M.M.A. Pires et al. 2017a. Evaluation of HBsAg and anti-HBc assays in saliva and dried blood spot samples according to HIV status. J. Virol. Methods 247: 32–37.

Flores, G.L., H.M. Cruz, V.A. Marques, C.A. Villela-Noguira, D.V. Potsch, S.B. May et al. 2017b. Performance of anti-HCV testing in dried blood spots and saliva according to HIV status. J. Med. Virol. 89: 1435–1441.

Flores, G.L., J.R. Barbosa, H.M. Cruz, J.C. Miguel, D.V. Potsch, J.H. Pilotto et al. 2021. Dried blood spot sampling as an alternative for the improvement of hepatitis B and C diagnosis in key populations. World J. Hepatol. 13: 504–514.

Fonseca, B.P.F., C.F.S. Marques, L.D Nascimento, M.B. Mello, L.B.R. Silva, N.M. Rubim et al. 2011. Development of a multiplex bead-based assay for detection of hepatitis C virus. Clin. Vacc. Immunol. 18: 802–806.

Food and Drug Administration (FDA). 2021. Premarket approval database. https://www.accessdata.fda.gov/scripts/cdrh/cfdocs/cfPMA/pma.cfm, accessed September 2021.

Freiman, J.M., T.M. Tran, S.G. Schumacher, L.F. White, S. Ongarello, J. Cohn et al. 2016. HCV core antigen testing for diagnosis of HCV infection: a systematic review and meta-analysis. Ann. Intern. Med. 165: 345–355.

Fung, J., C.L. Lai, W.K. Seto and M.F. Yuen. 2011. Nucleoside/nucleotide analogs in the treatment of chronic hepatitis B. J. Antimicrob. Chemother. 66: 2715–2725.

Gao, Y., Y. Li, Q. Meng, Z. Zhang, P. Zhao, Q. Shang et al. 2017. Serum hepatitis B virus DNA, RNA and HBsAg: which correlate better with intrahepatic covalently closed circular DNA before and after nucleos(t)ide analogue treatment? J. Clin. Microbiol. 55: 2972–2982.

Gaudy-Graffin, C., G. Lesage, I. Kousignian, S. Laperche, A. Girault, F. Dubois et al. 2010. Use of an anti-hepatitis C virus (HCV) IgG avidity assay to identify recent HCV infection. J. Clin. Microbiol. 48: 3281–3287.

Geretti, A.M., M. Patel, F.S. Sarfo, D. Chadwick, J. Verheyen, M. Fraune et al. 2010. Detection of highly prevalent hepatitis B virus coinfection among HIV-seropositive persons in Ghana. J. Clin. Microbiol. 48: 3223–3230.

Gerlich, W.H., A. Uy, F. Lembrecht and R. Thomssen. 1986. Cutoff levels of immunoglobulin M antibody against viral core antigen for differentiation of acute, chronic and past hepatitis B virus infection. J. Clin. Microbiol. 24: 288–293.

Giersch, K., L. Allweiss, T. Volz, M. Dandri and M. Lütgehetmann. 2017. Serum HBV pgRNA as a clinical marker for cccDNA activity. J. Hepatol. 66: 460–462.

Gish, R.G., B.D. Given, C.L. Lai, S.A. Locarnini, J.Y.N. Lau, D.L. Lewis et al. 2015. Chronic hepatitis B: virology, natural history, current management and a glimpse at future opportunities. Antiviral Res. 121: 47–58.

Gonçalves, P.L., C.B. Cunha, S.C.U. Busek, G.C. Oliveira, R. Ribeiro-Rodrigues and F.E.L. Pereira. 2005. Detection of HCV RNA in saliva samples from patients with seric anti-HCV antibodies. Braz. J. Infect. Dis. 9: 28–34.

Gootenberg, J.S., O.O. Abudayyeh, J.W. Lee, P. Essletzbichler, A.J. Dy, J. Joung et al. 2017. Nucleic acid detection with CRISPR-Cas13/C2c2. Science 356: 438–442.

Goto, M., E. Honda, A. Ogura, A. Nomoto and K.I. Hanaki. 2009. Colorimetric detection of loop-mediated isothermal amplification reaction by using hydroxy naphthol blue. BioTechniques 46: 167–172.

Gower, E., C. Estes, S. Blach, K. Razavi-Shearer and H. Razavi. 2014. Global epidemiology and genotype distribution of the hepatitis C virus infection. J. Hepatol. 61: S45–S57.

Granade, T.C., M. Kodani, S.K. Wells, A.S. Youngpairoj, S. Masciotra, K.A. Curtis et al. 2018. Characterization of real-time microarrays for simultaneous detection of HIV-1, HIV-2 and hepatitis viruses. J. Virol. Methods 259: 60–65.

Grossi, G., M. Vigano, A. Loglio and P. Lampertico. 2017. Hepatitis B virus long-term impact of antiviral therapy nucleos(t)ide analogues (NUCs). Liver Intl. 37: 45–51.

Grüner, N., O. Stambouli and R.S. Ross. 2015. Dried blood spots—preparing and processing for use in immunoassays and in molecular techniques. J. Vis. Exp. 97: e52619.

Guirgis, B.S.S., R.O. Abbas and H.M.E. Azzazy. 2010. Hepatitis B virus genotyping: current methods and clinical implications. Int. J. Inf. Dis. 14: e941–e953.

Hall, S., J. Howell, K. Visvanathan and A. Thompson. 2020. The yin and the yang of treatment for chronic hepatitis B—when to start, when to stop nucleos(t)ide analogue therapy. Viruses 12: 934.

Hadziyannis, E. and S. Hadziyannis. 2020. Current practice and contrasting views on discontinuation of nucleos(t)ide analog therapy in chronic hepatitis B. Expert Rev. Gastroenterol. Hepatol. 14: 243–251.

Hadziyannis, S.J. and G.V. Papatheodoridis. 2006. Hepatitis B e antigen-negative chronic hepatitis B: natural history and treatment. Semin. Liv. Dis. 26: 131–141.

Hermida, M., M.C. Ferreiro, S. Barral, R. Laredo, A. Castro and P. Diz Dios. 2002. Detection of HCV RNA in saliva of patients with hepatitis C virus infection by using a highly sensitive test. J. Virol. Methods 101: 29–35.

Hirzel, C., S. Pfister, M. Gergievski-Hrisoho, G. Wandeler and S. Zuercher. 2015. Performance of HBsAg point-of-care tests for detection of diagnostic escape-variants in clinical samples. J. Clin. Virol. 69: 33–35.

Ho, E., P. Michielsen, P. Van Damme, M. Ieven, I. Veldhuijzen and T. Vanwolleghem. 2020. Point-of-care tests for hepatitis B are associated with a higher linkage to care and lower cost compared to venepuncture sampling during outreach screenings in an Asian migrant population. Ann. Glob. Heal. 86: 1–8.

Hong, X., L. Luckenbaugh, M. Mendenhall, R. Walsh, L. Cabuang, S. Soppe et al. 2021. Characterization of hepatitis B precore/core-related antigens. J. Virol. 95: e01695–20.

Hongjaisee, S., N. Doungjinda, W. Khamduang, T.S. Carraway, J. Wipasa, J.D. Debes et al. 2021. Rapid visual detection of hepatitis C virus using a reverse transcription loop-mediated isothermal amplification assay. Int. J. Inf. Dis. 102: 440–445.

Huang, H., J. Wang, W. Li, R. Chen, X. Chen, F. Zhang et al. 2018. Serum HBV DNA plus RNA shows superiority in reflecting the activity of intrahepatic cccDNA in treatment-naïve HBV-infected individuals. J. Clin. Virol. 99-110: 71–78.

Ijaz, S., R. Szypulska, N. Andres and R.S. Tedder. 2012. Investigating the impact of hepatitis B virus surface gene polymorphism on antigenicity using *ex vivo* phenotyping. J. Gen. Virol. 93: 2473–2479.

Inoue, T. and Y. Tanaka. 2019. The role of hepatitis B core-related antigen. Genes 10: 357.

Irshad, M., P. Gupta, D.S. Mankotia and M.A. Ansari. 2016. Multiplex qPCR for serodetection and serotyping of hepatitis viruses: a brief review. World J. Gastroenterol. 22: 4824–4834.

Joung, J., A. Ladha, M. Saito, N.G. Kim, A.E. Wooley, M. Segel et al. 2020. Detection of SARS-CoV-2 with SHERLOCK one-pot testing. N. Engl. J. Med. 383: 15.

Kabamba, A.T., C.M. Mwamba, G. Dessilly, F. Dufrasne, B.M. Kabamba and A.O. Longanga. 2020. Evaluation of the analytical performance of six rapid diagnostic tests for the detection of viral hepatitis B and C in Lubumbashi, Democratic Republic of Congo. J. Virol. Methods 285: 113961.

Kamili, S., J. Drobeniuc, A.C. Araujo and T.M. Hayden. 2012. Laboratory diagnostics for hepatitis C virus infection. Clin. Infect. Dis. 55: S43–8.

Kimble, M.M., C. Stafylis, P. Treut, S. Saab and J.D. Klausner. 2019. Clinical evaluation of a hepatitis C antibody rapid immunoassay on self-collected oral fluid specimens. Diag. Micro. Inf. Dis. 95: 149–151.

Kimura, T., N. Ohno, N. Terada, A. Rokuhara, A. Matsumoto, S. Yagi et al. 2005. The hepatitis B virus DNA-negative Dane particles lack core protein but contain a 22-kDa precore protein without c-terminal arginine-rich domain. J. Biol. Chem. 280: 21713–21719.

Khodakov, D.A., N.V. Zakharova, D.A. Gryadunov, F.P. Filatov, A.S. Zasedatelev and V.M. Mikhailovich. 2008. An oligonucleotide microarray for multiplex real-time PCR identification of HIV-1, HBV and HCV. BioTechniques 44: 241–248.

Klimashevskaya, S., A. Obriadina, T. Ulanova, G. Bochkova, A. Burkov, A. Araujo et al. 2007. Distinguishing acute from chronic and resolved hepatitis C virus (HCV) infections by measurement of anti-HCV immunoglobulin G avidity index. J. Clin. Microbiol. 45: 3400–3403.

Kodani, M., T. Mixson-Hayden, J. Drobeniuc and S. Kamili. 2014. Rapid and sensitive approach to simultaneous detection of genomes of hepatitis A, B, C, D and E viruses. J. Clin. Virol. 61: 260–264.

Konerman, M.A. and A.S. Lok. 2016. Interferon treatment for hepatitis B. Clin. Liver Dis. 20: 645–665.

Kramvis, A., E.G. Kostaki, A. Hatzakis and D. Paraskevis. 2018. Immunoregulatoru function of HBeAg related to short-sighted evolution, transmissibility, and clinical manifestation of hepatitis B virus. Front. Microbiol. 9: 2521.

Lamoury, F.M., S. Bajis, B. Hajarizadeh, A.D. Marshall, M. Martinello, E. Ivanova et al. 2018. Evaluation of the Xpert HCV viral load finger-stick point-of-care assay. J. Inf. Dis. 217: 1889–1896.

Lange, B., J. Cohn, T. Roberts, J. Camp, J. Chauffour, N. Gummadi et al. 2017a. Diagnostic accuracy of serological diagnosis of hepatitis C and B using dried blood spot samples (DBS): two systematic reviews and meta-analysis. BMC Infect. Dis. 17: 87–106.

Lange, B., T. Roberts, J. Cohn, J. Greenman, J. Camp, A. Ishizaki et al. 2017b. Diagnostic accuracy of detection and quantification of HBV-DNA and HCV-RNA using dried blood spots (DBS) samples—a systemic review and meta-analysis. BMC Infect. Dis. 17: 71–86.

Laperche, S., N. Le Marrec, A. Girault, F. Bouchardeau, A. Servant-Delmas, M. Maniez-Montreuil et al. 2005. Simultaneous detection of hepatitis C virus (HCV) core antigen and anti-HCV antibodies improves the early detection of HCV infection. J. Clin. Microbiol. 43: 3877–3883.

Lara, J., M. Teka and Y. Khudyakov. 2017. Identification of recent cases of hepatitis C virus infection using physical-chemical properties of hypervariable region 1 and a radial basis function neural network classifier. BMC Genomics 18: 33–42.

Laury, J., L. Hiebert and J.W. Ward. 2021. Impact of COVID-19 response on hepatitis prevention care and treatment: results from global survey providers and program managers. Clin. Liver. Dis. 17: 41–46.

Law, M. 2021. Antibody responses in hepatitis C infection. Cold Spring Harb. Prospect. Med. 11: a036962.

Leathers, J.S., M.B. Pisano, V. Re, G. van Oord, A. Sultan, A. Boonstra et al. 2019. Evaluation of rapid diagnostic tests for assessment of hepatitis B in resource-limited settings. Ann. Glob. Heal. 85: 1–3.

Lee, H.W., S.H. Ahn and H.L.Y. Chan. 2021. Hepatitis B core-related antigen: from virology to clinical application. Semin. Liv. Dis. 41: 182–190.

Li, Y., L. He, Y. Li, M. Su, J. Su, J. Hou et al. 2018. Characterization of serum HBV RNA in patients with untreated HBeAg-positive and -negative chronic hepatitis B infection. Hepat. Mon. 18: e62079.

Liao, H., Y. Liu, X. Li, J. Wang, X. Chen, J. Zuo et al. 2019. Monitoring of serum HBV RNA, HBcrAg, HBsAg and anti-HBc levels in patients during long-term nucleoside/nucleotide analog therapy. Antiv. Ther. 24: 105–115.

Liem, K.S., A.J. Gehring, J.J. Reed and H.L.A. Janssen. 2020. Challenges with stopping long-term nucleos(t)ide analogue therapy in patients with chronic hepatitis B. Gastroenterol. 158: 1185–1190.

Lin, C.L. and J.H. Kao. 2011. The clinical implications of hepatitis B virus genotype: recent advances. J. Gastroenterol. Hepatol. 26: 123–130.

Liu, S., B. Zhou, J.D. Valdez, J. Sun and H. Guo. 2019. Serum hepatitis B virus RNA: a new potential biomarker for chronic hepatitis B virus infection. Hepatol. 69: 1816–1827.

Llibre, A., Y. Shimakawa, E. Mottez, S. Ainsworth, T.P. Buivan, R. Firth et al. 2018. Development and clinical validation of the Genedrive point-of-care test for qualitative detection of hepatitis C virus. Gut. 67: 2017–2024.

Lok, A.S.F. and B.J. McMahon. 2007. Chronic hepatitis B. Hepatol. 45: 507–539.

Maertens, G., F. Dekeyser, A. Van Geel, E. Sablon, F. Bosman, M. Zrein et al. 1999. Confirmation of HCV antibodies by the line immunoassay INNO-LIA HCV Ab III. Methods Mol. Med. 19: 11–25.

Margaritis, A.R., S.M. Brown, C.R. Seed, P. Kiely, B. D'Agostino and A.J. Keller. 2007. Comparison of two automated nucleic acid testing systems for simultaneous detection of human immunodeficiency virus and hepatitis C virus RNA and hepatitis B virus DNA. Transfusion 46: 1783–1793.

Martin, P., F. Fabrizi, V. Dixit, S. Quan, M. Brezina, E. Kaufman et al. 1998. Automated RIBA hepatitis C virus (HCV) strip immunoblot assay for reproducible HCV diagnosis. J. Clin. Microbiol. 36: 387–390.

Martinot-Peignoux, M., M. Lapalus, S. Maylin, N. Boyer, C. Castelnau, N. Giuily et al. 2016. Baseline HBsAg and HBcrAg titres allow peginterferon-based 'precision medicine' in HBeAg-negative chronic hepatitis B patients. J. Viral Hepat. 23: 905–911.

Mast, E.E., C.M. Weinbaum, A.E. Fiore, M.J. Alter, B.P. Bell, L. Finelli et al. 2006. A comprehensive immunization strategy to eliminate transmission of hepatitis B virus infection in the United States: recommendations of the advisory committee on immunization practices (ACIP) part II: immunization of adults. MMWR 55: 1–25.

McFall, S.M., R.L. Wagner, S.R. Jangam, D.H. Yamada, D. Hardie and D.M. Kelso. 2015. A simple and rapid DNA extraction method from whole blood for highly sensitive detection and quantitation of HIV-1 proviral DNA by real-time PCR. J. Virol. Methods 214: 37–42.

Mikawa, A.Y., S.A.T. Santos, F.R. Kenfe, F.H. da Silva and P.I. da Costa. 2009. Development of a rapid one-step immunochromatographic assay for HCV core antigen detection. J. Virol. Methods 158: 160–164.

Mixson-Hayden, T., G.J. Dawson, E. Teshale, T. Le, K. Cheng, J. Drobeniuc et al. 2015. Performance of ARCHITECT HCV core antigen test with specimens from US plasma donors and injection drug users. J. Clin. Virol. 66: 15–18.

Montoya, V., A.D. Olmstead, N.Z. Janjua, P. Tang, J. Grebely, D. Cook et al. 2015. Differentiation of acute from chronic hepatitis C virus infection by nonstructural 5B deep sequencing: a population-level tool for incidence estimation. Hepatol. 61: 1842–1850.

Mössner, B.K., B. Staugaard, J. Jensen, S.T. Lillevang, P.B. Christensen and D.K. Holm. 2016. Dried blood spots, valid screening for viral hepatitis and human immunodeficiency virus in real-life. World J. Gastroenterol. 22: 7604–7612.

Netski, D.M., T. Mosbruger, E. Depla, G. Maertens, S.C. Ray, R.G. Hamilton et al. 2005. Humoral immune response in acute hepatitis C virus infection. Clin. Inf. Dis. 41: 667–675.

Nikolaeva, L.I., N.P. Blakhina, N.N. Tsurikova, N.V. Veronkova, M.I. Miminoshvili, D.M. Braginsky. et al. 2002. Virus-specific antibody titres in different phases of hepatitis C virus infection. J. Viral Hepat. 9: 429–437.

Nguyen, T.T., V. Lemee, K. Bollore, H.V. Vu, K. Lacombe, X.L.T. Thi et al. 2018. Confirmation of HCV viremia using HCV RNA and core antigen testing on dried blood spot in HIV infected people who inject drugs in Vietnam. BMC Inf. Dis. 18: 622.

Notomi, T., H. Okayama, H. Masubuchi, T. Yonekawa, K. Watanabe, N. Amino et al. 2000. Loop-mediated isothermal amplification of DNA. Nuc. Acids. Res. 28: e63.

Nyan, D.C., L.E. Ulitzky, N. Cehan, P. Williamson, V. Winkelman, M. Rios et al. 2014. Rapid detection of hepatitis B virus in blood plasma by a specific and sensitive loop-mediated isothermal amplification assay. Clin. Inf. Dis. 59: 16–23.

Nyan, D.C. and K.L. Swinson. 2016. A method for rapid detection and genotype identification of hepatitis C virus 1-6 by one-step reverse transcription loop-mediated isothermal amplification. Intl. J. Inf. Dis. 43: 30–36.

Nyirenda, M., M.B.J. Beadsworth, P. Stephany, C.A. Hart, I.J. Hart, C. Munthali et al. 2008. Prevalence of infection with hepatitis B and C virus and coinfection with HIV in medical inpatients in Malawi. J. Infect. 57: 72–77.

Pallarés, C., A. Carvalho-Gomes, V. Hontangas, I. Conde, T. Di Maira, V. Aguilera et al. 2018. Performance of the OraQuick hepatitis C virus antibody test in oral fluid and fingerstick blood before and after treatment-induced viral clearance. J. Clin. Virol. 102: 77–83.

Papatheodoridi, M., E. Hadziyannis, F. Berby, K. Zachou, B. Testoni, E. Rigopoulou et al. 2020. Predictors of hepatitis B surface antigen loss, relapse and retreatment after discontinuation of effective oral antiviral therapy in noncirrhotic HBeAg-negative chronic hepatitis B. J. Viral Hepat. 27: 118–126.

Park, J.W., K.M. Kwak, S.M. Kim, M.K. Jang, D.J. Kim, M.S. Lee et al. 2015. Differentiation of acute and chronic hepatitis B in IgM anti-HBc positive patients. World J. Gastroenterol. 21: 3953–3959.

Patel, E.U., A.L. Cox, S.H. Mehta, D. Boon, C.E. Mullis, J. Astemborski et al. 2016. Use of hepatitis C virus (HCV) immunoglobulin G antibody avidity as a biomarker to estimate the population-level incidence of HCV infection. J. Inf. Dis. 214: 344–352.

Patel, J. and P. Sharma. 2020. Design of a novel rapid immunoassay for simultaneous detection of hepatitis C virus core antigen and antibodies. Arch. Virol. 165: 627–641.

Paul, R., E. Ostermann and Q. Wei. 2020. Advances in point-of-care nucleic acid extraction technologies for rapid diagnosis of human and plant diseases. Biosens. Bioelectron. 169: 112592.

Peeling, R.W., K.K. Holmes, D. Mabley and A. Ronald. 2006. Rapid tests for sexually transmitted infections (STIs): the way forward. Sex. Transm. Infect. 82: v1–v6.

Perrin, A., D. Duracher, M. Perret, P. Cleuziat and B. Mandrand. 2003. A combined oligonucleotide and protein microarray for the codetection of nucleic acids and antibodies associated with human immunodeficiency virus, hepatitis B virus and hepatitis C virus infections. Anal. Biochem. 322: 148–155.

Piepenburg, O., C.H. Williams, D.L. Stempie and N.A. Armes. 2006. DNA detection using recombination proteins. PLoS Biol. 4: 1115–1121.

Poiteau, L., A. Soulier, F. Raudot-Thoraval, C. Hezode, D. Challine, J.M. Pawlotsky et al. 2017. Performance of rapid diagnostic tests for the detection of anti-HBs in various patient populations. J. Clin Virol. 96: 64–66.

Poiteau, L., M. Wlassow, C. Hezode, J.M. Pawlotsky and S. Chevaliez. 2020. Evaluation of the Xpert HBV viral load for hepatitis B virus molecular testing. J. Clin. Virol. 129: 104481.

Pollock, K.G., S.A. McDonald, R. Gunson, A. McLeod, A. Went, D.J. Goldberg et al. 2020. Real-world utility of HCV core antigen as an alternative to HCV RNA testing: implications for viral load and genotype. J. Viral Hepat. 27: 996–1002.

Prakash, K., G.E. Rydell, S.B. Larsson, M. Anderson, G. Norkans, H. Norder et al. 2018. High serum levels of pregenomic RNA reflect frequently failing reverse transcription in hepatitis B virus particles. Virol. J. 15: 86.

Revill, P.A., T. Tu, H.J. Netter, L.K.W. Yuen, S.A. Locarnini and M. Littlejohn. 2020. The evolution and clinical impact of hepatitis B virus genome diversity. Nat. Rev. Gastroenterol. Hepatol. 17: 618–634.

Robin, L., R.S.M. Bouassa, Z.A. Nodjikouambaye, L. Charmant, M. Matta, S. Simon et al. 2018. Analytical performance of simultaneous detection of HIV-1, HIV-2 and hepatitis C-specific antibodies and hepatitis B surface antigen (HBsAg) by multiplex immunochromatographic rapid test with serum samples: a cross-sectional study. J. Virol. Methods 253: 1–4.

Rodella, A., C. Galli, L. Terlenghi, F. Perandin, C. Bonfanti and N. Manca. 2006. Quantitative analysis of HBsAg, IgM anti-HBc and anti-HBc avidity in acute and chronic hepatitis B. J. Clin Virol. 37: 206–212.

Sagnelli, E., N. Coppola, C. Marroco, G. Coviello, M. Battaglia, V. Messina et al. 2005. Diagnosis of hepatitis C virus related acute hepatitis by serial determination of IgM anti-HCV titees. J. Hepatol. 42: 646–651.

Saludes, V., A. Antuori, C. Folch, N. Gonzalez, N. Ibanez, X. Majo et al. 2019. Utility of a one-step screening and diagnosis strategy for viremic HCV infection among people who inject drugs in Catalonia. Int. J. Drug Policy 74: 236–245.

Sarrazin, C., V. Isakov, E.S. Svaravskaia, C. Hedskog, R. Martin, K. Chodavarapu et al. 2017. Late relapse versus hepatitis C virus reinfection in patients with sustained virologic response after Sofosbuvir-based therapies. Clin. Inf. Dis. 64: 44–52.

Scalioni, L.D.P., H.M. Cruz, V.S. de Paula, J.C. Miguel, V.A. Marques, C.A. Villela-Nogueira et al. 2014. Performance of rapid hepatitis C virus antibody assays among high- and low-risk populations. J. Clin. Virol. 60: 200–205.

Seck, A., F. Ndiaye, S. Maylin, B. Ndiaye, F. Simon, A.L. Funk et al. 2018. Poor sensitivity of commercial rapid diagnostic tests for hepatitis B e antigen in Senegal, West Africa. Am. J. Trop. Med. Hyg. 99: 428–434.

Seeger, C. and W.S. Mason. 2015. Molecular biology of hepatitis B virus infection. Virology 479: 672–686.

Seo, S., M.J. Silverberg, L.B. Hurley, J. Ready, V. Saxena, D. Witt et al. 2020. Prevalence of spontaneous clearance of hepatitis C virus infection doubled from 1998 to 2017. Clin. Gastroenterol. Hepatol. 18: 511–513.

Sharma, S., A. Kabir and W. Asghar. 2020. Lab-on-a-chip Zika detection with reverse transcription loop-mediated isothermal amplification-based assay for point-of-care settings. Arch. Pathol. Lab. Med. 144: 1335–1343.

Shen, X.X., F.Z. Qiu, L.P. Shen, T.F. Yan, M.C. Zhao, J.J. Qi et al. 2019. A rapid and sensitive recombinase aided amplification assay to detect hepatitis B virus without DNA extraction. BMC Inf. Dis. 19: 229.

Shepherd, S.J., J. Kean, S.J. Hutchinson, S.O. Cameron, D.J. Goldberg, W.F. Carman et al. 2013. A hepatitis C avidity test for determining recent and past infections in both plasma and dried blood spots. J. Clin Virol. 57: 29–35.

Shepherd, S.J., S.A. McDonald, N.E. Palmateer, R.N. Gunson, C. Aitken, G.J. Dore et al. 2018. HCV avidity as a tool for detection of recent HCV infection: sensitivity depends on HCV genotype. J. Med. Virol. 90: 120–130.

Singh, A., D.S. Mankotia and M. Irshad. 2017. A single-step multiplex quantitative real time polymerase chain reaction assay for hepatitis C virus genotypes. J. Trans. Int. Med. 5: 34–42.

Smith, D.B., J. Bukh, C. Kuiken, A.S. Muerhoff, C.M. Rice, J.T. Stapleton et al. 2014. Expanded classification of hepatitis C virus into 7 genotypes and 67 subtypes: updated criteria and genotype assignment web resource. Hepatol. 59: 318–327.

Song, J., M.G. Mauk, B.A. Hacket, S. Cherry, H.H. Bau and C. Liu. 2016. Instrument-free point-of-care molecular detection of Zika virus. Anal. Chem. 88: 7289–7294.

Song, Y., E. Dai, J. Wang, H. Liu, J. Zhai, C. Chen et al. 2006. Genotyping of hepatitis B virus (HBV) by oligonucleotide microarray. Molec. Cell. Probes 20: 121–127.

Sonneveld, M.J., G.W. van Oord, M.J. van Campenhout, R.A. De Man, H.L.A Janssen, R.J. de Knegt et al. 2019. Relationship between hepatitis B core-related antigen levels and sustained HBeAg seroconversion in patients treated with nucleos(t)ide analogues. J. Viral. Hepatol. 26: 828–834.

Soulier, A., L. Poiteau, I. Rosa, C. Hezode, F. Roudot-Thoraval, J.M. Pawlotsky et al. 2016. Dried blood spots: a tool to ensure broad access to hepatitis C screening, diagnosis and treatment monitoring. J. Infect. Dis. 213: 1087–1095.

Sullivan, R.P., J. Davies, P. Binks, R.G. Dhurrkay, G.G. Gurruwiwi, S.M. Bukulatjpi et al. 2019. Point-of-care and oral fluid hepatitis B testing in remote indigenous communities of Northern Australia. J. Viral Hepat. 27: 407–414.

Talha, S.M., P. Saviranta, L. Hattara, T. Vuorinen, J. Hytonen, N. Khanna et al. 2016. Array-In-well platform-based multiplex assay for the simultaneous detection of anti-HIV- and Treponemal-antibodies and hepatitis B surface antigen. J. Immuno. Methods 429: 21–27.

Tang, D., J. Tang, B. Su, J. Ren and G. Chen. 2010. Simultaneous determination of five-type hepatitis virus antigens in 5 min using an integrated automatic electrochemical immunosensor array. Biosens. Bioelectron. 25: 1658–1662.

Tang, L.S.Y., E. Covert, E. Wilson and S. Kottilil. 2018. Chronic hepatitis B infection. JAMA 319: 1802–1813.

Tang, W., W. Chen, A. Amini, D. Boeras, J. Falconer, H. Kelly et al. 2017. Diagnostic accuracy of tests to detect hepatitis C antibody: a meta-analysis and review of the literature. BMC Inf. Dis. 17: 41–56.

Tejada-Strop, A., J. Drobeniuc, T. Mixson-Hayden, J.C. Forbi, N.T. Le, L. Li et al. 2015. Disparate detection outcomes for anti-HCV IgG and HCV RNA in dried blood spots. J. Virol. Methods 212: 66–70.

Terrault, N.A., N.H. Bzowej, K.M. Chang, J.P. Hwang, M.M. Jonas and M.H. Murad. 2016. AASLD guidelines for treatment of chronic Hepatitis B. Hepatol. 63: 261–283.

Terrault, N.A., A.S.F. Lok, B.J. McMahon, K.M. Chang, J.P. Hwang, M.M. Jonas et al. 2018. Update on prevention, diagnosis and treatment of chronic hepatitis B: AASLD 2018 hepatitis B guidance. Hepatology 67: 1560–1599.

Testoni, B., F. Lebosse, C. Scholtes, F. Berby, C. Miaglia, M. Subic et al. 2019. Serum hepatitis B core-related antigen (HBcrAg) correlates with covalently closed circular DNA transcriptional activity in chronic hepatitis B patients. J. Hepatol. 70: 615–625.

Tsukuda, S. and K. Watashi. 2020. Hepatitis B virus biology and life cycle. Antivir. Res. 182: 104925.

Tuaillon, E., A.M. Mondain, F. Meroueh, L. Ottomani, M.C. Picot, N. Nagot et al. 2010. Dried blood spot for hepatitis C virus serology and molecular testing. Hepatology 51: 752–758.

Tuaillon, E., D. Kania, A. Pisoni, K. Bellore, F. Taleb, E.N.O. Ngoyi et al. 2020. Dried blood spot tests for the diagnosis and therapeutic monitoring of HIV and hepatitis B and C. Front. Microbiol. 11: 373.

UNAIDS/WHO Working Group on Global HIV/AIDS and STI Surveillance. 2011. When and how to use assays for recent infection to estimate HIV incidence at a population level. World Health Organization, Geneva, Switzerland.

Uusküla, A., A. Talu, J. Rannap, D.M. Barnes and D.D. Jarlais. 2021. Rapid point-of-care (POC) testing for hepatitis C antibodies in a very high prevalence setting: persons injecting drugs in Tallinn, Estonia. Harm Reduct. J. 18: 39.

van Bömmel, F., A. Bartens, A. Mysickova, J. Hofmann, D.H. Kruger, T. Berg et al. 2015. Serum hepatitis B virus RNA levels as an early predictor of hepatitis B envelope antigen seroconversion during treatment with polymerase inhibitors. Hepatol. 61: 66–76.

van Campenhout, M.J.H., W.P. Brouwer, G.W. van Oord, Q. Xie, Q. Zhang, N. Zhang et al. 2016. Hepatitis B core-related antigen levels are associated with response to Entecavir and peginterferon add-on therapy in hepatitis B e antigen-positive chronic hepatitis B patients. Clin. Microbiol. Inf. Dis. 22: 571.

van Loo, I.H.M., N.H.T.M. Dukers-Muijers, R. Heuts, M.A.B. van der Sande and C.J.P.A Hoebe. 2017. Screening for HIV, hepatitis B and syphilis on dried blood spots: a promising method to better reach hidden high-risk populations with self-collected sampling. PLoS ONE 12: e0186722.

Vanhomwegen, J., A. Kwasiborski, A. Diop, L. Boizeau, D. Hoinard, M. Vray et al. 2021. Development and clinical validation of loop-mediated isothermal amplification (LAMP) assay to diagnose high HBV DNA levels in resource-limited settings. Clin. Microbiol. Inf. 27: S1198.

Vázquez-Morón, S., P. Ryan, B. Ardizone-Jimenez, D. Martin, J. Troya, G. Cuevas et al. 2018. Evaluation of dried blood spot samples for screening of hepatitis C and human immunodeficiency virus in a real-world setting. Sci. Rep. 8: 1858.

Vázquez-Morón, S., B.A. Jimenez, M.A. Jimenez-Sousa, J.M. Bellon, P. Ryan and S. Resino. 2019. Evaluation of the diagnostic accuracy of laboratory-based screening for hepatitis C in dried blood spot samples: a systematic review and meta-analysis. Sci. Rep. 9: 7316.

Velkov, S., J.J. Ott, U. Protzer and T. Michler. 2018. The global hepatitis B virus genotype distribution approximated from available genotyping data. Genes 9: 495.

Villar, L.M., C.S. Bezerra, H.M. Cruz, M.M. Portilho and G.L. Flores. 2019. Applicability of oral fluids and dried blood spot for hepatitis B Virus diagnosis. Can. J. Gastroenterol. Hepatol. 2019: 5672795.

Villeneuve, J.P. 2005. The natural history of chronic hepatitis B virus infection. J. Clin. Virol. 34: S139–S142.

Wang, B., I. Carey, M. Bruce, S. Montague, G. Dusheiko and K. Agarwal. 2018. HBsAg and HBcrAg as predictors of HBsAg seroconversion in HBeAg-positive patients treated with nucleos(t)ide analogues. J. Viral Hepat. 25: 886–893.

Wang, J., T. Shen, X. Huang, G.R. Kumar, X. Chen, Z. Zeng et al. 2016. Serum hepatitis B virus RNA is encapsidated pregenome RNA that may be associated with persistence of viral infection and rebound. J. Hepatol. 65: 700–710.

Wang, J., Y. Yu, G. Li, C. Shen, J. Li, S. Chen et al. 2018a. Natural history of HBV-RNA in chronic HBV infection. J. Viral Hepatol. 25: 1038–1047.

Wang, J., Y. Yu, G. Li, C. Shen, Y. Meng, J. Zheng et al. 2018b. Relationship between serum HBV-RNA levels and intrahepatic viral as well as histologic activity markers in Entecavir-treated patients. J. Hepatol. 68: 16–24.

Ward, K.N., W. Dhaliwal, K.L. Ashworth, E.J. Clutterbuck and C.G. Teo. 1994. Measurement of antibody avidity for hepatitis C virus distinguishes primary antibody responses from passively acquired antibody. J. Med. Virol. 43: 367–372.

Warkad, S.D., S.B. Nimse, K.S. Song and T. Kim. 2020. Development of a method for screening and genotyping of HCV 1a, 1b, 2, 3, 4 and 6 genotypes. ACS Omega 5: 10794–10799.

Weber, J., M.K. Sahoo, N. Taylor, R.Z. Shi and B.A. Pinsky. 2019. Evaluation of the Aptima HCV Quant Dx assay using serum and dried blood spots. J. Clin. Microbiol. 57: e00030–19.

Westbrook, R.H. and G. Dusheiko. 2014. Natural history of hepatitis C. J. Hepatology 61: 558–568.

Wong, D.J.H., W.K. Seto, K.S. Cheung, C.K. Chong, F.Y. Huang, J. Fung et al. 2017. Hepatitis B virus core-related antigen as a surrogate marker for covalently closed circular DNA. Liver Intl. 37: 995–1001.

World Health Organization (WHO). 2016. Global health sector strategy on viral hepatitis 2016–2021. World Health Organization, Geneva, Switzerland.

World Health Organization (WHO). 2017. Guidelines on hepatitis B and C testing. World Health Organization, Geneva, Switzerland.

World Health Organization (WHO). 2021a. Global progress report on HIV, viral hepatitis and sexually transmitted infections, 2021. World Health Organization, Geneva, Switzerland.

World Health Organization (WHO). 2021b. Recommendations and guidance on hepatitis C virus self-testing. World Health Organization, Geneva, Switzerland.

World Health Organization (WHO), 2021c. WHO list of prequalified *in vitro* diagnostic products. https://extranet.who.int/pqweb/news/who-list-prequalified-vitro-diagnostic-products-updated-4, accessed September 2021.

Xiao, Y., A.J. Thompson and J. Howell. 2020. Point-of-care tests for hepatitis B: an overview. Cells 9: 2233.

Xie, C., S. Chen, L. Zhang, X. He, Y. Ma, H. Wu et al. 2021. Multiplex detection of bloodborne pathogens on a self-driven microfluidic chip using loop-mediated isothermal amplification. Anal. Bioanal. Chem. 413: 2923–2931.

Xie, Y., M. Li, X. Ou, S. Zheng, Y. Gao, X. Xu et al. 2021, HBeAg-positive patients with HBsAg < 100 IU/mL and negative HBV RNA have lower risk of virological relapse after nucleos(t)ide analogues cessation. J. Gastroenterol. 56: 856–867.

Xu, R., X. Gan, Y. Fang, S. Zheng and Q. Dong. 2007. A simple, rapid and sensitive integrated protein microarray for simultaneous detection of multiple antigens and antibodies of five human hepatitis viruses (HBV, HCV, HDV, HEV and HGV). Anal. Biochem. 362: 69–75.

Yamaguchi, N., K. Tokushige, K. Yamauchi and N. Hayashi. 2000. Humoral immune response in Japanese acute hepatitis patients with hepatitis C virus. Can. J. Gastroenterol. 14: 593–598.

Yang, Y.C., D.Y. Wang, H.F. Cheng, E.Y. Chuang and M.S. Tsai. 2015. A reliable multiplex genotyping assay for HCV using a suspension bead array. Microb. Biotech. 8: 93–102.

Yi, T.T., H.Y. Zhang, H. Liang, G.Z. Gong and Y. Cai. 2020. Betaine-assisted recombinase polymerase assay for rapid hepatitis B virus detection. Biotechnol. Appl. Biochem. 68: 469–475.

Yusrina, F., C.W. Chua, C.K. Lee, L. Chiu, T.S.Y. Png, M.J. Khoo et al. 2018. Comparison of Cobas HCV GT against Versant HCV Genotype 2.0 (LiPA) with confirmation by Sanger sequencing. J. Virol. Methods 255: 8–13.

Zhang, H., M. Ganova, Z.Q. Yan, H. Chang and P. Neuzil. 2020. PCR multiplexing based on a single fluorescent channel using dynamic melting curve analysis. ACS Omega 5: 30267–30273.

Zhang, J., X. Su, J. Xu, J. Wang, J. Zheng, C. Li et al. 2019. A point-of-care platform based on microfluidic chip for nucleic acid extraction in less than 1 minute. Biomicrofluidics 13: 034102.

Zhao, N., J. Liu and D. Sun. 2017. Detection of HCV genotypes 1b and 2a by a reverse transcription loop-mediated isothermal amplification assay. J. Med. Virol. 89: 1048–1054.

Part IV

Diagnosis of Human-Pathogenic Gastrointestinal Parasites

Chapter 6

Molecular Diagnosis of Intestinal Amebiasis#

Ibne Karim M. Ali

Introduction

Amebiasis is defined as the infection caused by a single-celled protozoan parasite called *Entamoeba histolytica* regardless of symptoms (WHO/PAHO/UNESCO 1997). *E. histolytica* belongs to the genus *Entamoeba* comprised of more than 40 species. Besides *E. histolytica*, seven other *Entamoeba* species can infect humans: *E. dispar, E. coli, E. moshkovskii, E. hartmanni, E. polecki, E. gingivalis*, and *E. bangladeshi*. Among these *E. histolytica* is universally recognized as a cause of intestinal and extraintestinal diseases in humans. It has two life cycle stages: a motile and actively feeding mono-nucleated trophozoite form and an environmentally resistant tetra-nucleated dormant cyst form. Some of these *Entamoeba* species can be differentiated from each other and *E. histolytica* based on their cyst morphology. *E. polecki* and *E. coli* cysts have one and eight nuclei, respectively. Although *E. hartmanni* cysts have four nuclei like those of *E. histolytica*, *E. hartmanni* is significantly smaller in size. *E. gingivalis*, a parasite of the human oral cavity, does not have a cyst form. However, *E. histolytica, E. dispar, E. moshkovskii,* and *E. bangladeshi* are morphologically indistinguishable in both cyst and trophozoite forms.

Entamoeba histolytica was responsible for a death toll of 55,500 people in 2010 worldwide (Lozano et al. 2012). However, about 90% of all *E. histolytica* infections remain asymptomatic while the remainder progresses to clinical illness (Gathiram and Jackson 1987). Asymptomatic individuals are thought to contribute to the

Waterborne Disease Prevention Branch, Division of Foodborne, Waterborne, and Environmental Diseases, National Center for Emerging and Zoonotic Infectious Diseases, Centers for Disease Control and Prevention, 1600 Clifton Road NE, Mailstop H23-9, Atlanta, GA 30329, USA.
Email: iali@cdc.gov

Disclaimer: Use of trade names is for identification only and does not imply endorsement of the Centers for Disease Control and Prevention (Prevention). The findings and conclusions in this report are those of the author and do not necessarily represent the official position of the CDC.

spread of the disease by excreting the environmentally resistant cyst form of the ameba in stools. Amebiasis remains a significant cause of morbidity and mortality in developing countries (Stanley 2003). *E. histolytica* infections are also linked to impaired cognitive development, growth stunting, and childhood morbidities (Tarleton et al. 2006; Petri et al. 2009; Mondal et al. 2012).

Entamoeba dispar is commonly accepted as a non-pathogenic commensal species. It appears to be ten times more common than *E. histolytica* worldwide (Gathiram and Jackson 1985), but local prevalence may vary significantly. Experimental and limited clinical data suggest that *E. moshkovskii* could be associated with human disease (Shimokawa et al. 2012; Ali et al. 2003). Similarly, the newest species of *Entamoeba*, *E. bangladeshi* has been detected in a handful of asymptomatic and symptomatic individuals in two studies. The true prevalence and pathogenicity of *E. bangladeshi* remain unknown. The purpose of this chapter is to discuss the tools commonly used for the diagnosis of intestinal amebiasis, their advantages and disadvantages, DNA-based tools, and recent advancements in the molecular detection of amebiasis.

Brief History of *E. histolytica*, *E. dispar*, *E. moshkovskii* and *E. bangladeshi*

Entamoeba histolytica

Fedor Aleksandrovich Lösch first detected intestinal amebiasis by identifying amebas in samples from a patient who died of dysentery in 1875. He reproduced the disease in dogs and suggested the name *Amoeba coli* (Petri 1996). Osler first detected a case of amebic liver abscess in 1890. Councilman and Lafleur at the Johns Hopkins Hospital studied the pathological role of amebas on patients with dysentery and liver abscesses and introduced the terms "amebic dysentery" and "amebic liver abscess." The organism was formally named *Entamoeba histolytica* by Schaudinn in 1903.

Entamoeba dispar

Emily Brumpt first suggested the existence of two morphologically identical species within *E. histolytica* in 1925, one being pathogenic and the other not. She proposed the name *Entamoeba dispar* for the non-pathogenic species. This proposal was ignored for decades. Eventually in 1993, biochemical, immunological, and genetic evidence emerged separating *E. histolytica* from *E. dispar* (Diamond and Clark 1993).

Entamoeba moshkovskii

Tshalaia first discovered *E. moshkovskii* in sewage in Moscow in 1941 (Tshalaia 1941). For the next 20 years, it was considered a free-living environmental ameba. In 1961, Dreyer reported the first human case of *E. moshkovskii* in a Texas resident in the USA, who presented with diarrhea, weight loss, and epigastric pain (Dreyer 1961). *E. moshkovskii* has unique features that are different from *E. dispar* or *E. histolytica*. For example, *E. moshkovskii* can be cultured at room temperature, is osmotolerant, and is resistant to emetine (Clark and Diamond 1997; Dreyer 1961; Entner and Most 1965; Richards et al. 1966).

Table 1. Countries with > 10% reported prevalence of infection with *Entamoeba moshkovskii*.

Country	Study population (size, N)	Prevalence of *E. moshkovskii*	*Detected in patients with diarrhea/ dysentery?	References
Australia	Patients with gastrointestinal disease (N = 110)	50.0%	Yes	(Fotedar et al. 2008)
Colombia	Asymptomatic children under 16 years old	25.4%	No	(Lopez et al. 2015)
Bangladesh	Preschool children 2–5 years old (N = 109)	21.1%	No	(Ali et al. 2003)
Kenya	Children and adults (N = 170)	19.5%	Yes	(Kyany'a et al. 2019)
Pakistan	Patients with chronic diarrhea (N = 161)	18.6%	Yes	(Yakoob et al. 2012)
Yemen	Children and adults (N = 800)	18.2%	No	(Al-Areeqi et al. 2017)
South Africa	Children and adults (N = 46)	15.9%	Yes	(Samie et al. 2020)
Tanzania	HIV-suspected or confirmed patients (N = 118)	13.0%	No	(Beck et al. 2008)
Egypt	Patients with gastrointestinal disease (N = 175)	11.8%	Yes	(Abozahra et al. 2020)

* These data should be interpreted with caution because a comprehensive workup to detect the etiology of diarrhea/dysentery was not performed.

Ali et al. (2003) were the first to detect a high > 20% prevalence of *E. moshkovskii* in children living in an urban slum in Dhaka, Bangladesh (Ali et al. 2003). Since then, human isolates of *E. moshkovskii* have been obtained from at least 19 other countries, including both resource-limited and resource-rich countries: USA, Italy, South Africa, India, Australia, Pakistan, Iran, Tanzania, Turkey, Malaysia, Thailand, Tunisia, United Arab Emirates, Iraq, Egypt, Kenya, Indonesia, Yemen, and Colombia. In most of these studies, only a few individuals were found to be infected. However, nine of these countries reported > 10.0% prevalence of *E. moshkovskii* (Fotedar et al. 2008; Parija and Khairnar 2005; Khairnar and Parija 2007; Ali et al. 2003; Beck et al. 2008; Yakoob et al. 2012; Abozahra et al. 2020; Al-Areeqi et al. 2017; Kyany'a et al. 2019; Lopez et al. 2015) (Table 1). In some of these countries, *E. moshkovskii* was detected in patients with diarrhea or dysentery. However, a complete work up to identify the pathogens responsible for diarrhea or dysentery was not performed, which raises the question of the role of *E. moshkovskii* in disease. In experimental mice, however, *E. moshkovskii* has been shown to cause diarrhea (Shimokawa et al. 2012).

Entamoeba bangladeshi

Royer et al. (Royer et al. 2012) first detected *E. bangladeshi* in 2010–2011 in stool samples from children living in an urban slum community in Dhaka, Bangladesh, while screening 2,039 stool samples from diarrheal and asymptomatic children by microscopy, culture, and species-specific *Entamoeba* PCRs for *E. histolytica*, *E. dispar* and *E. moshkovskii*. Forty-three of these stools were positive for *Entamoeba* by microscopy or culture, but species-specific *Entamoeba* PCRs were negative. Using the broad specificity primers for the *Entamoeba* small subunit ribosomal RNA (SSU rRNA) gene (Stensvold et al. 2011; Royer et al. 2012) followed by Sanger sequencing, a new species of *Entamoeba* was identified in two of the samples—one from a child with diarrhea and one from a child without symptoms. *E. bangladeshi* grows at both room temperature and 37°C like *E. moshkovskii*. The SSU rDNA phylogeny places *E. bangladeshi* closer to *E. histolytica* than *E. moshkovskii* but distant from *E. dispar*. Five years after its discovery, *E. bangladeshi* was detected outside of Bangladesh in individuals with and without symptoms in South Africa by Ngobeni et al. (2017). More investigations are required to understand (a) the epidemiology and potential pathogenicity of *E. bangladeshi*, (b) whether it can live outside a host like *E. moshkovskii*, (c) the potential host range, and (d) environmental reservoirs of this ameba.

Prevalence

The true prevalence of *E. histolytica* infection is not well understood as most studies used tools that were not *E. histolytica*-specific. Amebiasis is highly prevalent in the Indian subcontinent, Africa, the Far East, Mexico, and areas of South and Central America, where it is associated with cultural habits, crowding, age, level of sanitation, education, and socio-economic status. Amebiasis remains a major public health problem in countries with poor sanitation and hygiene and that lack safe drinking water (Atabati et al. 2020). In developed countries, *E. dispar* is about ten times more prevalent than *E. histolytica*. In Japan, however, *E. histolytica* is more common than *E. dispar* (Martínez-Palomo 1993). In developed countries, *E. histolytica* is primarily associated with immigrants from or travelers to endemic countries, men who have sex with other men, people living in institutionalized facilities, and patients infected with the human immunodeficiency virus (Petri 1996). *E. moshkovskii* is more common than *E. histolytica* or *E. dispar* in Australia (Fotedar et al. 2008). *E. bangladeshi* has been detected only in two countries, Bangladesh and South Africa.

Tegen et al. (2020) reviewed microscopy-based studies reported from Ethiopia to detect intestinal parasites and showed that 14.1% of the population had *Entamoeba* infection. Alasvand Javadi et al. (2019) detected only a 1.37% prevalence of *Entamoeba* species in stools by microscopy among 50,000 diarrheal patients attending the Naft Hospital of Ahvaz, Southwest Iran, during 2007–2017. Meyer et al. (2020) reviewed the data from 14 original publications reporting pathogens identified with the FilmArray GI panel and found that 39.7% of the patients had at least one pathogen in the panel. However, *E. histolytica* was identified in 0.3% of the patients. Samie et al. (2020) detected *E. histolytica*, *E. dispar*, and *E. moshkovskii* in stool samples from patients attending different rural clinics in northern South Africa and found 4.1%, 14.7%, and 15.9% prevalence of these species, respectively.

Mondal et al. (2006) investigated the effect of *E. histolytica* infection on growth and concluded that preschool children with a history of *E. histolytica*-associated diarrhea were more malnourished and stunted compared to the uninfected children. In a later study, Mondal et al. (2012) monitored 147 infants living in an urban slum of Dhaka, Bangladesh, for the first year of life and found that 10.9% of children developed at least one episode of *E. histolytica* diarrhea in their first year of life. Children born malnourished were more likely to be infected with *E. histolytica* during this period. An apparent vicious cycle exists where malnourished children are more likely to get *E. histolytica* infection, and *E. histolytica*-infected children are more likely to be malnourished.

Investigations with amebiasis patients in Natal, South Africa, showed that there was a peak incidence of *E. histolytica* infection among children < 14 years of age and a second peak in infection in adults > 40 years old (Gathiram and Jackson 1985). Acuno-Soto et al. (Acuna-Soto et al. 2000) reviewed the male-to-female ratios for invasive intestinal amebiasis and asymptomatic carriage in all the published reports from 1929 to 1997 and found that ratios were 3.2:1 and 1:1, respectively. A study in 1994 found that > 8% of the Mexican population was seropositive for *E. histolytica*. In 1996, 1.3 million cases of intestinal amebiasis were reported in Mexico (Caballero-Salcedo et al. 1994; PAHO 1998). In contrast, only 2,970 symptomatic cases of amebiasis were reported from the neighboring country, the USA, in 1993. About half of those cases in the USA were in immigrants from Mexico, South and Central America, Asia, and Pacific Islands (Prevention 1994).

Clinical Manifestations

About 10% of all *E. histolytica* infections are symptomatic (Gathiram and Jackson 1987). What determines the outcome of an *E. histolytica* infection is not clear. Parasite genotypes and host genetics are thought to play important roles in determining the outcome of an infection. Additionally, biological factors, such as the gut microbiome of the host, likely play a significant role [reviewed in (Burgess and Petri 2016)]. The majority of symptomatic *E. histolytica* infections progress to develop amebic colitis, which is marked by loose stools with mucus that may or may not contain visible blood. If not treated, amebic colitis may progress to perforation and ulcers in some cases that are associated with high mortality rates. Chronic ulceration may result in ameboma formation.

Amebic liver abscess is the most common form of extraintestinal amebiasis, which unless properly diagnosed and promptly treated is a potentially lethal disease. Patients with amebic liver abscess usually have a fever, weight loss, and right upper quadrant pain and tenderness (Haque et al. 2000). A small percentage of ALA patients may progress to develop brain abscess, which is highly fatal (Bauddh et al. 2020; Petri and Haque 2013; Victoria-Hernandez et al. 2020; Hughes et al. 1975; Solaymani-Mohammadi et al. 2007; Sundaram et al. 2004). Sporadic cases of *E. histolytica* infection of other organs have been reported: kidney (Saensiriphan et al. 2015; Ramakrishnan et al. 1971), heart (Gomersall et al. 1994; Mehta 1968), lung

(Liu et al. 2018; Hara et al. 2004; Lichtenstein et al. 2005), spleen (Goret and Goret 2019; Kruger et al. 2011), appendix (Ito et al. 2014; Singh et al. 2010; Ramdial et al. 2002), genital tract (Musthyala et al. 2019; Prasetyo 2015) and skin (Fernandez-Diez et al. 2012; Sasaki et al. 2016; Kroft et al. 2005; Magana et al. 2004).

Diagnosis of Intestinal Amebiasis

According to the recommendations of the World Health Organization, Pan American Health Organization and the United Nations Educational, Scientific and Cultural Organization, "Optimally, *E. histolytica* should be specifically identified and, if present, treated. If only *E. dispar* is identified, treatment is unnecessary. If the infected person has gastrointestinal symptoms, other causes should be sought" (WHO/PAHO/UNESCO 1997). However, these recommendations were made before the identification of *E. moshkovskii* and *E. bangladeshi* in human infections and may need an update.

For the diagnosis of *E. histolytica* various methods have been used such as microscopy, serology, antigen detection ELISA, and conventional and real-time PCRs (Figure 1). These methods have different levels of sensitivity and specificity. The advantages and disadvantages of some of the commonly used techniques for the diagnosis of intestinal amebiasis are provided in Table 2.

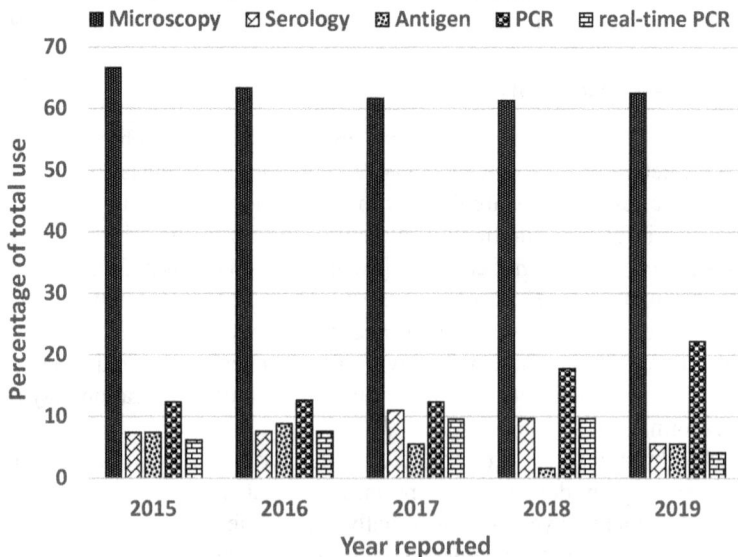

Figure 1. Tools used in the diagnosis of Entamoeba infections during 2015–2019. Percentages of total usage for each method are shown. Microscopy represents the sole method used in the diagnosis, and if microscopy was used in combination with other methods, it was excluded under microscopy. For the other methods, they might have been used in combination. Data were obtained from the PubMed search using the keywords: "Entamoeba," "diagnosis," and "detection." In most cases, only articles written in English were selected.

Table 2. Advantages and disadvantages of different methods used in the diagnosis of amebiasis [adapted from (Ali 2015)].

Methods	Advantages	Disadvantages
Microscopy	Simple, rapid, inexpensive. Available in most clinical laboratories.	Non-specific, insensitive; requires expertise, and cannot be performed on frozen specimens. Must be performed within 1–2 hours of collection if the stools are not preserved appropriately.
Culture	Visual observation of motile trophozoites; allows the expansion of ameba for downstream analysis, such as pathogenicity and virulence studies, and NGS.	Time-consuming, insensitive, and biased as it may facilitate the growth of the predominant species/strain, or other organisms. Usually not available in diagnostic laboratories.
Isoenzyme	Capable of differentiating between *E. histolytica* and other *Entamoeba* species.	Time-consuming, labor-intensive, and insensitive. It depends on the growth of a large number of trophozoites in a culture, which is not always successful.
Serology	Easy to perform; provides individual's past or present exposure to *E. histolytica*.	Unable to differentiate between an acute and a past infection; limited diagnostic value especially in the amebiasis-endemic areas. May give false-negative results in early infections, or in patients with immunocompromised status.
Antigen Detection	Simple, rapid, allows species-specific detection of *E. histolytica*; a well-suited diagnostic technique for resource-limited nations where amebiasis is endemic.	Does not work on formalin-fixed stools; works poorly with frozen stools; is less sensitive than PCR assays. Cannot detect mixed infections with more than one *Entamoeba* species. Species-specific antigen detection assays for *E. dispar*, *E. moshkovskii*, or *E. bangladeshi* are not available.
Conventional PCR	Sensitive, species-specific, rapid, works on broad specimen types.	Contamination-prone; sophisticated laboratory setup and expertise required - not suitable as a diagnostic test for resource-limited countries.
Real-time PCR	Highly sensitive; highly specific; rapid; works on broad specimen types; does not require post-analysis of PCR amplicons which minimizes contamination risks.	Sophisticated laboratory setup, high maintenance cost, and expertise required; not suitable for resource-limited countries.

Microscopy

The microscopic identification of intestinal parasites in stool samples is often referred to as the "ova and parasite (O&P)" examination. Three main microscopic techniques are used in clinical laboratories to identify *Entamoeba* species in stool samples: wet preparation, concentration, and permanent staining of smears. Wet preparation is most widely used, which works best if motile trophozoites of *Entamoeba* are seen. A motile *Entamoeba* trophozoite with the ingested red blood cell (RBC) is often considered a confirmation of the presence of *E. histolytica,* especially in diarrheal stools. The formol-ether concentration technique is used to concentrate parasites in stool samples, and it works best if the cyst form of *E. histolytica* is present, especially in asymptomatic cyst carriers. Stool smears may be stained with a permanent dye such as trichrome if the delay is anticipated in the evaluation of slides. For several reasons, microscopy should not be the method of choice for the diagnosis of amebiasis. Microscopy requires expertise with knowledge of the morphology of *Entamoeba* species. Still then, a knowledgeable microscopist may not be able to differentiate a pathogenic *E. histolytica* from non-pathogenic *E. dispar, E. moshkovskii,* or *E. bangladeshi* with full confidence because they all look the same under a microscope. Microscopy has a low sensitivity (Huston et al. 1999). Because *E. histolytica* trophozoites disintegrate within 1–2 hours of stool production if not properly stored, it requires immediate processing of fresh stool samples where trophozoites are the predominant forms of the parasite. Also, since the amebas are not excreted in every stool uniformly, at least three consecutive stool samples collected within 10 days are recommended, which increases the chance of ameba detection by 23% (Hiatt et al. 1995). Unfortunately, microscopy remains the most widely used method of amebiasis diagnosis, especially in endemic countries. A PubMed search on the methods used for the diagnosis of *E. histolytica* during 2015–2019 shows that microscopy has been used overwhelmingly in 61.3 to 66.7% of studies during this period (Figure 1).

Culture and Isoenzyme Analysis

Unpreserved and fresh stool samples may be used to grow amebas in culture. However, culture is a difficult, labor-intensive, time-consuming, and yet insensitive and non-specific process. Most clinical laboratories do not have the capacity to perform *Entamoeba* culture processes. *Entamoeba* culture is performed with clinical (usually stool) samples in a few research laboratories. Three main cultivation methods used for *Entamoeba* culture are xenic cultivation (where the ameba is grown in the presence of an undefined flora, mostly bacteria), monoxenic cultivation (where the ameba is grown in the presence of another single species), and axenic cultivation (where the ameba is grown in the absence of any other living organisms).

Culture has poor diagnostic value. Amebic culture has limited diagnostic value because several *Entamoeba* species can be found in human stools that are too morphologically similar to distinguish but can grow in the same culture media. However, culture can be an important research tool for other investigations such as ameba virulence, pathogenicity, and whole genome sequencing. It was the culture-derived isoenzyme analysis that provided the first glimpse of the evidence that what used to be considered "non-pathogenic *E. histolytica*" was indeed a separate species, *E. dispar* (Sargeaunt et al. 1978).

Isoenzyme analysis was first used by Reeves and Bischoff (1968) to classify *Entamoeba* species. However, Sargeaunt et al. (1978) were the pioneers in establishing the unique isoenzyme profiles (known as zymodemes) for *E. histolytica* and *E. dispar*. Prior to the development of DNA-based techniques, zymodemes of cultured amebas were considered the gold standard for species-specific identification of *Entamoeba* infection. A major disadvantage of isoenzyme profiling is that it depends on the growth of the amebas in culture, which is often unsuccessful. Also, it is time-consuming and labor-intensive.

Antibody Detection Tests

Serum anti-amebic antibodies against an *E. histolytica* infection can persist for several years (Abd-Alla et al. 1998). ELISA-based assays to detect *E. histolytica*-specific antibodies have been developed and some are commercially available. These tests demonstrated > 90% sensitivity and specificity in different studies (reviewed in (Fotedar et al. 2007)). However, serologic tests have little diagnostic value in endemic countries because a high percentage of asymptomatic individuals may have antibodies against *E. histolytica* due to past exposure (Caballero-Salcedo et al. 1994; Gathiram and Jackson 1987). In contrast, serologic tests are useful in non-endemic countries where people do not have high baseline antibody levels against *E. histolytica* (Ohnishi and Murata 1997; Weinke et al. 1989). Recently Moss et al. (2014) developed a high throughput multiplex, bead-based assay to detect serologic responses to multiple diarrhea-causing intestinal pathogens including *E. histolytica, Giardia intestinalis*, and *Cryptosporidium parvum*. Serologic tests specific for *E. moshkovskii* (could be an emerging pathogen), *E. bangladeshi* (of unknown pathogenicity), or *E. dispar* are not available.

Antigen Detection Tests

The amebic antigen can be detected directly in stool samples. Detection of amebic antigen is more sensitive than microscopy or culture simpler than the isoenzyme analysis, and superior to antibody detection as it detects acute infections. Antigen detection ELISAs are rapid and do not require expertise or sophisticated instruments to interpret the results. All these criteria make it suitable for diagnostic use in resource-limited endemic countries.

Some of the commercially available antigen detection ELISA tests are specific for *E. histolytica*, while the others are common for both *E. histolytica* and *E. dispar* (Table 3). These tests showed 55 to 100% sensitivity and 93 to 100% specificity compared to other established tests as detected in various investigations [reviewed in (Fotedar et al. 2007). Among these, the TechLab *E. histolytica* II ELISA assay is approved by the US Food and Drug Administration (FDA) for diagnostic use. Two major limitations of the current ELISA kits are one, they do not work on formalin-fixed stool samples; two, they work poorly on frozen stool samples. Commercial ELISA kits for the species-specific detection of *E. dispar*, *E. moshkovskii,* or *E. bangladeshi* do not exist.

ELISA-Based Point-of-Care (POC) Tests

An *E. histolytica*-specific POC test was first developed by TechLab in 2006 (Leo et al. 2006). This test is based on the amebic adherence lectin. Compared to the TechLab *E. histolytica* II antigen detection ELISA, the POC test showed 97% sensitivity and 100% specificity, respectively. However, compared to real-time

Table 3. Commercially available antigen detection ELISA kits for diagnosis of *E. histolytica* [adapted from (Ali 2015)].

Name	Target	Ameba specificity	Manufacturers	Detection limit
TechLab *E. histolytica* II ELISA	Gal/GalNac lectin	*E. histolytica*	TechLab, Blacksburg, VA	0.2–0.4 ng of lectin per well
Entamoeba CELISA-PATH	Gal/GalNac lectin	*E. histolytica*	Cellabs Pty Ltd., Brookvale, Australia	0.2–0.4 ng of lectin per well
Optimum S *Entamoeba histolytica* antigen ELISA	*SREHP	*E. histolytica*	Merlin Diagnostika, Berheim-Hersel, Germany	Unknown
Triage parasite panel	29-kDa surface antigen	**E. histolytica/E. dispar*	BIOSITE Diagnostics, San Diego, CA	Unknown
ProSpecT *Entamoeba histolytica* microplate assay	$EHSA	*E. histolytica/E. dispar*	REMEL Inc., Lenexa, KS	40 ng/ml of *E. histolytica*-specific antigen
Entamoeba histolytica/ Entamoeba dispar	Unknown	*E. histolytica/E. dispar*	IVD Research, Carlsbad, CA, USA	Unknown
R-Biopharm Ridascreen *Entamoeba* test	Gal/GalNac lectin	*E. histolytica/E. dispar*	Darmstadt, Germany	#17 *E. histolytica* or 595 *E. dispar* per well
TechLab *E. histolytica* II ELISA	Gal/GalNac lectin	*E. histolytica*	TechLab, Blacksburg, VA	0.2–0.4 ng of lectin per well
Entamoeba CELISA-PATH	Gal/GalNac lectin	*E. histolytica*	Cellabs Pty Ltd., Brookvale, Australia	0.2–0.4 ng of lectin per well
Optimum S *Entamoeba histolytica* antigen ELISA	*SREHP	*E. histolytica*	Merlin Diagnostika, Berheim-Hersel, Germany	Unknown
Triage parasite panel	29-kDa surface antigen	**E. histolytica/E. dispar*	BIOSITE Diagnostics, San Diego, CA	Unknown
ProSpecT *Entamoeba histolytica* microplate assay	$EHSA	*E. histolytica/E. dispar*	REMEL Inc., Lenexa, KS	40 ng/ml of *E. histolytica*-specific antigen
Entamoeba histolytica/ Entamoeba dispar	Unknown	*E. histolytica/E. dispar*	IVD Research, Carlsbad, CA, USA	Unknown
R-Biopharm Ridascreen *Entamoeba* test	Gal/GalNac lectin	*E. histolytica/E. dispar*	Darmstadt, Germany	#17 *E. histolytica* or 595 *E. dispar* per well

*   SREHP = Serine-rich *Entamoeba histolytica* protein.
** The Triage parasite panel also detects *G. lamblia* (target: alpha-1-giardin) and *C. parvum* (target: disulfide isomerase) in addition to *E. histolytica/E. dispar*.
$ EHSA = *E. histolytica*-specific antigen (targeted using polyclonal antibodies).
#   According to the manufacturer.

PCR, the POC test was only 80% sensitive. Verke et al. (2015) evaluated a newer version of TechLab's POC test, the *E. histolytica* QUIK CHEK test to detect *E. histolytica*-specific antigen in fecal samples from South Africa and Bangladesh. The performance of this newer POC test was compared to those of the commercially available ProSpecT *Entamoeba histolytica* microplate assay (Remel) and TechLab's *E. histolytica* II ELISA using un-fixed frozen stool samples. Compared to the *E. histolytica* II ELISA and ProSpecT microplate assay, the *E. histolytica* QUIK CHEK exhibited 98% and 97% sensitivity, respectively. Compared to *E. histolytica* II ELISA and ProSpecT microplate assay, the *E. histolytica* QUIK CHEK exhibited 100% specificity with both assays. Yanagawa et al. (2020) conducted a multicenter cross-sectional study in Japan to evaluate the utility of the *E. histolytica* QUIK CHEK in comparison with a PCR assay. They used stool samples that had been submitted for O&P examination. The overall sensitivity and specificity of the *E. histolytica* QUIK CHEK was 44.7% and 99.8%, respectively. The sensitivity of the *E. histolytica* QUIK CHEK was higher for diarrheal cases (60.0%) than non-diarrheal cases (27.8%). The *E. histolytica* QUIK CHEK assay sensitivity was lower for cyst-containing stools than for trophozoite-containing stools. Perhaps this may be explained by the fact that this assay is directed toward trophozoite-specific proteins.

PCR-Based Diagnostic Testing

Since the recognition of the "pathogenic *E. histolytica*" (i.e., true *E. histolytica*) and the "non-pathogenic *E. histolytica*" (i.e., *E. dispar*) as two separate species, several PCR-based species-specific tests have been developed for the diagnosis of amebiasis. These tests are highly sensitive and specific. Some of these tests are sensitive enough to detect a single ameba in the clinical specimens. However, one caveat is that these PCR-based assays require appropriate laboratory facilities, expertise and costly equipment which limits their use to resource-rich countries only.

A variety of clinical specimens have been used for the detection of species-specific *Entamoeba* by PCR assays including stools, liver aspirated pus, blood, saliva, urine, tissue, and formalin-preserved biopsy or autopsy samples. Among these, stool is the most complex and often most difficult specimen type for the extraction of DNA prior to performing PCR assays.

Complexity of Stool Samples

Intestinal amebiasis is diagnosed with stool samples. Although it is easy to collect, the stool is considered one of the most complex specimen types for PCR-based diagnosis. For two reasons, stool samples must be handled with appropriate caution when used in the diagnosis of amebiasis. First, infected individuals may not excrete amebas in every stool requiring checking of at least three consecutive stools collected within 10 days. Second, stool samples may contain PCR inhibitors that can be co-purified during DNA extraction if additional steps are not implemented. Components such as heme, bile salts, bilirubin, or complex carbohydrates in stools may act as

PCR inhibitors (Holland et al. 2000). DNA extraction should include steps to remove PCR inhibitors in order to achieve success in PCR and avoid false-negative results.

Storage and Transportation of Stool Samples

Appropriate storage of stool samples is critical to stabilize DNA and prevent it from degradation. Several PCR-compatible buffers or preservatives may be used to increase the shelf-life of stools prior to DNA extraction. The trophozoite form of *Entamoeba* species is not very stable outside the host environment and may spontaneously disintegrate rapidly. If only trophozoites are excreted in the stools, the processing of a sample should begin within 1–2 hours of production. If this is not possible, stools should either be kept frozen or stored at room temperature after adding PCR-compatible preservatives such as TotalFix, UniFix, EcoFix, and modified PVA (Zn- or Cu-based). Alternatively, stools can be mixed in potassium dichromate 2.5% (1:1 dilution) or in absolute ethanol (1:1 dilution). Stools kept with preservatives such as sodium acetate-acetic acid-formalin (SAF) or formalin may be ideal to preserve the intact morphology of ameba for wet mount microscopy, but formalin may negatively impact the DNA quality and subsequent PCR detection. Formalin-fixed stools, however, have been used by some investigators for the detection of *E. histolytica* and *E. dispar* by PCR (Ogren et al. 2020; Rivera et al. 1998; Sanuki et al. 1997). The longer a stool sample remains with formalin the greater DNA damage it may cause. Therefore, it is advisable to avoid formalin preservatives if the stool is to be used for DNA extraction and PCR. Frozen stool samples may be shipped to a diagnostic laboratory in frozen form using either dry-ice or plenty of icepacks. Stools preserved in PCR-compatible preservatives may be shipped at room temperature. Also, fresh stools preserved in Cary Blair medium may be shipped to a diagnostic laboratory at room temperature within 3 days of storage.

Considerations for DNA Extraction

DNA extraction is a crucial first step for achieving success in PCR. A stool sample suspected of *Entamoeba* infection may contain either trophozoite or cyst or both forms of the parasite. Stool samples may also contain PCR inhibitors. Therefore, two important factors to consider in DNA extraction. One, the extraction procedure should include a step to break open the rigid cyst wall of *Entamoeba*. Extraction procedures that work for the *Entamoeba* cysts should also work for the trophozoites. For cysts, a chemical, enzymatic, or mechanical shearing step is needed in the extraction procedure. Two, the extraction procedure should include steps to remove PCR inhibitors from stools. PCR inhibitors can be removed from stool samples in a variety of ways. Heating stool samples at 90°C for a few minutes may inactivate the PCR inhibitors. PCR inhibitors can also be removed by using absorbent substances such as polyvinylpolypyrrolidone. Some also used inhibition-factor-binding substances such as BSA to remove PCR inhibitors. An alternative way to eliminate the inhibitory effect of PCR inhibitors is by using inhibitor-resistance DNA polymerases during PCR.

Controls in DNA Extraction

DNA extraction procedures must include appropriate controls such as negative and positive extraction controls. A negative extraction control ensures that sample-to-sample or vial-to-vial contamination has not occurred during extraction. A previously characterized *Entamoeba*-negative stool can be used as the NEC. If the NEC gives a positive amplification in PCR (or fluorescence signal in real-time PCR) this will invalidate the DNA extraction and test results, and a repeat extraction will be warranted. A positive extraction control (PEC) or internal control should be included in DNA extraction steps too. A PEC will ensure that the extraction procedure has worked, and a negative PCR result is not because of extraction failure. A previously characterized, well-preserved *E. histolytica*-positive stool sample may serve as a PEC. An *Entamoeba*-negative stool may also be spiked with culture lysates of *E. histolytica* (containing 10^2 amebas, for example) and may serve as a PEC too. Additionally, to ensure that the DNA quality in the stool sample is ideal for PCR, one or more human house-keeping genes with varying sizes may be included as an internal control. This internal control will ensure that a negative PCR result is not due to the degradation of DNA in stool samples. Finally, the test stool sample may be spiked with a known target (which may or may not be of ameba origin) that can serve as a control for PCR inhibitors. If the known target does not give an amplification, this would suggest that the stool sample contains PCR inhibitors. In this case, the DNA extraction must be repeated with a more effective protocol to remove PCR inhibitors.

DNA Extraction Kits

Commercially available DNA extraction kits can isolate PCR-quality DNA directly from stool samples. The QIAamp Stool DNA Mini Kit (Qiagen) is the most widely used extraction method in amebiasis (Evangelopoulos et al. 2000; Freitas et al. 2004; Gonin and Trudel 2003; Heckendorn et al. 2002; Paglia and Visca 2004; Verweij et al. 2000a; Verweij et al. 2000b). It is rapid and can be performed within an hour. It includes both chemical (using alkaline lysis buffer) and enzymatic (using proteinase-K) degradation steps that work well on stool samples containing either a cyst or trophozoite form of *Entamoeba* species. Further improvements in the QIAamp DNA extraction procedure have been reported. These include (a) prewashing stools with PBS followed by resuspension in polyvinylpolypyrrolidone prior to proteinase-K treatment (Kebede et al. 2004), (b) an overnight incubation with proteinase-K and sodium dodecyl sulfate, and (c) freezing of stool samples prior to DNA extraction (Cnops and Esbroeck 2010). Other commercially available kits that have been used in *Entamoeba* DNA isolation are the XTRAX DNA extraction kit (Gull Laboratories, USA) (Evangelopoulos et al. 2000), the Extract MasterFaecal DNA extraction kit (Epicenter Biotechnologies, USA), and the Genomic DNA Prep Plus kit (A&A Biotechnology, Poland) (Myjak et al. 1997).

Several automated DNA extraction methods are commercially available. Qiagen has automated the DNA extraction process using additional robotics equipment such as QIAcube or QIAsymphony. One caveat is the high price of these machines and associated maintenance costs limiting their use to resource-rich countries. Several other automated DNA isolation systems are available, including magLEAD® 6gC & magLEAD® 12gC (Precision System Science, Japan), Freedom EVO VacS (Tecan, Switzerland), MagNAPure Compact (Roche, Switzerland), BioRobot EZ1 (Qiagen, Germany), and NucliSens easyMAG (bioMérieux, USA). However, generally, these have not been evaluated for extraction of DNA from stool samples, and none have been used for *Entamoeba* DNA isolation.

Design of PCR Primers

For both conventional and real-time PCRs, primer design is important to achieve success. For diagnostic purposes, primer targets should be chosen in the conserved DNA regions of the pathogen of interest. For *E. histolytica*, the small subunit ribosomal RNA gene (SSU rDNA) is an optimal primer target for two main reasons. First, SSU rDNA is maintained at several hundred copies per genome in the extrachromosomal episomes. Second, it is highly conserved among all the different genotypes of *E. histolytica*; yet it is distinct from other closely related *Entamoeba* species.

Several primer design software tools are available with some of these available online for free. Alternatively, if primers are designed manually, one should follow basic primer design principles. For example, (i) each primer should comprise about 20–30 nucleotides with about 50% G+C content. (ii) For primers with low G+C content, a longer primer should be chosen to avoid a low melting temperature. Primers with a low melting temperature will likely give non-specific amplification. (iii) Sequences with long runs (i.e., more than three or four) of a single nucleotide should be avoided, if possible. (iv) Primers with secondary structure should be avoided. (v) Complementarity between the two primers of a pair and self-complementarity should be avoided. (vi) The specificity of the primer should be checked using basic local alignment search tool (BLAST).

Conventional PCR

In 1990, Edman et al. (Edman et al. 1990) first used PCR to characterize a gene that encodes an immuno-dominant variable surface antigen from the pathogenic *E. histolytica* and non-pathogenic *E. histolytica* (now known as *E. dispar*). This work led to the development of first PCR to detect and differentiate between *E. histolytica* and *E. dispar* in 1991 (Tannich and Burchard 1991). In 1992, at least three groups reported the development of PCRs for the detection of *E. histolytica* and *E. dispar* using culture-derived DNA (Romero et al. 1992; Tachibana et al. 1992; Cruz-Reyes et al. 1992). Romero et al. (Romero et al. 1992) used repetitive sequences to detect and differentiate between *E. histolytica* and *E. dispar*. They verified PCR products using non-radioactive probes. Cruz-Reyes et al. (Cruz-Reyes et al. 1992) used extrachromosomal ribosomal gene sequences in the PCR. ALA pus fluid was first used in amebiasis PCR by Tachibana et al. (Tachibana et al. 1992) using primers specific for a gene encoding a 30 kDa protein of *E. histolytica*.

In 1993, PCR was first used in the epidemiological studies of amebiasis by Acuna-Soto et al. (Acuna-Soto et al. 1993). They used 201 formalin-fixed stool samples, 25 (12%) of which were microscopy positive for *E. histolytica*. PCR positively detected 21 of the 25, plus an additional three microscopy negative samples. They were the first to report mixed infections of both *E. histolytica* and *E. dispar* in 14 out of 24 PCR positive samples. Ali et al. (Ali et al. 2003) first developed a simple nested PCR followed by verification by restriction endonuclease digestion to detect *E. moshkovskii* directly in the stool samples. No conventional PCR is known to detect *E. bangladeshi*.

Different genomic regions have been used in various conventional PCRs. The success of PCR in clinical samples depends on several factors.

i) *Copy number of targets*: A PCR with a multi-copy target is generally more sensitive than that with a single-copy target. For this reason, the multi-copy SSU rDNA has been the most widely used target for *Entamoeba* PCRs (Ito et al. 2014).

ii) *Size of amplicons*: A shorter target is easier to amplify than a larger one (Ito et al. 2014). In clinical samples, especially in stools, unless appropriately stored, DNA may undergo spontaneous degradation over time. Therefore, a PCR designed to amplify a shorter target would have more chance of success than that with a larger target.

iii) *Cyclic conditions*: Optimization of the thermal cyclic conditions must occur to reduce the production of non-specific amplicons.

iv) *Concentration of PCR reagents*: Concentrations of major PCR components such as primers, $MgCl_2$, and DNA polymerase should be optimized with known control samples. Overall, the success of PCR will also depend on the quality of extracted DNA and the correct design of primers as described above.

Since 2017, there appears to be an upward trend in the use of conventional PCR suggesting that it is becoming more accessible in the diagnosis of amebiasis (Figure 1). A list of different conventional PCRs used in the detection of *E. histolytica*, *E. dispar*, and *E. moshkovskii* is provided in Table 4.

Real-Time PCR

Real-time PCR is an advancement of conventional PCR. It uses fluorescence-labeled probes in addition to primers that provide additional specificity. The fluorescence signals from the probes can be monitored as the amplicons are being produced during PCR allowing real-time monitoring of the PCR results. Real-time PCR has several advantages over conventional PCR: (i) it has a faster turnaround time (Ito et al. 2014); (ii) it is more sensitive and specific; and (iii) it does not require post-PCR processing of amplicons to interpret the data, which eliminates post-processing amplicon contamination.

Blessmann et al. first developed a real-time PCR for the detection of *E. histolytica* and *E. dispar* based on the SSU rDNA sequences using the LightCycler probes (Blessmann et al. 2002). The real-time PCR was highly sensitive and could detect 0.1 trophozoite-equivalent per gram of feces for both *E. histolytica* and *E. dispar*. Since then several other real-time PCRs have been developed using primers (and probes) located almost exclusively in the SSU rDNA genes (Table 5).

Table 4. Primers used for conventional PCR for *E. histolytica, E. dispar*, and *E. moshkovskii* [adapted from (Ali 2015)].

Assay type	Gene target or name	Primer name	*Primer sequence (5'→3')	**References
Triplex	Cysteine protease-8	EHCP8-S1[a]	ATTTGTTAAGTATTGTAAATGGG	(Bahrami et al. 2019)
		EHCP8-As1[a]	ATTGTAACCTTTCATTGTAACAT	
Singleplex	Conserved sequences	P1-S17[a]	GCAACTAGTGTTAGTTA	(Tannich and Burchard 1991)
		P1-AS20[a]	CCTCCAAGATATGTTTTAAC	
	30-kDa protein	P11[a]	GGAGGAGTAGGAAAGTTGAC	(Tachibana et al. 1991)
		P12[a]	TTCTTGCAATTCCTGCTTCGA	
		P13[b]	AGGAGGAGTAGGAAAATTAGG	
		P14[b]	TTCTTGAAACTCCTGTTTCTAC	
	DNA highly repetitive sequences	EHP1[a]	TCAAAATGGTCGTCGTCTAGGC	(Romero et al. 1992)
		EHP2[a]	CAGTTAGAAAATTATTGTACTTTGTA	
		EHNP1[b]	GGATCCTCCAAAAAATAAAGT	
		EHNP2[b]	CCACAGAAACGATATTGGATACC	
	SSU rDNA	Psp F[a]	GGCCAATTCATTCAATGAATTGAG	(Clark and Diamond 1992)
		Psp R[a]	CTCAGATCTAGAAACAATGCTTCTC	
		NPspF[b]	GGCCAATTTATGTAAGTAAAATTGAG	
		NPspR[b]	CTTGGATTTAGAAACAATGTTTCTTC	
		P1[a]	TCAAAATGGTCGTCGTCTAGGC	(Acuna-Soto et al. 1993)
		P2[a]	CAGTTAGAAAATTATTGTACTTTGTA	
		NP1[b]	GGATCCTCCAAAAAATAAAGTTT	
		NP2[b]	ATGATCCCATAGGTTATAGCAAGACA	
		RD5[c]	GGAAGCTTATCTGGTTGATCCTGCCAGTA	(Zaman et al. 2000)
		RD3[c]	GGGGATCCTGATCCTTCCGCAGGTTCACCTAC	

Method	Gene	Primer	Sequence	Reference
		Eh5[a]	GTACAAAAATGGCCAATTCATTCAATG	(Troll et al. 1997)
		Eh3[a]	CTCAGATCTAGAAACAATGCTTCTCT	
		Ed5[b]	GTACAAAGTGGCCAATTTATGTAAGT	
		Ed5[b]	ACTTGGATTTAGAAACAATGTTTCTTC	
		EH1[a]	GTACAAAAATGGCCAATTCATTCAATG	(Gonin and Trudel 2003)
		ED1[b]	TACAAAGTGGCCAATTTATGTAAGTA	
		EHD2[c]	ACTACCAACTGATTGATAGATCAG	
	Hemolysin gene (HLY6) LSU rRNA	EH6F[a]	GACCTCTCCTAATATCCTCGT	(Zindrou et al. 2001)
		Eh6R[a]	GCAGAGAAGTACTGTGAAGG	
	30-kDa protein	HF[c]	AAGAAATTGATATTAATGAATATA	(Hooshyar et al. 2004)
		HR[c]	ATCTTCCAATTCCATCATCAT	
Duplex	Cysteine proteinase	Ehcp6F[a]	GTTGCTGCTGAAGAAACTTG	(Freitas et al. 2004)
		Ehcp6R[a]	GTACCATAACCAACTACTGC	
	Actin gene	Act3F[c]	GGGACGATATGGAAAAGATC	(Gonin and Trudel 2003)
		Act5R[c]	CAAGTCTAAGAATAGCA TGTG	
Nested	SSU rDNA	EH1[a]	GTACAAAAATGGCCAATTCATTCAATG	
		ED1[b]	TACAAAGTGGCCAATTTATGTAAGTA	
		EHD 2[c]	ACTACCAACTGATTGATAGATCAG	
	SSU rDNA	EH-1[c]	TTTGTATTAGTACAAA	(Katzwinkel-Wladarsch et al. 1994)
		EH-2[c]	GTA(A/G)TATTGATATACT	
		EHP-1[a]	AATGGCCAATTCATTCAATG	
		EHP-2[a]	TCTAGAAACAATGCTTCTCT	
		EHN-1[b]	AGTGGCCAATTTATGTAAGT	
		EHN-2[b]	TTTAGAAACAATGTTTCTTC	

Table 4 contd. ...

...Table 4 contd.

Assay type	Gene target or name	Primer name	*Primer sequence (5'→3')	**References
	SSU rDNA	E1[c]	TGCTGTGATTAAAAACGCT	(Evangelopoulos et al. 2000)
		E2[c]	TTAACTATTTCAATCTCGG	
		Eh-L[a]	ACATTTTGAAGACTTTATGTAAGTA	
		Eh-R[a]	CAGATCTAGAAACAATGCTTCTCT	
		Ed-L[b]	GTTAGTTATCTAATTTCGATTAGAA	
		Ed-R[b]	ACACCACTTACTATCCCTACC	
	SSU rDNA	Outer 1F[c]	GAAATTCAGATGTACAAAGA	(Hung et al. 2005)
		Outer 1R[c]	CAGAATCCTAGAATTTCAC	
		Eh1[a]	AAGCAATTGTTTCTAGATCTG	
		Eh2[a]	CACGTTAAAAGAGGTCTAAC	
		Ed1[b]	AAACAATGTTTCTAAATCCA	
		Ed2[b]	ACCACTTACTATCCCTACC	
	SSU rDNA	Em-1[d]	CTCTTCACGGGGAGTGCG	(Ali et al. 2003)
		Em-2[d]	TCGTTAGTTTCATTACCT	
		nEm-1[d]	GAATAAGGATGGTATGAC	
		nEm-2[d]	AAGTGGAGTTAACCACCT	
Multiplex	SSU rDNA	EntaF[e]	ATGCACGAGAGGCGAAAGCAT	(Hamzah et al. 2006)
		Eh-Ra	GATCTAGAAACAATGCTTCTCT	
		Ed-R[b]	CACCACTTACTATCCCTACC	
		EmR[d]	TGACCGGAGCCAGAGACAT	
	SSU rDNA	EhP1[a]	CGATTTTCCCAGTTAGAAATTA	(Nunez et al. 2001)
		EhP2[a]	CAAAATGGTCGTCGTCTAGGC	
		EdP1[b]	ATGGTGAGGTTGTAGCAGAGA	
		EdP2[b]	CGATATTGGATACCTAGTACT	

Target	Primer	Sequence	Reference
SSU rDNA	EH-1[a]	AAGCATTGTTTCTAGATCTGAG	(Khairnar and Parija 2007)
	EH-2[a]	AAGAGGTCTAACCGAAATTAG	
	Mos-1[d]	GAAACCAAGAGTTTCACAAC	
	Mos-2[d]	CAATATAAGGCTTGGATGAT	
	ED-1[b]	TCTAATTTCGATTAGAACTCT	
	ED-2[b]	TCCCTACCTATTAGACATAGC	
SSU rDNA	Eg-SS-F1[c]	TGTGATTAAAACGCTCGTAGTTGAA	(Foo et al. 2012)
	Eg-SS-CR1[e]	CTCGTTCGTTACCGGAATTAACC	
	Eh-SS-F1[a]	GAAGCATTGTTTCTAGATCTGA	
	Ed-SS-F7[b]	AATGCTGAGGAGATGTCAGTT	
SSU rDNA	P1 5'[a]	ATGCACGAGAGCGAAAGCAT	(Singh et al. 2011)
	P2 5'[a]	GATCTAGAAACAATGCTTCTCT	
SSU rDNA	E1[c]	TAGGATGAAACTGCGGACGGT	(Intarapuk et al. 2009)
	E2[c]	AGCCTTGTGACCATACTCCC	
SSU rDNA	Eh-fa	AACAGTAATAGTTTCTTTGGTTAGTAAAA	(Solaymani-Mohammadi et al. 2007)
	Ehr[a]	CTTAGAAATGTCATTTCTCAATTCAT	
SSU rDNA	EntaF2[a]	CGATCAGATACCGTCGTAGTCC	(Lamien-Meda et al. 2020)
	Eh-Ra	GATCTAGAAACAATGCTTCTCT	

a = specific for *E. histolytica*; b = specific for *E. dispar*; c = common for *E. histolytica* and *E. dispar*; d = specific for *E. moshkovskii*; e = *Entamoeba* species broad-spectrum.

* In some of the multiplex PCR, besides *E. histolytica*, *E. dispar*, or *E. moshkovskii* other pathogens were included in the PCR panel. However, for simplicity primers specific to these *Entamoeba* species are included in this Table. ** Only the original references describing the development of the respective PCRs are shown.

Table 5. Real-time PCR primer and probe information [adapted from (Ali 2015)].

Assay type	Gene target	Primer or probe	*Sequence (5'→3')	**References
LightCycler	SSU rDNA	Eh-S26C[a]	GTACAAAATGGCCAATTCATTCAACG	(Blessmann et al. 2002)
		Ed-27 C[b]	GTACAAAGTGGCCAATTTATGTAAGCA	
		Eh-Ed-AS25[c]	GAATTGATTTTACTCAACTCTAGAG	
		Eh/Ed-24LC-Red 640[c]	LC-Red-640-TCGAACCCAATTCCTCGTTATCCp	
		Eh-Ed-25-F[c]	FL-GCCATCTGTAAAGCTCCCTCTCCGAX	
TaqMan	SSU rDNA	Eh-196F[a]	AAATGGCCAATTCATTCAATGA	(Desoubeaux et al. 2014)
		Eh-294R[a]	CATTGGTTACTTGTTAAACACTGTGTG	
		Eh-245[a]	FAM-AGGATGCCACGACAA-NFQ	
TaqMan	SSU rDNA	F_ehis_02[a]	AGACGATCCAGTTTGTATTAG	(Mero et al. 2017)
		R_ehis_02[a]	GGCATCCTAACTCACTTAG	
		P_ehis_02[a]	JOEN/ACAAAATGGCCAATTCATTCAATGAA/3IABkFQ	
TaqMan	SSU rDNA	E. histolytica forward[a]	GCGGACGGCTCATTATAACA	(Won et al. 2016)
		E. histolytica reverse[a]	TGTCGTGGCATCCTAACTCA	
		E. histolytica probe[a]	VIC-AAATGGCCAATTCATTCAATG-non-fluorescent quencher MGB	
TaqMan (Duplex)	SSU rDNA	Ehd-239F[c]	ATTGTCGTGGCATCCTAACTCA	(Verweij et al. 2003)
		Ehd-88R[c]	GCGGACGGCTCATTATAACA	
		Histolytica-96T[a]	VIC-TCATTGAATGAATTGGCCATTT- nonfluorescent quencher	
		dispar-96T[b]	FAM—TTA CTT ACA TAA ATT GGC CAC TTTG-non-fluorescent quencher	
TaqMan	SSU rDNA	E. histolytica forward[a]	AACAGTAATAGTTTCTTTGGTTAGTAAAA	(Verweij and van Lieshout 2011)
		E. histolytica reverse[a]	CTTAGAATGTCATTTCTCAATTCAT	
		E. histolytica probe[a]	ROX—ATTAGTACAAAATGGCCAATTCATTCA—IBRQ	

SYBER green	SSU rDNA	PSP5[a]	GGCCAATTCATTCAATGAATTGAG	(Qvarnstrom et al. 2005)
		PSP3[a]	CTCAGATCTAGAAACAATGCTTCTC	
		NPSP5[b]	GGCCAATTTATGTAAGTAAATTGAG	
		NPSP3[b]	CTTGGATTTAGAAACAATGTTTCTTC	
Molecular beacon	SSU rDNA	Eh-fa	AACAGTAATAGTTTCTTTGGTTAGTAAAA	(Roy et al. 2005)
		Ehr[a]	CTTAGAATGTCATTTCTCAATTCAT	
		Molecular beacon[a]	Texas Red-GCGAGC-ATTAGTACAAAATGGCCAATTCATTCA-GCTCGC-dR Elle	
LightCycler (Multiplex real-time PCR)	SSU rDNA	EhdmF[f]	CGAAAGCATTTCACTCAACTG	(Hamzah et al. 2010)
		EhdmR[f]	TCCCCCTGAAGTCCATAAACTC	
		Ehdm-FL[f]	5'FluoresceinLabel-ACT ATA AAC gAT gTC AAC CAA ggA TTg gAT gAAA-FITC-3'	
		Ehd-640[c]	5'LCRed640-TCA gAT gTA CAA AgA TAg AgA AgC ATT gTT TCTA-phosphate-3'	
		Em-705[e]	5'LCRed705-AAg AAA TTC gCg gAT gAA gAA ACA TTg TTT-phosphate-3'	
TaqMan	SSU rDNA	Eh-f[a]	AACAGTAATAGTTTCTTTGGTTAGTAAAA	(Haque et al. 2010)
		Eh-r[a]	CTTAGAATGTCATTTCTCAATTCAT	
		Eh-YYT[a]	5'YYT-ATT AGT ACA AAC TGG CCA ATT CAT TCA-Eclipse3'	
TaqMan	Episomal repeats	Histolytica-50F[a]	CATTAAAAATGGTGAGGTTCTTAGGAA	(Verweij et al. 2003)
		Histolytica-132R[a]	TGGTCGTCGTCTAGGCAAAATATT	
		Histolytica-78T[a]	FAM-TTGACCAATTTACACCGTTGATTTTCGGA-Eclipse Dark quencher	
		Dispar-1F[b]	GGATCCTCCAAAAAATAAAGTTTTATCA	
		Dispar-137R[b]	ATCCACAGAAACGATATTGGATACCTAGTA	
		Dispar-33[b]	HEX-UGGUGAGGUUGUAGCAGAGAUAUUAAUU-TAMRA	

Table 5 contd. ...

...Table 5 contd.

Assay type	Gene target	Primer or probe	*Sequence (5'→3')	**References
Multiplex SYBR Green	Tandem repeats in circular rDNA episome	EhP1[a]	CGATTTTCCCAGTTAGAAATTA	(Gomes Tdos et al. 2014)
		EhP2[a]	CAAAATGGTCGTCGTCTAGGC	
		EdP1[b]	ATGGTGAGGTTGTAGCAGAGA	
		EdP2[b]	CGATATTGGATACCTAGTACT	
Tetraplex TaqMan	SSU rDNA	Ehd-88R[h]	GCGGACGGCTCATTATAACA	(Ngobeni et al. 2017)
		EM-RT-F2[b]	GTCCTCGATACTACCAAC	
		E. histolytica[a]	FAM-TCATT+GAATGAATTGGCCATTT[5]	
		E. dispar[b]	HEX-ACTTA+CATAAATTGGCCAACTTT[5]	
		E. moshkovskii[c]	Quasar670-CCGTGAAGAGAGTGGCCGA[i]	
		E. bangladeshi[g]	Texas Red-CCTTACAGAG+TATGGCCAATTT[5]	
Tetraplex TaqMan	SSU rDNA	EhF3[a]	CAGTAATAGTTTCTTTGGTTAGTAAAA	(Ali and Roy 2020)
		EhR3[a]	CTTAGAAATGTCATTTCTCAATTCAT	
		EhP3[a]	HEX-GTTTGTATTAGTACAAAATGGC-BHQ1	
		EdF3[b]	CAGTAATAGTTTCTTTGGTTAGTAAAG	
		EdR3[b]	CTTAGAAATGTCATTTCTCAATTTAC	
		EdP3n1[b]	Cy5-GTATTAGTACAAAGTGGCCAA-BHQ3	
		EmF4[c]	CAGATGGCTACCACTTCTAC	
		EmR4[c]	GATTTCGTAAGAGTATTTACTTCT	
		EmP4[c]	FAM-CTCGAGGTGGTTAACTCCAC-BHQ1	
		EbF2[g]	GTTTCTAGAGATGTGATAATGG	
		EbR2[g]	CAATATTGTCCCATGCTTGAATATC	
		EbP2[g]	TAMRA-GGGTGTTTAAAGCAAAACATTAA-BHQ2	

Multiplex tandem real-time PCR (MT-rtPCR)	#Pxr	Not available	Not available	(Stark et al. 2011)
Artus (Hamburg, Germany) real-time LC-PCR kit[d]	#Unknown	Not available	Not available	(Furrows et al. 2004)

5' and 3' modifications indicate a probe sequence. Sequences that do not have any modifications are primers.

a = Specific for *E. histolytica*. b = Specific for *E. dispar*. c = Common for *E. histolytica* and *E. dispar*. d = Discontinued. e = Specific for *E. moshkovskii*. f = Common for *E. histolytica, E. dispar,* and *E. moshkovskii*. g = Specific for *E. bangladeshi*. h = common for *E. histolytica, E. dispar, E. moshkovskii,* and *E. bangladeshi*. i = Each "+" indicates the location of a "locked" nucleotide [for details, see (Kumar et al. 1998)]. SSU rDNA = Small subunit ribosomal RNA gene. #Gene targets or primer/probe sequences were not revealed due to licensing issues.

* In some of the multiplex real-time PCR, besides *E. histolytica, E. dispar,* or *E. moshkovskii* other pathogens were included in the PCR panel. However, for simplicity primers specific to these Entamoeba species are included in this Table.

** Only the original references describing the development of the respective PCRs are shown.

Most of the real-time PCRs target only *E. histolytica*, or *E. histolytica* plus *E. dispar*. Few real-time PCRs target three *Entamoeba* species: *E. histolytica, E. dispar,* and *E. moshkovskii* in stool specimens (Hamzah et al. 2010; Lau et al. 2013). Only two real-time PCRs have been reported to detect four *Entamoeba* species simultaneously: *E. histolytica, E. dispar, E. moshkovskii,* and *E. bangladeshi*. The first tetraplex real-time was developed by Ngobeni et al. in 2017, which used a common pair of primers located in the conserved regions of *Entamoeba* SSU rDNA, but used species-specific probes (Ngobeni et al. 2017). The second tetraplex real-time PCR was developed by Ali and Roy in 2020, which used four sets of species-specific primers and probes located in the SSU rDNA (Ali and Roy 2020). It could detect *Entamoeba* DNA originating from 0.1 trophozoite-equivalent per reaction. Detection of DNA originating from just 0.1 trophozoite is not surprising given the hundreds of copies of target SSU rDNA molecules per ameba genome (Bhattacharya et al. 1988). In mixed infection scenarios, this real-time PCR could detect *E. histolytica* DNA in the excess of up to 10-fold more DNA from another *Entamoeba* species. Another advantage of this tetraplex real-time PCR is the smaller amplicon sizes (132–145 bp) compared to those of Ngobeni et al. (250 bp each), which gives it increased sensitivity compared to the other tetraplex real-time PCR (Varga and James 2006). Perhaps the most important advantage of the Ali and Roy tetraplex real-time PCR over the Ngobeni et al. is that it can be used in conventional PCR format in the absence of real-time PCR equipment and expertise as it uses species-specific primers. This would be useful in many resource-limited countries where real-time PCR is not feasible.

Several groups developed real-time PCR to detect *E. histolytica* and two other diarrhea-causing parasites, *Cryptosporidium* and *Giardia* species (Verweij et al. 2004; McAuliffe et al. 2013; Soonawala et al. 2014; Taniuchi et al. 2011; Van Lint et al. 2013). It is also commercially available in kit format from Fast-Track Diagnostics Ltd., Malta. Stark et al. (2014) evaluated a commercially available EasyScreen™ enteric parasite detection real-time PCR kit (Genetic Signatures, Sydney, Australia) for the detection of *Entamoeba* species, *Blastocystis* species, *Cryptosporidium* species, *D. fragilis,* and *G. intestinalis* from clinical stool samples. The kit was rapid and exhibited 92–100% sensitivity and 100% specificity in detecting these five clinically important human parasites compared to individual PCRs. However, one major limitation of this kit is that it uses a broad-spectrum *Entamoeba* primer which does not differentiate between pathogenic *E. histolytica* and non-pathogenic *Entamoeba* species. A list of real-time PCRs used in the detection of *E. histolytica, E. dispar, E. moshkovskii* and *E. bangladeshi* with primer and probe sequences is provided in Table 5.

Recent Advancement in *E. histolytica* Diagnostics

Several novel techniques have been developed recently for the diagnosis of *E. histolytica*. Some of the most promising ones are briefly discussed below.

BIOFIRE® FILMARRAY® Gastrointestinal (GI) Panel

The BIOFIRE® FILMARRAY® GI Panel is a closed, multiplexed, and real-time PCR-based nucleic acid detection system that allows rapid and automated identification of 22 common gastrointestinal pathogens that cause diarrhea including viruses (both DNA and RNA viruses), bacteria, and parasites. The parasites included in the GI panel are *E. histolytica*, *Giardia lamblia*, *Cryptosporidium*, and *Cyclospora cayetanensis*. It detects nucleic acids from these pathogens directly in stool samples transported in Cary Blair medium from symptomatic GI patients.

The NanoCHIP® Gastrointestinal Panels (GIP)

This assay is performed on the NanoCHIP® platform, which is an automated and qualitative *in vitro* diagnostic test for the direct detection and differentiation of human diarrheal bacteria and parasites in stool specimens from symptomatic patients. The parasites in the panel include *E. histolytica*, *E. dispar*, *Giardia lamblia*, *Cryptosporidium* spp., *Dientamoeba fragilis*, and *Blastocystis hominis*. The test is performed directly on extracted DNA from stool specimens. It detects species-specific DNA to characterize an organism in the panel. The test is intended to be used in the clinical laboratory in healthcare settings.

ImmunoCardSTAT CGE Rapid Antigen Detection

ImmunoCardSTAT CGE (Meridian Bioscence, Milan, Italy) is a rapid immunochromatographic assay for the qualitative detection of *Cryptosporidium parvum*, *Giardia intestinalis*, and *E. histolytica*. One study (Formenti et al. 2015) compared the ImmunoCardSTAT CGE test results with those of a real-time PCR that was specific for *E. histolytica* and *E. dispar*. The ImmunoCard rapid antigen detection test exhibited 88% sensitivity and 92% specificity compared to the real-time PCR but cross-reacted with *E. dispar*.

Single Chain Fragment Variable (scFv) Probes

One group has developed an assay using the scFv probes for *E. histolytica* (Gray et al. 2012). The scFv probes are proteins that are antibody-like molecules expressed on the surface of *Saccharomyces cerevisiae*. This assay was directed toward *E. histolytica* cyst antigen and had comparable sensitivity to that of a monoclonal antibody-based ELISA (Lozano et al. 2012). However, one caveat is that the scFv molecules are insoluble and often too large for diagnostic application. Additionally, they need a labeled secondary antibody to detect the specific amebic antigen. The same group has further improved the previous system by using cell-wall fragments of selected scFv clones (Grewal et al. 2013) and made it a label-free antigen detection system (Grewal et al. 2014). The new method is rapid, more cost-effective than that of the mouse mAb production, and shows promise as an effective diagnostic tool.

Luminex Assay

Luminex is a high throughput multiplex assay, where fluorescently labeled beads are used to produce a specific spectral identifier. Pathogen-specific oligonucleotides are conjugated to the surface of beads in PCR-based Luminex assays to capture pathogen DNA during PCR. A major advantage of the Luminex technology is that it can detect up to several hundred proteins or genes in a single assay using a very small volume of samples.

Several Luminex assays have been developed to detect *E. histolytica*. In 2011, Taniuchi et al. developed a real-time PCR-based Luminex assay to simultaneously detect seven intestinal parasites, including *E. histolytica* (Taniuchi et al. 2011). The Luminex assay could detect as low as 10 *E. histolytica* trophozoites in 200 mg of stool like that of the parent real-time PCR assays (Taniuchi et al. 2011). Santos et al. (2013) developed a Luminex test for the simultaneous detection of five *Entamoeba* species that infect humans—*E. histolytica*, *E. dispar*, *E. moshkovskii*, *E. coli*, and *E. hartmanni* (Santos et al. 2013). Wessels et al. (Wessels et al. 2014) reported the successful development of a multiplex Luminex Gastrointestinal Pathogen Panel (xTAG GPP) that can detect 15 of the most common gastrointestinal pathogens or toxins including *E. histolytica*. The advantage of the Luminex assay is that it is a highly sensitive method and allows simultaneous screening of a large panel of pathogens. However, Luminex instruments are expensive and require expertise to run them, which is often absent in amebiasis-endemic countries, limiting its use in research laboratories in resource-rich countries.

Loop-Mediated Isothermal Amplification (LAMP)

LAMP is a simple, closed-tube, DNA amplification technique for the detection of a specific DNA at a constant temperature (Notomi et al. 2000). Unlike PCR which requires just two primers, the LAMP technique requires four to six primers, which provides increased sensitivity and productivity. As a byproduct of amplification in LAMP, a large quantity of magnesium pyrophosphate is produced which causes a drop in pH and introduces turbidity in the solution. The turbidity of the solution can be determined via photometry or can be seen with the naked eye. There are various ways to detect LAMP amplicons as well. These include the use of colorimetric dyes such as malachite green or hydroxy naphthalene that are sensitive to specific pH ranges. SYBR green dye can be used that binds DNA molecules. Real-time detection of LAMP products is also possible using an intercalating fluorescence dye such as SYTO 9 (Njiru et al. 2008).

LAMP is simple and inexpensive, and it does not require sophisticated instrumentation or expertise. LAMP assay is sensitive and specific. It does not require post-amplification manipulation to detect the results minimizing contamination issues. Also, LAMP assay is more tolerant to sample matrix inhibitors than PCR assay. All these criteria make the LAMP assay suitable for diagnostic use in resource-limited countries. However, a major disadvantage of LAMP is that the amplification products are not suitable for cloning or other molecular biology applications.

Liang et al. (2009) developed the first LAMP assay specific for *E. histolytica* based on the SSU rDNA gene and compared its performance with that of a nested PCR based also on the SSU rDNA gene. Both tests could detect as few as one amebae per reaction. Rivera and Ong (2013) developed an *E. histolytica*-specific LAMP assay based on the hemolysin gene HLY6, which is a single-copy gene in the genome. This LAMP assay could detect as few as five amebas per reaction. Foo et al. (2017) developed the first triplex LAMP assay to detect *E. histolytica*, *E. dispar,* and *E. moshkovskii.* The LAMP primers are designed from the Serine-rich *E. histolytica* protein (SREHP) gene for *E. histolytica*, and SSU rDNA sequences of *E. dispar* and *E. moshkovskii.* It is a "dry-reagent LAMP" assay that does not require its reagents to be maintained in frozen form by using a deglycerolized form of the *Bst* DNA polymerase prior to lyophilizing all the LAMP reagents in the master mix. They further utilized the nitrocellulose membrane-based lateral flow immunoassay strip technology for the simultaneous detection of *E. histolytica*, *E. dispar,* and *E. moshkovskii.* While *E. histolytica* produces a unique line in the lateral flow strip, both *E. dispar* and *E. moshkovskii* produce a line that is different from *E. histolytica* but identical to each other (i.e., *E. dispar* and *E. moshkovskii* cannot be differentiated by this assay). A major criticism of this assay is the use of a single-copy SREHP gene as a target for *E. histolytica*, resulting in reduced sensitivity—the limit of detection is 10 amebas per reaction. Nevertheless, this LAMP assay has added advantages for use in resource-limited endemic countries because of its simple results interpretation and increased reagent stability.

TaqMan Array Card

TaqMan Array Card (TAC) is a high throughput platform developed by Life Technologies that enables spatial multiplexing of up to 384 targets to perform simultaneous real-time PCR reactions. Liu et al. (2013) first utilized the TAC platform for the simultaneous detection of *E. histolytica* and 18 other most common diarrhea-causing enteropathogens including 5 viruses, 9 bacteria, and 4 parasites. Using clinical stool samples from Haydon, Tanzania and Mirpur, Bangladesh, TAC showed 98% sensitivity and 96% specificity for the detection of *E. histolytica* when compared to the PCR-Luminex assays (Taniuchi et al. 2011). Overall, TAC is faster than conventional PCR. It is accurate, and it allows the quantitative detection of multiple pathogens. TAC is thus well suited for clinical or epidemiological investigations. The caveats of TAC are that it requires expensive setup and maintenance costs and, therefore, is not feasible in resource-limited countries.

Broad-Spectrum Entamoeba PCR Coupled With Pyrosequencing

Stensvold et al. (2010) utilized a single-round PCR designed to amplify a 252 bp region of *E. histolytica*, *E. dispar*, and *E. moshkovskii* based on the SSU rDNAs. Although the primers were in the conserved regions, species-specific single nucleotide polymorphisms (SNPs) were present in the amplified products. PCR was coupled to pyrosequencing for the detection of members of the *Entamoeba* complex. This technique was used on 102 stool samples from patients from Sweden, Denmark, and the Netherlands. Results of pyrosequencing were compared to that of a duplex real-time PCR for the diagnosis of *E. histolytica* and *E. dispar*, and a conventional PCR for the diagnosis of *E. moshkovskii*. Overall, pyrosequencing, real-time PCR, and conventional PCR showed excellent agreement in the detection of 17 *E. histolytica*, 86 *E. dispar*, and one mixed infection of both species. No *E. moshkovskii* infection could be detected in those stool samples by either of these tools. By choosing primers in the conserved regions of SSU rDNA from a broad range *Entamoeba* species, where species-specific SNPs would be present internally in the amplicons, a similar technique can be used to detect potentially novel species of *Entamoeba* directly in stool-derived DNA without prior need of cultivation.

Conclusions and Future Direction

Stool O&P (microscopy) examination remains a primary means of diagnosis for the detection of *E. histolytica* and other intestinal parasites in most endemic countries despite its major limitations in sensitivity and specificity. Molecular diagnostic tools capable of species-specific detection of *Entamoeba* species, as well as other intestinal parasites commonly identified by microscopy, are urgently needed for clinical laboratories. The FDA-approved, highly multiplexed Luminex xTAG GPP has the potential to meet this need. The xTAG GPP detects common diarrhea-causing parasites *E. histolytica*, *Giardia*, and *Cryptosporidium* and numerous other bacterial and viral pathogens. However, molecular assays like this would require a laboratory with proficiency in molecular testing, limiting their use to major academic hospitals and reference laboratories. Alternatively, tools that provide a direct diagnosis from unprocessed samples, such as the BioFire Diagnostics in FilmArray platform can be used in virtually any laboratory setting although there would be initial setup expenses and subsequent maintenance costs. Another alternative to these advanced multiplex PCR-based assays or even conventional or real-time PCR would be the LAMP assay, especially for resource-limited countries.

Antigen detection ELISAs or lateral flow immunological POC assays can be good alternatives to PCR-based tests in resource-limited settings. The existing *E. histolytica* antigen detection ELISA or POC tests are designed to detect proteins predominantly expressed in the trophozoite form of the ameba. Therefore, a cyst-directed ELISA is still needed. This is important because a majority of the *E. histolytica*-infected individuals are asymptomatic and excrete predominantly the cyst form of the ameba in their stools. From a clinical point of view, a stage-neutral ELISA or POC test (i.e., one that works for both ameba stages) would be ideal. Currently, species-specific antigen detection ELISA or POC tests are not available for *E. dispar*, *E. moshkovskii*, or *E. bangladeshi*. These tests would be useful in understanding the true epidemiology of these amebae separately and the pathogenicity of *E. moshkovskii* and *E. bangladeshi*. Development of these assays in traditional ways would need production of monoclonal antibodies in mice, which is expensive and time-consuming. However, the development cost could be substantially reduced if yeast-scFv affinity reagents can be used.

Most of the current diagnostic methods are based on species-specific targets, which makes them inherently unsuitable for detection of novel pathogens. Next generation sequencing (NGS) approaches are needed to discover novel pathogens directly in the clinical samples without needing to culture them in selective media. The NGS approach will have two major advantages. First, it would eliminate the bias introduced by growing a selective pathogen in culture. Secondly, it would identify strains and species that are non-culturable at present. With significant improvement in the sensitivity and coverage, and substantial decrease in the sequencing cost, NGS will play a central part in future diagnostic needs and research priorities.

Acknowledgment

The effort of Ibne K. M. Ali was funded by HHS | Centers for Disease Control and Prevention (Prevention).

References

Abd-Alla, M.D., T.G. Jackson and J.I. Ravdin. 1998. Serum IgM antibody response to the galactose-inhibitable adherence lectin of *Entameoba histolytica*. Am. J. Trop. Med. Hyg. 59(3): 431–4.

Abozahra, R., M. Mokhles and K. Baraka. 2020. Prevalence and molecular differentiation of *Entamoeba histolytica*, *Entamoeba dispar*, *Entamoeba moshkovskii*, and *Entamoeba hartmanni* in Egypt. Acta Parasitol.

Acuna-Soto, R., J. Samuelson, P. De Girolami, L. Zarate, F. Millan-Velasco, G. Schoolnick and D. Wirth. 1993. Application of the polymerase chain reaction to the epidemiology of pathogenic and nonpathogenic *Entamoeba histolytica*. Am. J. Trop. Med. Hyg. 48(1): 58–70.

Acuna-Soto, R., J.H. Maguire and D.F. Wirth. 2000. Gender distribution in asymptomatic and invasive amebiasis. Am. J. Gastroenterol. 95(5): 1277–83.

Al-Areeqi, M.A., H. Sady, H.M. Al-Mekhlafi, T.S. Anuar, A.H. Al-Adhroey, W.M. Atroosh, S. Dawaki, F.N. Elyana, N.A. Nasr, I. Ithoi, Y.L. Lau and J. Surin. 2017. First molecular epidemiology of *Entamoeba histolytica*, *E. dispar* and *E. moshkovskii* infections in Yemen: different species-specific associated risk factors. Trop. Med. Int. Health 22(4): 493–504.

Alasvand Javadi, R., F. Kazemi, S. Fallahizadeh and R. Arjmand. 2019. The prevalence of intestinal parasitic infections in Ahvaz, Southwest of Iran, during 2007–2017. Iran J. Public Health 48(11): 2070–2073.

Ali, I.K., M.B. Hossain, S. Roy, P.F. Ayeh-Kumi, W.A. Petri, Jr., R. Haque and C.G. Clark. 2003. *Entamoeba moshkovskii* infections in children, Bangladesh. Emerg. Infect. Dis. 9(5): 580–4.

Ali, I.K. 2015. Intestinal amebae. Clin. Lab Med. 35(2): 393–422.

Ali, I.K.M. and S. Roy. 2020. A real-time PCR assay for simultaneous detection and differentiation of four common *Entamoeba* species that infect humans. J. Clin. Microbiol.

Atabati, H., H. Kassiri, E. Shamloo, M. Akbari, A. Atamaleki, F. Sahlabadi, N.T.T. Linh, A. Rostami, Y. Fakhri and A.M. Khaneghah. 2020. The association between the lack of safe drinking water and sanitation facilities with intestinal *Entamoeba* spp. infection risk: A systematic review and meta-analysis. PLoS One 15(11): e0237102.

Bahrami, F., A. Haghighi, G. Zamini and M. Khademerfan. 2019. Differential detection of *Entamoeba histolytica*, *Entamoeba dispar* and *Entamoeba moshkovskii* in faecal samples using nested multiplex PCR in west of Iran. Epidemiol. Infect. 147: e96.

Bauddh, N.K., R.S. Jadon, P. Ranjan and N.K. Vikram. 2020. Metastatic amebic brain abscess: A rare presentation. Trop. Parasitol. 10(1): 47–49.

Beck, D.L., N. Dogan, V. Maro, N.E. Sam, J. Shao and E.R. Houpt. 2008. High prevalence of *Entamoeba moshkovskii* in a Tanzanian HIV population. Acta Trop. 107(1): 48–9.

Bhattacharya, S., A. Bhattacharya and L.S. Diamond. 1988. Comparison of repeated DNA from strains of *Entamoeba histolytica* and other *Entamoeba*. Mol. Biochem. Parasitol. 27(2-3): 257–62.

Blessmann, J., H. Buss, P.A. Nu, B.T. Dinh, Q.T. Ngo, A.L. Van, M.D. Alla, T.F. Jackson, J.I. Ravdin, and E. Tannich. 2002. Real-time PCR for detection and differentiation of *Entamoeba histolytica* and *Entamoeba dispar* in fecal samples. J. Clin. Microbiol. 40(12): 4413–7.

Burgess, S.L. and W.A. Petri, Jr. 2016. The Intestinal Bacterial Microbiome and *E. histolytica* Infection. Curr. Trop. Med. Rep. 3: 71–74.

Caballero-Salcedo, A., M. Viveros-Rogel, B. Salvatierra, R. Tapia-Conyer, J. Sepulveda-Amor, G. Gutierrez and L. Ortiz-Ortiz. 1994. Seroepidemiology of amebiasis in Mexico. Am. J. Trop. Med. Hyg. 50(4): 412–9.

Centers for Disease Control and Prevention. 1994. Summary of notifiable diseases, United States, 1993. MMWR Morb Mortal Wkly Rep. 42((53):i-xvii): 1–73.

Clark, C.G. and L.S. Diamond. 1992. Differentiation of pathogenic *Entamoeba histolytica* from other intestinal protozoa by riboprinting. Arch. Med. Res. 23(2): 15–6.

Clark, C.G. 1997. Intraspecific variation and phylogenetic relationships in the genus *Entamoeba* as revealed by riboprinting. J. Eukaryot. Microbiol. 44(2): 142–54.

Cnops, L. and M.V. Esbroeck. 2010. Freezing of stool samples improves real-time PCR detection of *Entamoeba dispar* and *Entamoeba histolytica*. J. Microbiol. Methods 80(3): 310–2.

Cruz-Reyes, J.A., W.M. Spice, T. Rehman, E. Gisborne and J.P. Ackers. 1992. Ribosomal DNA sequences in the differentiation of pathogenic and non-pathogenic isolates of *Entamoeba histolytica*. Parasitology 104(Pt 2): 239–46.

Desoubeaux, G., H. Chaussade, M. Thellier, S. Poussing, F. Bastides, E. Bailly, P. Lanotte, D. Alison, L. Brunereau, L. Bernard and J. Chandenier. 2014. Unusual multiple large abscesses of the liver: interest of the radiological features and the real-time PCR to distinguish between bacterial and amebic etiologies. Pathog. Glob. Health 108(1): 53–7.

Diamond, L.S. and C.G. Clark. 1993. A redescription of *Entamoeba histolytica* Schaudinn, 1903 (Emended Walker, 1911) separating it from *Entamoeba dispar* Brumpt, 1925. J. Eukaryot. Microbiol. 40(3): 340–4.

Dreyer, D.A. 1961. Growth of a strain of *Entamoeba histolytica* at room temperature. Tex Rep. Biol. Med. 19: 393–6.

Edman, U., M.A. Meraz, S. Rausser, N. Agabian and I. Meza. 1990. Characterization of an immuno-dominant variable surface antigen from pathogenic and nonpathogenic *Entamoeba histolytica*. J. Exp. Med. 172(3): 879–88.

Entner, N. and H. Most. 1965. Genetics of *Entamoeba*: Characterization of two new parasitic strains which grow at room temperature (and at 37 Degrees C). J. Protozool. 12: 10–3.

Evangelopoulos, A., G. Spanakos, E. Patsoula, N. Vakalis and N. Legakis. 2000. A nested, multiplex, PCR assay for the simultaneous detection and differentiation of *Entamoeba histolytica* and *Entamoeba dispar* in faeces. Ann. Trop. Med. Parasitol. 94(3): 233–40.

Fernandez-Diez, J., M. Magana and M.L. Magana. 2012. Cutaneous amebiasis: 50 years of experience. Cutis 90(6): 310–4.

Foo, P.C., Y.Y. Chan, W.C. See Too, Z.N. Tan, W.K. Wong, P. Lalitha and B.H. Lim. 2012. Development of a thermostabilized, one-step, nested, tetraplex PCR assay for simultaneous identification and differentiation of *Entamoeba* species, *Entamoeba histolytica* and *Entamoeba dispar* from stool samples. J. Med. Microbiol. 61(Pt 9): 1219–1225.

Foo, P.C., Y.Y. Chan, M. Mohamed, W.K. Wong, A.B. Nurul Najian and B.H. Lim. 2017. Development of a thermostabilised triplex LAMP assay with dry-reagent four target lateral flow dipstick for detection of *Entamoeba histolytica* and non-pathogenic *Entamoeba* spp. Anal. Chim. Acta 966: 71–80.

Formenti, F., F. Perandin, S. Bonafini, M. Degani and Z. Bisoffi. 2015. [Evaluation of the new ImmunoCard STAT!(R) CGE test for the diagnosis of Amebiasis]. Bull Soc. Pathol. Exot. 108(3): 171–4.

Fotedar, R., D. Stark, N. Beebe, D. Marriott, J. Ellis and J. Harkness. 2007. Laboratory diagnostic techniques for *Entamoeba* species. Clin. Microbiol. Rev. 20(3): 511–32, table of contents.

Fotedar, R., D. Stark, D. Marriott, J. Ellis and J. Harkness. 2008. *Entamoeba moshkovskii* infections in Sydney, Australia. Eur. J. Clin. Microbiol. Infect. Dis. 27(2): 133–7.

Freitas, M.A., E.N. Vianna, A.S. Martins, E.F. Silva, J.L. Pesquero and M.A. Gomes. 2004. A single step duplex PCR to distinguish *Entamoeba histolytica* from *Entamoeba dispar*. Parasitology 128(Pt 6): 625–8.

Furrows, S.J., A.H. Moody and P.L. Chiodini. 2004. Comparison of PCR and antigen detection methods for diagnosis of *Entamoeba histolytica* infection. J. Clin. Pathol. 57(12): 1264–6.

Gathiram, V. and T.F. Jackson. 1985. Frequency distribution of *Entamoeba histolytica* zymodemes in a rural South African population. Lancet 1(8431): 719–21.

Gathiram, V. 1987. A longitudinal study of asymptomatic carriers of pathogenic zymodemes of *Entamoeba histolytica*. S Afr. Med. J. 72(10): 669–72.

Gomersall, L.N., J. Currie and R. Jeffrey. 1994. Amoebiasis: a rare cause of cardiac tamponade. Br Heart J. 71(4): 368–9.

Gomes Tdos, S., M.C. Garcia, F. de Souza Cunha, H. Werneck de Macedo, J.M. Peralta and R.H. Peralta. 2014. Differential diagnosis of *Entamoeba* spp. in clinical stool samples using SYBR green real-time polymerase chain reaction. ScientificWorldJournal 2014: 645084.

Gonin, P. and L. Trudel. 2003. Detection and differentiation of *Entamoeba histolytica* and *Entamoeba dispar* isolates in clinical samples by PCR and enzyme-linked immunosorbent assay. J. Clin. Microbiol. 41(1): 237–41.

Goret, N.E. and C.C. Goret. 2019. Splenic abscess is extremely rare after amoebic dysentery. A case report and review of the literature. Ann. Ital. Chir. 8.

Gray, S.A., K.M. Weigel, I.K. Ali, A.A. Lakey, J. Capalungan, G.J. Domingo and G.A. Cangelosi. 2012. Toward low-cost affinity reagents: lyophilized yeast-scFv probes specific for pathogen antigens. PLoS One 7(2): e32042.

Grewal, Y.S., M.J. Shiddiky, S.A. Gray, K.M. Weigel, G.A. Cangelosi and M. Trau. 2013. Label-free electrochemical detection of an *Entamoeba histolytica* antigen using cell-free yeast-scFv probes. Chem. Commun. (Camb) 49(15): 1551–3.

Grewal, Y.S., M.J. Shiddiky, L.J. Spadafora, G.A. Cangelosi and M. Trau. 2014. Nano-yeast-scFv probes on screen-printed gold electrodes for detection of *Entamoeba histolytica* antigens in a biological matrix. Biosens Bioelectron 55: 417–22.

Hamzah, Z., S. Petmitr, M. Mungthin, S. Leelayoova and P. Chavalitshewinkoon-Petmitr. 2006. Differential detection of *Entamoeba histolytica*, *Entamoeba dispar*, and *Entamoeba moshkovskii* by a single-round PCR assay. J. Clin. Microbiol. 44(9): 3196–200.

Hamzah, Z. 2010. Development of multiplex real-time polymerase chain reaction for detection of *Entamoeba histolytica*, *Entamoeba dispar*, and *Entamoeba moshkovskii* in clinical specimens. Am. J. Trop. Med. Hyg. 83(4): 909–13.

Haque, R., N.U. Mollah, I.K. Ali, K. Alam, A. Eubanks, D. Lyerly and W.A. Petri, Jr. 2000. Diagnosis of amebic liver abscess and intestinal infection with the TechLab *Entamoeba histolytica* II antigen detection and antibody tests. J. Clin. Microbiol. 38(9): 3235–9.

Haque, R., M. Kabir, Z. Noor, S.M. Rahman, D. Mondal, F. Alam, I. Rahman, A. Al Mahmood, N. Ahmed and W.A. Petri, Jr. 2010. Diagnosis of amebic liver abscess and amebic colitis by detection of *Entamoeba histolytica* DNA in blood, urine, and saliva by a real-time PCR assay. J. Clin. Microbiol. 48(8): 2798–801.

Hara, A., Y. Hirose, H. Mori, H. Iwao, T. Kato and Y. Kusuhara. 2004. Cytopathologic and genetic diagnosis of pulmonary amebiasis: a case report. Acta Cytol. 48(4): 547–50.

Heckendorn, F., E.K. N'Goran, I. Felger, P. Vounatsou, A. Yapi, A. Oettli, H.P. Marti, M. Dobler, M. Traore, K.L. Lohourignon and C. Lengeler. 2002. Species-specific field testing of *Entamoeba* spp. in an area of high endemicity. Trans. R Soc. Trop. Med. Hyg. 96(5): 521–8.

Hiatt, R.A., E.K. Markell and E. Ng. 1995. How many stool examinations are necessary to detect pathogenic intestinal protozoa? Am. J. Trop. Med. Hyg. 53(1): 36–9.

Holland, J.L., L. Louie, A.E. Simor and M. Louie. 2000. PCR detection of *Escherichia coli* O157:H7 directly from stools: evaluation of commercial extraction methods for purifying fecal DNA. J. Clin. Microbiol. 38(11): 4108–13.

Hooshyar, H., M. Rezaian, B. Kazemi, M. Jeddi-Tehrani and S. Solaymani-Mohammadi. 2004. The distribution of *Entamoeba histolytica* and *Entamoeba dispar* in northern, central, and southern Iran. Parasitol. Res. 94(2): 96–100.

Hughes, F.B., S.T. Faehnle and J.L. Simon. 1975. Multiple cerebral abscesses complicating hepatopulmonary amebiasis. J. Pediatr. 86(1): 95–6.

Hung, C.C., H.Y. Deng, W.H. Hsiao, S.M. Hsieh, C.F. Hsiao, M.Y. Chen, S.C. Chang and K.E. Su. 2005. Invasive amebiasis as an emerging parasitic disease in patients with human immunodeficiency virus type 1 infection in Taiwan. Arch. Intern. Med. 165(4): 409–15.

Huston, C.D., R. Haque and W.A. Petri, Jr. 1999. Molecular-based diagnosis of *Entamoeba histolytica* infection. Expert Rev. Mol. Med. 1999: 1–11.

Intarapuk, A., T. Kalambaheti, N. Thammapalerd, P. Mahannop, P. Kaewsatien, A. Bhumiratana and D. Nityasuddhi. 2009. Identification of *Entamoeba histolytica* and *Entamoeba dispar* by PCR assay of fecal specimens obtained from Thai/Myanmar border region. Southeast Asian J. Trop. Med. Public Health 40(3): 425–34.

Ito, D., S. Hata, S. Seiichiro, K. Kobayashi, M. Teruya and M. Kaminishi. 2014. Amebiasis presenting as acute appendicitis: Report of a case and review of Japanese literature. Int. J. Surg. Case Rep. 5(12): 1054–7.

Katzwinkel-Wladarsch, S., T. Loscher and H. Rinder. 1994. Direct amplification and differentiation of pathogenic and nonpathogenic *Entamoeba histolytica* DNA from stool specimens. Am. J. Trop. Med. Hyg. 51(1): 115–8.

Kebede, A., J.J. Verweij, T. Endeshaw, T. Messele, G. Tasew, B. Petros and A.M. Polderman. 2004. The use of real-time PCR to identify *Entamoeba histolytica* and *E. dispar* infections in prisoners and primary-school children in Ethiopia. Ann. Trop. Med. Parasitol. 98(1): 43–8.

Khairnar, K. and S.C. Parija. 2007. A novel nested multiplex polymerase chain reaction (PCR) assay for differential detection of *Entamoeba histolytica*, *E. moshkovskii* and *E. dispar* DNA in stool samples. BMC Microbiol. 7: 47.

Kroft, E.B., A. Warris, L.E. Jansen and R. van Crevel. 2005. [A Dutchman from Mali with a perianal ulcer caused by cutaneous amebiasis]. Ned Tijdschr Geneeskd 149(6): 308–11.

Kruger, C., I. Malleyeck and N. Naman. 2011. Amoebic abscess of the spleen and fatal colonic perforation. Pediatr. Infect. Dis. J. 30(1): 91–2.

Kumar, R., S.K. Singh, A.A. Koshkin, V.K. Rajwanshi, M. Meldgaard and J. Wengel. 1998. The first analogues of LNA (locked nucleic acids): phosphorothioate-LNA and 2'-thio-LNA. Bioorg. Med. Chem. Lett. 8(16): 2219–22.

Kyany'a, C., F. Eyase, E. Odundo, E. Kipkirui, N. Kipkemoi, R. Kirera, C. Philip, J. Ndonye, M. Kirui, A. Ombogo, M. Koech, W. Bulimo and C.E. Hulseberg. 2019. First report of *Entamoeba moshkovskii* in human stool samples from symptomatic and asymptomatic participants in Kenya. Trop. Dis. Travel Med. Vaccines 5: 23.

Lamien-Meda, A., R. Schneider, J. Walochnik, H. Auer, U. Wiedermann and D. Leitsch. 2020. A novel 5-Plex qPCR-HRM assay detecting human diarrheal parasites. Gut Pathog. 12: 27.

Lau, Y.L., C. Anthony, S.A. Fakhrurrazi, J. Ibrahim, I. Ithoi and R. Mahmud. 2013. Real-time PCR assay in differentiating *Entamoeba histolytica*, *Entamoeba dispar*, and *Entamoeba moshkovskii* infections in Orang Asli settlements in Malaysia. Parasit Vectors 6(1): 250.

Leo, M., R. Haque, M. Kabir, S. Roy, R.M. Lahlou, D. Mondal, E. Tannich and W.A. Petri, Jr. 2006. Evaluation of *Entamoeba histolytica* antigen and antibody point-of-care tests for the rapid diagnosis of amebiasis. J. Clin. Microbiol. 44(12): 4569–71.

Liang, S.Y., Y.H. Chan, K.T. Hsia, J.L. Lee, M.C. Kuo, K.Y. Hwa, C.W. Chan, T.Y. Chiang, J.S. Chen, F.T. Wu and D.D. Ji. 2009. Development of loop-mediated isothermal amplification assay for detection of *Entamoeba histolytica*. J. Clin. Microbiol. 47(6): 1892–5.

Lichtenstein, A., A.T. Kondo, G.S. Visvesvara, A. Fernandez, E.F. Paiva, T. Mauad, M. Dolhnikoff and M.A. Martins. 2005. Pulmonary amoebiasis presenting as superior vena cava syndrome. Thorax 60(4): 350–2.

Liu, J., J. Gratz, C. Amour, G. Kibiki, S. Becker, L. Janaki, J.J. Verweij, M. Taniuchi, S.U. Sobuz, R. Haque, D.M. Haverstick and E.R. Houpt. 2013. A laboratory-developed TaqMan Array Card for simultaneous detection of 19 enteropathogens. J. Clin. Microbiol. 51(2): 472–80.

Liu, Y.Y., Y. Ying, C. Chen, Y.K. Hu, F.F. Yang, L.Y. Shao, X.J. Cheng and Y.X. Huang. 2018. Primary pulmonary amebic abscess in a patient with pulmonary adenocarcinoma: a case report. Infect. Dis. Poverty 7(1): 34.

Lopez, M.C., C.M. Leon, J. Fonseca, P. Reyes, L. Moncada, M.J. Olivera and J.D. Ramirez. 2015. Molecular epidemiology of *Entamoeba*: First description of *Entamoeba moshkovskii* in a rural area from Central Colombia. PLoS One 10(10): e0140302.

Lozano, R., M. Naghavi, K. Foreman, S. Lim, K. Shibuya, V. Aboyans, J. Abraham, T. Adair, R. Aggarwal, S.Y. Ahn, M. Alvarado, H.R. Anderson, L.M. Anderson, K.G. Andrews, C. Atkinson, L.M. Baddour, S. Barker-Collo, D.H. Bartels, M.L. Bell, E.J. Benjamin, D. Bennett, K. Bhalla, B. Bikbov, A. Bin Abdulhak, G. Birbeck, F. Blyth, I. Bolliger, S. Boufous, C. Bucello, M. Burch, P. Burney, J. Carapetis, H. Chen, D. Chou, S.S. Chugh, L.E. Coffeng, S.D. Colan, S. Colquhoun, K.E. Colson, J. Condon, M.D. Connor, L.T. Cooper, M. Corriere, M. Cortinovis, K.C. de Vaccaro, W. Couser, B.C. Cowie, M.H. Criqui, M. Cross, K.C. Dabhadkar, N. Dahodwala, D. De Leo, L. Degenhardt, A. Delossantos, J. Denenberg, D.C. Des Jarlais, S.D. Dharmaratne, E.R. Dorsey, T. Driscoll, H. Duber, B. Ebel, P.J. Erwin, P. Espindola, M. Ezzati, V. Feigin, A.D. Flaxman, M.H. Forouzanfar, F.G. Fowkes, R. Franklin, M. Fransen, M.K. Freeman, S.E. Gabriel, E. Gakidou, F. Gaspari, R.F. Gillum, D. Gonzalez-Medina, Y.A. Halasa, D. Haring, J.E. Harrison, R. Havmoeller, R.J. Hay, B. Hoen, P.J. Hotez, D. Hoy, K.H. Jacobsen, S.L. James, R. Jasrasaria, S. Jayaraman, N. Johns, G. Karthikeyan, N. Kassebaum, A. Keren, J.P. Khoo, L.M. Knowlton, O. Kobusingye, A. Koranteng, R. Krishnamurthi, M. Lipnick, S.E. Lipshultz, S.L. Ohno, J. Mabweijano, M.F. MacIntyre, L. Mallinger, L. March, G.B. Marks, R. Marks, A. Matsumori, R. Matzopoulos, B.M. Mayosi, J.H. McAnulty, M.M. McDermott, J. McGrath, G.A. Mensah, T.R. Merriman, C. Michaud, M. Miller, T.R. Miller, C. Mock, A.O. Mocumbi, A.A. Mokdad, A. Moran, K. Mulholland, M.N. Nair, L. Naldi, K.M. Narayan, K. Nasseri, P. Norman, M. O'Donnell, S.B. Omer, K. Ortblad, R. Osborne, D. Ozgediz, B. Pahari, J.D. Pandian, A.P. Rivero, R.P. Padilla, F. Perez-Ruiz, N. Perico, D. Phillips, K. Pierce, C.A. Pope, 3rd, E. Porrini, F. Pourmalek, M. Raju, D. Ranganathan, J.T. Rehm, D.B. Rein, G. Remuzzi, F.P. Rivara, T. Roberts, F.R. De Leon, L.C. Rosenfeld, L. Rushton, R.L. Sacco, J.A. Salomon, U. Sampson, E. Sanman, D.C. Schwebel, M. Segui-Gomez, D.S. Shepard, D. Singh, J. Singleton, K. Sliwa, E. Smith, A. Steer, J.A. Taylor, B. Thomas, I.M. Tleyjeh, J.A. Towbin, T. Truelsen, E.A. Undurraga, N. Venketasubramanian, L. Vijayakumar, T. Vos, G.R. Wagner, M. Wang, W. Wang, K. Watt, M.A. Weinstock, R. Weintraub, J.D. Wilkinson, A.D. Woolf, S. Wulf, P.H. Yeh, P. Yip, A. Zabetian, Z.J. Zheng, A.D. Lopez, C.J. Murray, M.A. AlMazroa and Z.A. Memish. 2012. Global and regional mortality from 235 causes of death for 20 age groups in 1990 and 2010: a systematic analysis for the Global Burden of Disease Study 2010. Lancet 380(9859): 2095–128.

Magana, M., M.L. Magana, A. Alcantara and M.A. Perez-Martin. 2004. Histopathology of cutaneous amebiasis. Am. J. Dermatopathol. 26(4): 280–4.

Martínez-Palomo, A. 1993. Parasitic amebas of the intestinal tract. Edited by J. P. K. a. J. R. B. (ed.). 2nd ed. Vol. 3, Parasitic Protozoa. San Diego, CA: Academic Press.

McAuliffe, G.N., T.P. Anderson, M. Stevens, J. Adams, R. Coleman, P. Mahagamasekera, S. Young, T. Henderson, M. Hofmann, L.C. Jennings and D.R. Murdoch. 2013. Systematic application of multiplex PCR enhances the detection of bacteria, parasites, and viruses in stool samples. J. Infect. 67(2): 122–9.

Mehta, A.B., B.C. Mehta, S.L. Balse and J.C. Patel. 1968. Amebic abscess of myocardium; a case report. Ind. J. Med. Sci. 22: 720–722.

Mero, S., J. Kirveskari, J. Antikainen, J. Ursing, L. Rombo, P.E. Kofoed and A. Kantele. 2017. Multiplex PCR detection of *Cryptosporidium* sp, *Giardia lamblia* and *Entamoeba histolytica* directly from dried stool samples from Guinea-Bissauan children with diarrhoea. Infect. Dis. (Lond) 49(9): 655–663.

Meyer, J., E. Roos, C. Combescure, N.C. Buchs, J.L. Frossard, F. Ris, C. Toso and J. Schrenzel. 2020. Mapping of aetiologies of gastroenteritis: a systematic review and meta-analysis of pathogens identified using a multiplex screening array. Scand. J. Gastroenterol. 1–6.

Mondal, D., W.A. Petri, Jr., R.B. Sack, B.D. Kirkpatrick and R. Haque. 2006. *Entamoeba histolytica*-associated diarrheal illness is negatively associated with the growth of preschool children: evidence from a prospective study. Trans R Soc. Trop. Med. Hyg. 100(11): 1032–8.

Mondal, D., J. Minak, M. Alam, Y. Liu, J. Dai, P. Korpe, L. Liu, R. Haque and W.A. Petri, Jr. 2012. Contribution of enteric infection, altered intestinal barrier function, and maternal malnutrition to infant malnutrition in Bangladesh. Clin. Infect. Dis. 54(2): 185–92.

Moss, D.M., J.W. Priest, K. Hamlin, G. Derado, J. Herbein, W.A. Petri, Jr. and P.J. Lammie. 2014. Longitudinal evaluation of enteric protozoa in Haitian children by stool exam and multiplex serologic assay. Am. J. Trop. Med. Hyg. 90(4): 653–60.

Musthyala, N.B., S. Indulkar, V.R. Palwai, M. Babaiah, M.A. Ali and P. Marriapam. 2019. Amebic infection of the female genital tract: a report of three cases. J. Midlife Health 10(2): 96–98.

Myjak, P., J. Kur and H. Pietkiewicz. 1997. Usefulness of new DNA extraction procedure for PCR technique in species identification of *Entamoeba* isolates. Wiad Parazytol. 43(2): 163–70.

Ngobeni, R., A. Samie, S. Moonah, K. Watanabe, W.A. Petri, Jr. and C. Gilchrist. 2017. *Entamoeba* species in South Africa: Correlations with the host microbiome, parasite burdens, and first description of *Entamoeba bangladeshi* outside of Asia. J. Infect. Dis. 216(12): 1592–1600.

Njiru, Z.K., A.S. Mikosza, T. Armstrong, J.C. Enyaru, J.M. Ndung'u and A.R. Thompson. 2008. Loop-mediated isothermal amplification (LAMP) method for rapid detection of *Trypanosoma brucei* rhodesiense. PLoS Negl Trop. Dis. 2(1): e147.

Notomi, T., H. Okayama, H. Masubuchi, T. Yonekawa, K. Watanabe, N. Amino and T. Hase. 2000. Loop-mediated isothermal amplification of DNA. Nucleic Acids Res. 28(12): E63.

Nunez, Y.O., M.A. Fernandez, D. Torres-Nunez, J.A. Silva, I. Montano, J.L. Maestre and L. Fonte. 2001. Multiplex polymerase chain reaction amplification and differentiation of *Entamoeba histolytica* and *Entamoeba dispar* DNA from stool samples. Am. J. Trop. Med. Hyg. 64(5-6): 293–7.

Ogren, J., O. Dienus and A. Matussek. 2020. Optimization of routine microscopic and molecular detection of parasitic protozoa in SAF-fixed faecal samples in Sweden. Infect. Dis. (Lond) 52(2): 87–96.

Ohnishi, K. and M. Murata. 1997. Present characteristics of symptomatic amebiasis due to *Entamoeba histolytica* in the east-southeast area of Tokyo. Epidemiol. Infect. 119(3): 363–7.

Paglia, M.G. and P. Visca. 2004. An improved PCR-based method for detection and differentiation of *Entamoeba histolytica* and *Entamoeba dispar* in formalin-fixed stools. Acta Trop. 92(3): 273–7.

PAHO. 1998. Health in the Americas. Washington DC. Pan American Health Organization, Mexico 357–378.

Parija, S.C. and K. Khairnar. 2005. *Entamoeba moshkovskii* and *Entamoeba dispar*-associated infections in pondicherry, India. J. Health Popul. Nutr. 23(3): 292–5.

Petri, W.A., Jr. 1996. Recent advances in amebiasis. Crit. Rev. Clin. Lab. Sci. 33(1): 1–37.

Petri, W.A., Jr., D. Mondal, K.M. Peterson, P. Duggal and R. Haque. 2009. Association of malnutrition with amebiasis. Nutr. Rev. 67 Suppl 2: S207–15.

Petri, W.A. and R. Haque. 2013. *Entamoeba histolytica* brain abscess. Handb. Clin. Neurol. 114: 147–52.

Prasetyo, R.H. 2015. Scrotal abscess, a rare case of extra intestinal amoebiasis. Trop. Biomed. 32(3): 494–6.

Qvarnstrom, Y., C. James, M. Xayavong, B.P. Holloway, G.S. Visvesvara, R. Sriram and A.J. da Silva. 2005. Comparison of real-time PCR protocols for differential laboratory diagnosis of amebiasis. J. Clin. Microbiol. 43(11): 5491–7.

Ramakrishnan, A.S., P.M. Ratnasabapathy, R. Natanasabapathy, L. Ananthakrishnan and A.R. Balakrishnan. 1971. An amoebic liver abscess presenting as a space occupying lesion of the right kidney. Am. Surg. 37(12): 756–8.

Ramdial, P.K., T.E. Madiba, S. Kharwa, B. Clarke and B. Zulu. 2002. Isolated amoebic appendicitis. Virchows Arch. 441(1): 63–8.

Reeves, R.E. and J.M. Bischoff. 1968. Classification of *Entamoeba* species by means of electrophoretic properties of amebal enzymes. J. Parasitol. 54(3): 594–600.

Richards, C.S., M. Goldman and L.T. Cannon. 1966. Cultivation of *Entamoeba histolytica* and *Entamoeba histolytica*-like strains at reduced temperature and behavior of the amebae in diluted media. Am. J. Trop. Med. Hyg. 15(4): 648–55.

Rivera, W.L., H. Tachibana and H. Kanbara. 1998. Field study on the distribution of *Entamoeba histolytica* and *Entamoeba dispar* in the northern Philippines as detected by the polymerase chain reaction. Am. J. Trop. Med. Hyg. 59(6): 916–21.

Rivera, W.L. and V.A. Ong. 2013. Development of loop-mediated isothermal amplification for rapid detection of *Entamoeba histolytica*. Asian Pac. J. Trop. Med. 6(6): 457–61.

Romero, J.L., S. Descoteaux, S. Reed, E. Orozco, J. Santos and J. Samuelson. 1992. Use of polymerase chain reaction and nonradioactive DNA probes to diagnose *Entamoeba histolytica* in clinical samples. Arch. Med. Res. 23(2): 277–9.

Roy, S., M. Kabir, D. Mondal, I.K. Ali, W.A. Petri, Jr. and R. Haque. 2005. Real-time-PCR assay for diagnosis of *Entamoeba histolytica* infection. J. Clin. Microbiol. 43(5): 2168–72.

Royer, T.L., C. Gilchrist, M. Kabir, T. Arju, K.S. Ralston, R. Haque, C.G. Clark and W.A. Petri, Jr. 2012. *Entamoeba bangladeshi* nov. sp., Bangladesh. Emerg. Infect. Dis. 18(9): 1543–5.

Saensiriphan, S., L. Rungmuenporn, P. Phiromnak, S. Yingyeun, S. Klayjunteuk and T. Pengsakul. 2015. First report of genitourinary amoebiasis in Thailand. Trop. Biomed. 32(3): 551–3.

Samie, A., L. Mahlaule, P. Mbati, T. Nozaki and A. ElBakri. 2020. Prevalence and distribution of *Entamoeba* species in a rural community in northern South Africa. Food Waterborne Parasitol. 18: e00076.

Santos, H.L., K. Bandyopadhyay, R. Bandea, R.H. Peralta, J.M. Peralta and A.J. Da Silva. 2013. LUMINEX(R): a new technology for the simultaneous identification of five *Entamoeba* spp. commonly found in human stools. Parasit Vectors 6: 69.

Sanuki, J., T. Asai, E. Okuzawa, S. Kobayashi and T. Takeuchi. 1997. Identification of *Entamoeba histolytica* and *E. dispar* cysts in stool by polymerase chain reaction. Parasitol. Res. 83(1): 96–8.

Sargeaunt, P.G., J.E. Williams and J.D. Grene. 1978. The differentiation of invasive and non-invasive *Entamoeba histolytica* by isoenzyme electrophoresis. Trans R Soc. Trop. Med. Hyg. 72(5): 519–21.

Sasaki, Y., T. Yoshida, J. Suzuki, S. Kobayashi and T. Sato. 2016. [A Case of Peristomal Cutaneous Ulcer Following Amebic Colitis Caused by *Entamoeba histolytica*]. Kansenshogaku Zasshi 90(1): 73–6.

Shimokawa, C., M. Kabir, M. Taniuchi, D. Mondal, S. Kobayashi, I.K. Ali, S.U. Sobuz, M. Senba, E. Houpt, R. Haque, W.A. Petri, Jr. and S. Hamano. 2012. *Entamoeba moshkovskii* is associated with diarrhea in infants and causes diarrhea and colitis in mice. J. Infect. Dis. 206(5): 744–51.

Singh, N.G., A.A. Mannan and M. Kahvic. 2010. Acute amebic appendicitis: report of a rare case. Indian J. Pathol. Microbiol. 53(4): 767–8.

Singh, P., B.R. Mirdha, V. Ahuja and S. Singh. 2011. Evaluation of small-subunit rRNA touchdown polymerase chain reaction for direct detection of *Entamoeba histolytica* in human pus samples from patients with amoebic liver abscess. Indian J. Med. Microbiol. 29(2): 141–6.

Solaymani-Mohammadi, S., M.M. Lam, J.R. Zunt and W.A. Petri, Jr. 2007. *Entamoeba histolytica* encephalitis diagnosed by PCR of cerebrospinal fluid. Trans R Soc. Trop. Med. Hyg. 101(3): 311–3.

Soonawala, D., L. van Lieshout, M.A. den Boer, E.C. Claas, J.J. Verweij, A. Godkewitsch, M. Ratering and L.G. Visser. 2014. Post-travel screening of asymptomatic long-term travelers to the tropics for intestinal parasites using molecular diagnostics. Am. J. Trop. Med. Hyg. 90(5): 835–9.

Stanley, S.L., Jr. 2003. Amebiasis. Lancet 361(9362): 1025–34.

Stark, D., S.E. Al-Qassab, J.L. Barratt, K. Stanley, T. Roberts, D. Marriott, J. Harkness and J.T. Ellis. 2011. Evaluation of multiplex tandem real-time PCR for detection of *Cryptosporidium* spp., *Dientamoeba fragilis*, *Entamoeba histolytica*, and *Giardia intestinalis* in clinical stool samples. J. Clin. Microbiol. 49(1): 257–62.

Stark, D., T. Roberts, J.T. Ellis, D. Marriott and J. Harkness. 2014. Evaluation of the EasyScreen enteric parasite detection kit for the detection of *Blastocystis* spp., *Cryptosporidium* spp., *Dientamoeba fragilis*, *Entamoeba* complex, and Giardia intestinalis from clinical stool samples. Diagn. Microbiol. Infect. Dis. 78(2): 149–52.

Stensvold, C.R., M. Lebbad, J.J. Verweij, C. Jespersgaard, G. von Samson-Himmelstjerna, S.S. Nielsen and H.V. Nielsen. 2010. Identification and delineation of members of the *Entamoeba* complex by pyrosequencing. Mol. Cell Probes 24(6): 403–6.

Stensvold, C.R., M. Lebbad, E.L. Victory, J.J. Verweij, E. Tannich, M. Alfellani, P. Legarraga and C.G. Clark. 2011. Increased sampling reveals novel lineages of *Entamoeba*: consequences of genetic diversity and host specificity for taxonomy and molecular detection. Protist 162(3): 525–41.

Sundaram, C., B.C. Prasad, G. Bhaskar, V. Lakshmi and J.M. Murthy. 2004. Brain abscess due to *Entamoeba histolytica*. J. Assoc. Physicians India 52: 251–2.

Tachibana, H., S. Kobayashi, M. Takekoshi and S. Ihara. 1991. Distinguishing pathogenic isolates of *Entamoeba histolytica* by polymerase chain reaction. J. Infect. Dis. 164(4): 825–6.

Tachibana, H., S. Kobayashi, E. Okuzawa and G. Masuda. 1992. Detection of pathogenic *Entamoeba histolytica* DNA in liver abscess fluid by polymerase chain reaction. Int. J. Parasitol. 22(8): 1193–6.

Taniuchi, M., J.J. Verweij, Z. Noor, S.U. Sobuz, Lv Lieshout, W.A. Petri, Jr., R. Haque and E.R. Houpt. 2011. High throughput multiplex PCR and probe-based detection with Luminex beads for seven intestinal parasites. Am. J. Trop. Med. Hyg. 84(2): 332–7.

Tannich, E. and G.D. Burchard. 1991. Differentiation of pathogenic from nonpathogenic *Entamoeba histolytica* by restriction fragment analysis of a single gene amplified *in vitro*. J. Clin. Microbiol. 29(2): 250–5.

Tarleton, J.L., R. Haque, D. Mondal, J. Shu, B.M. Farr and W.A. Petri, Jr. 2006. Cognitive effects of diarrhea, malnutrition, and *Entamoeba histolytica* infection on school age children in Dhaka, Bangladesh. Am. J. Trop. Med. Hyg. 74(3): 475–81.

Tegen, D., D. Damtie and T. Hailegebriel. 2020. Prevalence and associated risk factors of human intestinal protozoan parasitic infections in Ethiopia: a systematic review and meta-analysis. J. Parasitol. Res. 2020: 8884064.

Troll, H., H. Marti and N. Weiss. 1997. Simple differential detection of *Entamoeba histolytica* and *Entamoeba dispar* in fresh stool specimens by sodium acetate-acetic acid-formalin concentration and PCR. J. Clin. Microbiol. 35(7): 1701–5.

Tshalaia, L.E. 1941. On a species of *Entamoeba* detected in sewage effluents. Med. Parazit (Moscow) 10: 244–252.

Van Lint, P., J.W. Rossen, S. Vermeiren, K. Ver Elst, S. Weekx, J. Van Schaeren and A. Jeurissen. 2013. Detection of *Giardia lamblia, Cryptosporidium* spp. and *Entamoeba histolytica* in clinical stool samples by using multiplex real-time PCR after automated DNA isolation. Acta Clin. Belg 68(3): 188–92.

Varga, A. and D. James. 2006. Real-time RT-PCR and SYBR Green I melting curve analysis for the identification of Plum pox virus strains C, EA, and W: effect of amplicon size, melt rate, and dye translocation. J. Virol. Methods 132(1-2): 146–53.

Verkerke, H.P., B. Hanbury, A. Siddique, A. Samie, R. Haque, J. Herbein and W.A. Petri, Jr. 2015. Multisite clinical evaluation of a rapid test for *Entamoeba histolytica* in Stool. J. Clin. Microbiol. 53(2): 493–7.

Verweij, J.J., J. Blotkamp, E.A. Brienen, A. Aguirre and A.M. Polderman. 2000a. Differentiation of *Entamoeba histolytica* and *Entamoeba dispar* cysts using polymerase chain reaction on DNA isolated from faeces with spin columns. Eur. J. Clin. Microbiol. Infect. Dis. 19(5): 358–61.

Verweij, J.J., L. van Lieshout, C. Blotkamp, E.A. Brienen, S. van Duivenvoorden, M. van Esbroeck and A.M. Polderman. 2000b. Differentiation of *Entamoeba histolytica* and *Entamoeba dispar* using PCR-SHELA and comparison of antibody response. Arch. Med. Res. 31(4 Suppl): S44–6.

Verweij, J.J., F. Oostvogel, E.A. Brienen, A. Nang-Beifubah, J. Ziem and A.M. Polderman. 2003. Short communication: Prevalence of *Entamoeba histolytica* and *Entamoeba dispar* in northern Ghana. Trop. Med. Int. Health 8(12): 1153–6.

Verweij, J.J., R.A. Blange, K. Templeton, J. Schinkel, E.A. Brienen, M.A. van Rooyen, L. van Lieshout and A.M. Polderman. 2004. Simultaneous detection of *Entamoeba histolytica, Giardia lamblia,* and *Cryptosporidium parvum* in fecal samples by using multiplex real-time PCR. J. Clin. Microbiol. 42(3): 1220–3.

Verweij, J.J. and L. van Lieshout. 2011. Intestinal parasitic infections in an industrialized country; a new focus on children with better DNA-based diagnostics. Parasitology 138(12): 1492–8.

Victoria-Hernandez, J.A., A. Ventura-Saucedo, A. Lopez-Morones, S.L. Martinez-Hernandez, M.N. Medina-Rosales, M. Munoz-Ortega, M.E. Avila-Blanco, D. Cervantes-Garcia, L.F. Barba-Gallardo and J. Ventura-Juarez. 2020. Case report: multiple and atypical amoebic cerebral abscesses resistant to treatment. BMC Infect. Dis. 20(1): 669.

Weinke, T., W. Scherer, U. Neuber and M. Trautmann. 1989. Clinical features and management of amebic liver abscess. Experience from 29 patients. Klin Wochenschr 67(8): 415–20.

Wessels, E., L.G. Rusman, M.J. van Bussel and E.C. Claas. 2014. Added value of multiplex Luminex Gastrointestinal Pathogen Panel (xTAG(R) GPP) testing in the diagnosis of infectious gastroenteritis. Clin. Microbiol. Infect. 20(3): O182–7.

WHO/PAHO/UNESCO. 1997. World Health Organization/Pan American Health Organization/UNESCO report of a consultation of experts on amebiasis. Wkly Epidemiol. Rec. 72(14): 97–100.

Won, E.J., S.H. Kim, S.J. Kee, J.H. Shin, S.P. Suh, J.Y. Chai, D.W. Ryang and M.G. Shin. 2016. Multiplex real-time PCR assay targeting eight parasites customized to the Korean population: potential use for detection in diarrheal stool samples from gastroenteritis patients. PLoS One 11(11): e0166957.

Yakoob, J., Z. Abbas, M.A. Beg, S. Naz, R. Khan and W. Jafri. 2012. *Entamoeba* species associated with chronic diarrhoea in Pakistan. Epidemiol. Infect. 140(2): 323–8.

Yanagawa, Y., R. Shimogawara, T. Endo, R. Fukushima, H. Gatanaga, K. Hayasaka, Y. Kikuchi, T. Kobayashi, M. Koga, T. Koibuchi, T. Miyagawa, A. Nagata, H. Nakata, S. Oka, R. Otsuka, K. Sakai, M. Shibuya, H. Shingyochi, E. Tsuchihashi, K. Watanabe and K. Yagita. 2020. Utility of the rapid antigen detection test *E. histolytica* Quik Chek for the diagnosis of *Entamoeba histolytica* infection in nonendemic situations. J. Clin. Microbiol. 58(11).

Zaman, S., J. Khoo, S.W. Ng, R. Ahmed, M.A. Khan, R. Hussain and V. Zaman. 2000. Direct amplification of *Entamoeba histolytica* DNA from amoebic liver abscess pus using polymerase chain reaction. Parasitol. Res. 86(9): 724–8.

Zindrou, S., E. Orozco, E. Linder, A. Tellez and A. Bjorkman. 2001. Specific detection of *Entamoeba histolytica* DNA by hemolysin gene targeted PCR. Acta Trop. 78(2): 117–25.

Chapter 7

Recent Developments in the Diagnosis and Detection of the Zoonotic Parasite *Toxoplasma gondii* Causing Foodborne Disease and Outbreaks[#]

S. Almeria[1],* and J.P. Dubey[2]

Introduction: Importance of *Toxoplasma gondii* Infection and Ways of Transmission

Toxoplasmosis is a zoonotic disease of global distribution caused by the protozoan parasite *Toxoplasma gondii,* the only species in the *Toxoplasma* genus. *Toxoplasma* remarkably can infect virtually every warm-blooded animal, including humans and livestock, as intermediate hosts (IH) (Dubey 2010). Felids (domestic and wild) are the definitive hosts (DH), and the only hosts able to eliminate oocysts

[1] Department of Health and Human Services, U.S. Food and Drug Administration, Center for Food Safety and Nutrition, Office of Applied Research and Safety Assessment, Division of Virulence Assessment, Laurel, Maryland 20708, USA.
[2] U.S. Department of Agriculture, Agricultural Research Service, Animal Parasitic Disease Laboratory, Beltsville, Maryland 20705, USA.
* Corresponding author: maria.almeria@fda.hhs.gov
[#] Disclaimer: The views expressed are those of the authors and should not be construed as the U.S. Food and Drug Administration or U.S. Department of Agriculture views or policies. The mention of commercial products, their sources, or their use in connection with material reported herein is not to be construed as either an actual or implied endorsement or promotion of such products by the Department of Health and Human Services or the U.S. Department of Agriculture.

from the environment. *Toxoplasma gondii* is considered one of the most important foodborne and waterborne parasites of veterinary and medical infections (Jones and Dubey 2010; Havelaar et al. 2012; Hoffmann et al. 2012; Jones and Dubey 2012; FAO/WHO 2014; Torgerson et al. 2015; Opsteegh et al. 2016; Gisbert Algaba et al. 2017). Based on the disease burden expressed in Quality or Disability Adjusted Life Years (DALY), *T. gondii* ranked second out of 14 foodborne pathogens in the USA (Batz et al. 2012), and first in the Netherlands (Havelaar et al. 2012). Furthermore, an FAO/WHO report considered *T. gondii* as the fourth more important parasite in the world (FAO/WHO 2014). In addition, it has been estimated that about one-third of human beings are chronically infected by this pathogen (Moncada and Montoya 2012; Rostami et al. 2020).

The transmission of *T. gondii* in the IH may result from the ingestion of tissue cysts in raw or undercooked meat (by carnivorous and/or omnivorous species), ingestion of contaminated raw vegetables or water with *T. gondii* oocysts from cat feces (the main way of transmission in herbivores), and by vertical or transplacental transmission in all IH species (Table 1; Figure 1) (Dubey 2010; Yan et al. 2016). This parasite has been incriminated as one of the most fatal foodborne pathogens in the USA (Scallan et al. 2011). In a European multicenter case-control study, 30 to 63% of infections in pregnant women were attributed to ingestion of meat, whereas 6 to 17% were most likely soil-borne (Cook et al. 2000); however, this study is more than two decades old.

Oocysts in the environment are excreted only by felids, domestic and wild cats. However, a single cat may excrete millions of oocysts into the environment in a matter of a few days (Dubey 2010). Felids eliminate oocysts unsporulated (noninfectious) and oocysts sporulate in the environment and become infectious for all warm-blooded hosts. Sporulated oocysts can persist for months or years in the environment (Dubey 2010).

Consumption of contaminated non-properly or undercooked meat, particularly pork and lamb, has been ascribed as the major risk factor for the acquisition of toxoplasmosis in humans (Weiss and Dubey 2009; Dubey et al. 2020a, b, c) with

Table 1. *Toxoplasma gondii* infective stages and role in the transmission of infection for humans and animals.

Infective Stage	Type of Multiplication Localization in Host	Role in Transmission
Tachyzoite	Fast asexual multiplication (acute infection) intracellular in multiple organs and tissues	Vertical transmission (from mother to fetus). In rare occasions ingestion, via seminal or ocular
Bradyzoite (in tissue cysts)	Slow asexual multiplication (subacute infection) in cysts and pseudocysts in tissues including brain, muscle, and liver	Horizontal transmission by food ingestion of muscle tissues raw or undercooked
Sporulated oocysts	Sexual multiplication of the parasite in the small intestine epithelial cells of felines	Resistance forms in the environment. Transmission to Intermediate host by ingestion of water, fruits, vegetables, and/or plants contaminated with sporulated oocysts

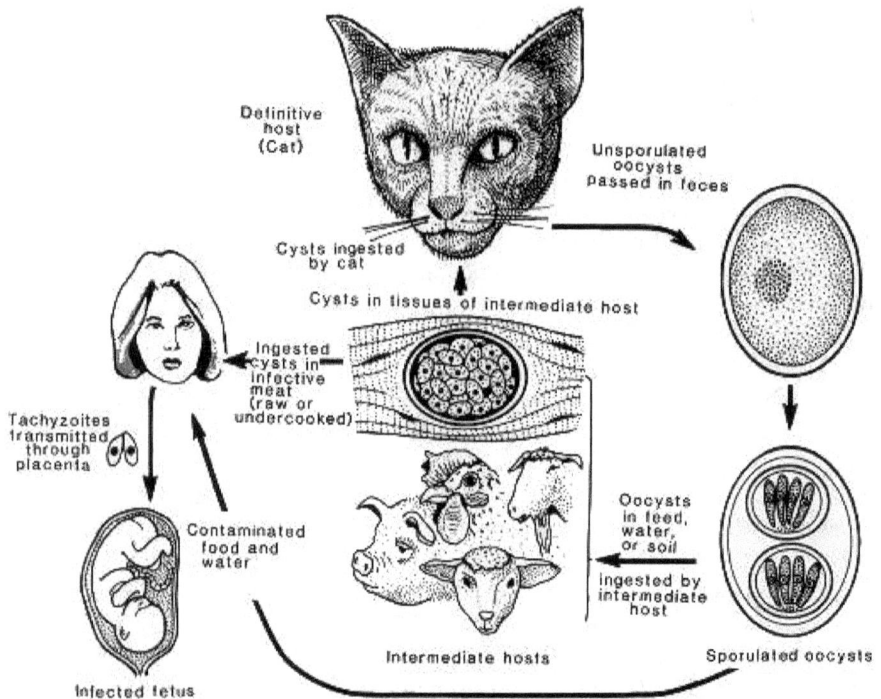

Figure 1. *Toxoplasma gondii* life cycle (Courtesy of Dr. Dubey).

unpasteurized goat's milk, vegetables, plants, and water, thus being linked to both food and water-borne transmission (Jones and Dubey 2010; Jones and Dubey 2012; Torgerson et al. 2015; Hill and Dubey 2016). Improved animal husbandry practices as well as increased awareness of the risks of consuming undercooked meat have resulted in a decreased prevalence of toxoplasmosis worldwide in the last decades. Transplacental or congenital transmission from the mother to the fetus occurs when tachyzoites pass through the placenta during pregnancy. In women and livestock species, the vertical transmission risk during pregnancy increases with gestational age. However, fetal damage is more severe in mothers who become infected early in pregnancy (Robert-Gangneux and Darde 2012; Fricker-Hidalgo et al. 2017). Primary infections early in the pregnancy can lead to abortion, fetal death, stillbirths, and congenitally infected progeny (McLeod et al. 2020; Peyron et al. 2016). The severity of congenital toxoplasmosis in alive infected progeny ranges from asymptomatic or mild infection to severe defects, including neurological problems such as hydrocephalus, microcephaly, intracranial calcifications, encephalitis, and psychomotor or mental retardation, deafness, and/or ocular lesions (blindness, visual impairment retinochoroiditis, and unilateral microphthalmia), particularly in untreated infants (Montoya and Liesenfeld 2004; Weiss and Dubey 2009; Fallahi et al. 2018; Khan and Khan 2018).

A recent systematic review and meta-analysis of pregnant women's data according to their geographic regions, as defined by the World Health Organization

(WHO) (Rostami et al. 2020), estimated that the global prevalence of latent toxoplasmosis in pregnant women is 33.8% worldwide (95% CI, 31.8–35.9%; 345,870/1, 148,677), with South America having the highest prevalence (56.2%; 50.5–62.8%), whereas the Western Pacific region had the lowest prevalence (11.8%; 8.1–16.0%). A significantly higher prevalence of latent toxoplasmosis was associated with countries with low income and low human development indices (Rostami et al. 2020).

In immunocompetent individuals, infection with *T. gondii* is usually asymptomatic. However, there have been some reports of severe toxoplasmosis in healthy individuals (Demar et al. 2007). Recently, a 31-year-old immunocompetent man suffered from spinal cord toxoplasmosis as a primary *T. gondii* infection (Martinot et al. 2020). Toxoplasmosis can also cause retinochoroiditis in adults, especially in individuals with an impaired immune system (McLeod et al. 2020). There is vast literature associating *T. gondii* infection with psychiatric conditions, such as bipolar disorder, epilepsy, depression, suicidal behavior and/or Alzheimer's disease, and automobile fatalities (Flegr et al. 2014; Fuglewicz et al. 2017; Abo-Al-Ela 2019). A recent meta-analysis study observed a positive relationship between toxoplasmosis and obsessive-compulsive disease (OCD) (Nayeri Chegeni et al. 2019). The most reported behavioral deviations are related to greater impulsivity and aggressiveness (Martinez et al. 2018). However, these results are not universally accepted (Fuglewicz et al. 2017; Abo-Al-Ela 2019; Sutterland et al. 2019).

In immunocompromised individuals, particularly HIV-positive patients and immunosuppressed patients, the parasite can cause life-threatening toxoplasmosis, usually presented as encephalitis, severe myocarditis, pneumonitis, and ophthalmitis (Eza and Lucas 2006). Cerebral toxoplasmosis is considered the third most prevalent opportunistic infection in HIV-infected populations (Bowen et al. 2016). In most HIV-patients, toxoplasmosis results from the reactivation of chronic asymptomatic infection, but severe toxoplasmosis has also been reported in HIV patients that recently acquired infection (McLeod et al. 2020). *T. gondii* infection seems to be also a severe problem in cancer patients (Cong et al. 2015). Huang et al. (2016) suggested that *T. gondii* infection might be a risk factor for leukemia, and seroprevalence of *T. gondii* infection was reported to be higher in children with leukemia than that in healthy children in Eastern China (Zhou et al. 2019).

T. gondii is also an important cause of abortion in sheep and goats, causing economic losses and a risk to public health. Toxoplasmosis abortion also occurs in pigs, but it is not a major abortifacient for this host (Dubey 2010; Opsteegh et al. 2016; Dubey et al. 2020a, b). *T. gondii* infection in susceptible animal species may cause early embryonic death and resorption, fetal death and mummification, abortion, stillbirth, and neonatal death (Dubey 2010). As in humans, the severity of infection is associated with the stage of pregnancy at which the ewe/dam becomes infected, the earlier in gestation, the more severe the consequences. Infected meat from farm animals is a source of *T. gondii* infection for humans and carnivorous animals (Dubey 2009; Opsteegh et al. 2016). Toxoplasmosis causes heavy economic losses to the sheep industry worldwide (Buxton et al. 2007; Dubey 2009; Dubey et al. 2020a, b). In addition, the parasite is also a cause of concern for wildlife and zoo species. New

World primates, pro-simians of Madagascar, and marsupials from Australasia are all highly susceptible species to *T. gondii* infection, and severe cases of toxoplasmosis have been reported in these species worldwide with high susceptibility in marsupials and primates (Fernandez-Aguilar et al. 2013). Certain felids (Pallas cats; Sand cats) are highly susceptible to clinical toxoplasmosis and these captive cats in zoos are a public health concern because they can excrete oocysts (Dubey 2010; Dubey et al. 2020d). Understanding the roles of animal reservoirs in the spread of *T. gondii* infection is important to control the dissemination of the parasite.

The parasite is, therefore, an important concern for public health, particularly, in immunocompromised people and pregnant women, and for animal health, especially pertaining to farm animals which caused important economic losses. Diagnosis of active infections (primary infection or reactivation of bradyzoites in tissue cysts to tachyzoites) in humans is of crucial importance for optimal treatment.

Laboratory Diagnosis

The symptoms and clinical signs of toxoplasmosis are non-specific, in fact, the symptoms are like those in several other infectious diseases and therefore, the differential diagnosis must be considered (Montoya 2002; Hill and Dubey 2016). In addition, *T. gondii* infection presents different clinical manifestations depending on the type of person infected and the clinical course of the infection, and methods of diagnosis and their interpretations may differ for each clinical category (Montoya 2002). Human clinical manifestations of *T. gondii* can include: (1) toxoplasmosis acquired by immunocompetent patients, (2) toxoplasmosis acquired by or reactivated in immunodeficient patients, (3) toxoplasmosis acquired during pregnancy or congenitally in the fetus, and (4) ocular infections. In animals, detection in meat-producing animals is important for their zoonotic role, and detection of oocysts in the environment, water, and vegetables is important to avoid other foodborne routes of transmission.

Laboratory diagnosis is therefore crucial for the correct diagnosis of the disease in humans and animals and for the detection of the parasite in the environment. Methods for the diagnosis of toxoplasmosis can be categorized as direct methods and indirect methods. Detection of DNA of *T. gondii* from samples of body fluids by PCR, demonstration of the parasite through bioassay in mice or cell culture, ophthalmic testing, and radiological studies come under the direct methods of diagnosis, while detection of anti-*Toxoplasma* antibodies by many different techniques is included under the indirect methods of diagnosis. Table 2 includes a summary of the main techniques used for *T. gondii* diagnosis.

In humans and animals, diagnosis of toxoplasmosis is made by biological, histological, serological, or molecular methods, or by some combination of the above. For example, for diagnosis of infections acquired congenitally in the fetus in humans, serological tests are first-line strategies in the diagnosis, while after confirmed serological diagnosis, ultrasound imaging, amniocentesis and polymerase chain reaction (PCR) are very sensitive and useful (Murat et al. 2013a).

There are also methods for the recovery and detection of oocysts in fecal samples from cats as well as in the environment, fresh produce, soil, and water. In addition,

Table 2. Summary of the methods described to date for *Toxoplasma gondii* diagnosis.

Type of Method	Name of Method	Antigen or Antibodies Used	Isotypes Tested (if Serology) or Target (if Molecular Methods) (Selected References)	Purpose or Indication in Clinical Settings. Notes
Serology	Sabin-Feldman (dye test or DT test)	Whole live tachyzoites in presence of complement	Total specific immunoglobulins (Sabin and Feldman 1948)	Acute and chronic infection. Reference methods for many studies. High sensitivity and specificity
Serology	Modified agglutination tests (MAT)	Formalin-fixed tachyzoites in U-shaped microwells	If added 2-mercaptoethanol: the presence of IgG (Dubey and Desmonts 1989)	Mainly used in veterinary studies of epidemiology and/or infection in animals
Serology	Latex agglutination test (LAT)	Soluble antigen coated in latex or polystyrene particles	IgG, IgM	Mainly used in veterinary studies of epidemiology and or infection in animals
Serology	Indirect hemagglutination (IHA)	Erythrocytes sensitized with *T. gondii* soluble antigens	IgG. Not very specific, not very sensitive (Dubey 2010)	Agglutination. Screening in epidemiological settings
Serology	immunosorbent agglutination assay (ISAGA)	Anti-Ig antibodies coating microwells incubated with formalin-fixed tachyzoites	IgM (Desmonts et al. 1981), IgE, IgA	Congenital toxoplasmosis in infants. Early detection for acute infection
Serology	Indirect fluorescent antibody techniques (IFAT)	Formalin-fixed whole tachyzoites. Need specific conjugate anti-host species	IgM, IgG	Chronic and acute infection in humans. Epidemiological studies in veterinary science. Requires fluorescent microscopy.
Serology	Enzyme-linked immunoabsorbent assay (ELISA)	*T. gondii* lysate antigens (TLA) or recombinant antigens, or specific antibodies in microwells	IgM, IgG. Acute or chronic infection. Can be used as kinetic measure of antibodies	Routine screening of pregnant women. Automated and commercial (need a lector). May need confirmatory methods

Table 2 contd. ...

...Table 2 contd.

Type of Method	Name of Method	Antigen or Antibodies Used	Isotypes Tested (if Serology) or Target (if Molecular Methods) (Selected References)	Purpose or Indication in Clinical Settings. Notes
Serology	Avidity ELISA	TLA or recombinant antigens. Compares standard ELISA with a modified ELISA using a dissociating IgG-antigen agent (i.e., urea). Ratio or index calculation	IgG (Hedman et al. 1989), IgM, IgA, IgE	Measure high or low-avidity antibodies. If high-avidity recent infection (3–5 months) can be excluded: chronic infection. Many commercial platforms
Serology	Chemiluminescent immunoassays (CLIA)	Similar to ELISA but using chemiluminescence or electrochemiluminescence as detection systems	SAG2-GRA1-ROP1L-Chimeric antigens (Ferra et al. 2015; Holec-Gąsior et al. 2018)	Automated. Differentiation of healthy versus infected humans.
Serology	Serotyping	Synthetic recombinant antigen or recombinant chimeric *T. gondii* antigens	IgG, IgM. GRA2, GRA3, GRA4, GRA5, GRA6, GRA7, GRA8, GRA14, MIC1, MIC2, MAG1, SAG1, SAG2A, ROP1, ROP6, MIC12, SRS29A SRS13	Human serotyping. Also used in some animal meat-producing studies
Serology	Immunochromatographic test (ICT), Rapid detection tests, and port-of-care (POC)	Antigens or antibodies labeled with colloidal gold	Excretory/secretory antigens, SAG2, SAG1, GRA1, GRA7, GRA14, SAG3-3A7 and 4D5 [Huang et al. 2004; Wang et al. 2011; Terkawi et al. 2013]	*T. gondii* anti-excretory/secretory circulating antigens. Also used in cats (Jiang et al. 2015)
Serology	Western blotting or immune blotting (IB)	On nitrocellulose strips with TLA or recombinant antigens	IgM, IgG (Rilling et al. 2003)	Early detection of IgG. Confirmatory test for congenital toxoplasmosis in mother and child or for ocular toxoplasmosis

Serology	Enzyme-linked aptamer assay (ELAA)	DNA and RNA aptamers	Aptamers against *Toxoplasma* ROP18 protein [Ospina-Villa et al. 2018; Varga-Montes et al. 2019]	Congenital infection
Serology	Piezoelectric immunoagglutination assay (PIA)	Antigen-coated gold nanoparticles	IgG (Wang et al. 2004)	Agglutination test with detection by a piezoelectric device
Serology	Luciferase immunoprecipitation (LIPS)	Recombinant *Toxoplasma* antigens fused with nanoluciferase	GRA6, GRA7, GRA8 and BAG1 (Aye et al. 2018)	Shorter overall assay time, and less sample volume than ELISA
Serology	Laser induced fluorescence (LIF)	Zinc oxide nanoparticles conjugated with *T-gondii* antigens	IgG (Medawar-Aguilar et al. 2019)	microfluidic immunosensor
Immunological-T-cell based response	Interferon-gamma release assay (IGRA)	Whole blood or T cells	GST-GRA1 (Ciardelli et al. 2008; Mahmoudi et al. 2017)	Congenital infection
Direct detection	Direct autofluorescence	Useful but low sensitivity	Oocysts detection (Lindquist et al. 2003)	Oocysts detection in fecal and produce samples
Direct detection	Immunomagnetic separation assay (IMS *Toxo*)	Using a specific purified *T. gondii* monoclonal antibody	Raspberries and basil (Hohweyer et al. 2016)	Detection and quantification in produce by IMS *Toxo*, coupled to microscopic and qPCR
Direct detection	magnetic capture and qPCR (MC-qPCR)	Crude DNA extract from meat and magnetic capture of *T. gondii* DNA followed by qPCR targeting the 29-bp repeat element	Meat for human consumption from several species (Opsteegh et al. 2010)	Detection and quantification in meat samples
Direct detection	Improved MC-qPCR	Inclusion of a non-competitive PCR inhibition control, and release of the target DNA from the streptavidin-coated paramagnetic beads UV-dependent	Meat for human consumption from several species (Gisbert Algaba et al. 2017)	Detection and quantification in meat samples

Table 2 contd. ...

...Table 2 contd.

Type of Method	Name of Method	Antigen or Antibodies Used	Isotypes Tested (if Serology) or Target (if Molecular Methods) (Selected References)	Purpose or Indication in Clinical Settings. Notes
Direct-Molecular detection	Conventional PCR	DNA	B1 (Burg et al. 1989) 529 bp repeat element (Homan et al. 2000; Reischl et al. 2003), 18S rDNA, ITS-1, P30	Species detection. Tissue biopsies, cerebrospinal fluid, bronchoalveolar lavage, peripheral blood, vitreous and aqueous humor, fetal blood, AF Produce
Direct-Molecular detection	Real-Time PCR	DNA. Fluorescent resonance energy transfer (FRET), hydrolysis probes, SYBR Green	B1 gene, 529 bp repeat element, 18S rDNA, SAG1	Species detection, same as conventional PCR. Also used in produce
Direct-Molecular detection	Loop-mediated isothermal amplification (LAMP)	DNA. Constant temperature	529 bp (Lin et al. 2012; Kong et al. 2012); B1 gene (Valian et al. 2019) 18S rDNA, SAG1, SAG2, GRA1, oocyst wall protein antigens	Needs less equipment than PCR but the contamination rate of LAMP is higher than PCR Species detection. Environmental (Lalle et al. 2018)
Direct-Molecular detection	Recombinase polymerase amplification (RPA) RPA	Uses a combination of enzymes such as recombinase, polymerase, and single-stranded DNA-binding (SSB) protein. Amplification between 37 and 42°C	B1 gene (Wu et al. 2017)	Environment from soil and water samples
Direct-Molecular detection	PCR-RFLP	Nested PCR Currently 10 markers	SAG1, SAG2, SAG3, BTUB, GRA6, c22-8, c29-2, L358, PK1, and Apico (Howe and Sibley 1995; Su et al. 2010)	Genotyping
Direct-Molecular detection	Microsatellite analysis (MS)	Very useful for genotyping from DNA. Multiplex PCR	TUB2, W35, TgM-A, B17, B18, IV.1, XI.1, M48, M102, N60, N82, AA, N61, N83 (Ajzenberg et al. 2010)	Genotyping

Direct-Molecular detection	Multilocus sequence typing (MLST)	Need large amounts of DNA	BTUB, SAG2, SAG3, GRA6 (Khan et al. 2007)	Genotyping. Identify DNA polymorphisms without predetermined genetic data limitations to be used in clinical samples because need large amounts of DNA. Also used to produce.
Direct-Molecular detection	RAPD-PCR	Patron of bands after restriction enzyme digestion of PCR products	Genomic DNA (Guo et al. 1997)	Genotyping
Direct-Molecular detection	HRM analysis	Post-PCR method for genetic variation based on their melting temperatures related to their sequences	B1 gene (Costa et al. 2011, Liu et al. 2015)	Genotyping. It is more informative than MS and can be used as a supplementary test in genotyping
Direct detection. Metabolomics	Matrix-assisted laser desorption ionization time-of-flight mass spectrometry (MALDI-TOF MS)	Detection of mass spectral fingerprints of proteins	*T. gondii* tachyzoites in human foreskin fibroblasts (King et al. 2020)	Metabolite profiling (metabolomic)

the parasite can also be detected, directly or indirectly, in living farm animals and in meat products intended for human consumption, including many epidemiological studies (Opsteegh et al. 2010). Many research methodologies have also been applied to the study of the parasite itself as a parasite model and will be briefly included in this review. In the past century, serological methods, mainly based on ELISA and avidity tests in humans and agglutination, IFAT and ELISA in animals, and molecular methods based on PCR were the most used method. More recently advanced ELISA-based methods including chemiluminescence (CLIA), Immunochromatographic tests (ICT), and rapid tests, together with new molecular methods, such as real-time PCR, and loop-mediated isothermal amplification (LAMP) have been developed, with high sensitivity and specificity, and certainly new methods will continue to be developed in the future.

There have been several reviews on the subject of analytical methods, including among others: Montoya 2002; Wei and Dubey 2009; Dubey 2010; Murat et al. 2013a; Dard et al. 2016; Peyron et al. 2016; Pomares and Montoya 2016; Ozgonul and Besirli 2017; Fallahi et al. 2018; Rostami et al. 2018; Khan and Noordin 2020; McLeod et al. 2020. Diagnosis of acute *T. gondii* infection in women and their fetuses is complex and only an overview is provided here; this subject is reviewed extensively in two book chapters (Peyron et al. 2016; McLeod et al. 2020).

Indirect Methods: Serological Analysis of Antibodies in Serum and Fluids

The most common diagnostic tools for the detection of *T. gondii* infection are indirect detection methods based on serological tests. These tests play an important role in the initial step in the diagnosis of the parasite in humans and animals. The serological methods allow the detection of antibodies and circulating antigens in a serum sample, but they only establish previous exposure to the parasite and do not imply the actual presence of the parasite in the infected person/animal. In addition, antibodies can persist in infected people after their clinical recovery, which can make interpretation of the results of serological tests difficult. *Toxoplasma*-specific immunoglobulins IgG and IgM detection tests are the most frequently used (Murat et al. 2013a). IgA shows similar kinetics to IgM, but it is less used in routine testing (Pinon et al. 2001), while IgE tests are rarely assessed, although they are considered specific for acute infection (Murat et al. 2013a). In a retrospective study of 690 pregnant women with positive *T. gondii*, IgG antibody test results that also had *T. gondii* IgA and IgM antibody tests performed, pregnant women with *T. gondii* IgA antibodies were more likely than pregnant women without *T. gondii* IgA antibodies to have had a recent infection with *T. gondii* (Olariu et al. 2019).

The presence or absence of IgM antibodies, titers of IgG antibodies, and antibody kinetics may help diagnosis whether the *T. gondii* infection is acute (recent) or latent (acquired in the past). Classically, *Toxoplasma* infection is associated with high levels of *Toxoplasma*-specific IgM antibodies usually produced within the first three weeks after infection (Marcolino et al. 2000; Murat et al. 2013a) and a rise in specific IgG levels 1 to 3 weeks later (Fricker-Hidalgo et al. 2013). IgM antibodies, generally, disappear faster than IgG antibodies after recovery and become undetectable after

6–9 months. However, low titers of IgM might persist long after the acute phase of the disease, so IgM positivity cannot be directly interpreted as a sign of acute infection (Villard et al. 2013). The detection of IgM can help diagnose congenital infection because maternal IgM does not cross the placenta. The passage of IgG antibodies from the dam to the fetus depends on the host species; in humans, IgG can cross the placenta but not in domestic animals (sheep, goats, and pigs) (Dubey 2010).

As elevated levels of IgM antibodies alone should not be considered evidence of recent infection, similarly, low serum IgG levels should not be considered as an inactive disease (Murat et al. 2013a). If the laboratory testing is unequivocal, serological tests should be repeated within 15–21 days (Marcolino et al. 2000; Suzuki et al. 2001). This kinetic analysis is important in cases of toxoplasmosis in pregnant women. It is recommended to analyze two samples from the same individual with the second sample being collected 2–4 weeks after the first. If there are higher antibody titers in the second sample (more than a 16-fold increase), it will indicate an acute infection. Another problem is that IgG seroconversion, without IgM detection or with transient IgM levels, has been described during serologic follow-ups of previously seronegative pregnant women and raises difficulties in interpreting the results (Fricker-Hidalgo et al. 2013).

There are numerous serological tests used currently, these are listed in Table 2. Several of these serological tests use whole parasites (tachyzoites) such as Sabin-Feldman, direct agglutination, MAT, ISAGA, or IFAT. Other serological methods, including many commercially available ELISA, are based on antigenic preparations isolated from tachyzoites obtained from infected mice or *in vitro* tissue culture to obtain *T. gondii* lysate antigen (TLA), instead of whole parasites.

Sabin-Feldman Test (Dye Test or DT Test)

The development of the dye test for *T. gondii* by Sabin and Feldman in 1948 was a major advance in the study of toxoplasmosis. It demonstrated the widespread prevalence of this infection worldwide and assisted in the differential diagnosis of congenital infections (Sabin and Feldman 1949; Weiss and Dubey 2009). This test is both highly sensitive and specific in humans (Weiss and Dubey 2009; Dubey 2020) even in the undiluted serum, and it has been considered the gold standard serological test for several validation studies for definitive human toxoplasmosis (Petersen et al. 2005; Sickinger et al. 2008; Gay-Andrieu et al. 2009; Prusa et al. 2010). However, it uses live *T. gondii* tachyzoites (highly virulent RH strain) to detect total specific anti-*Toxoplasma* antibodies and currently is only available in some reference laboratories in the world due to the possible risk for laboratory-acquired infection. It is essentially a complement-mediated neutralizing type of antigen-antibody reaction (Dubey 2010) and involves the addition of an accessory factor (complement-like factor from human serum that is heat labile) to the serum samples in the analysis which makes it complex (Ozgonul and Besirli 2017). Most hosts develop DT antibodies within four weeks and titers may remain stable for months or years (Dubey 2010). It is specific in many animal species, but in ruminants can give false results unless sera are inactivated at 60°C; it also does not work with some avian sera (Dubey 2010).

Agglutination Tests

Agglutination tests are currently used extensively, mainly in epidemiological animal studies, because they do not require special equipment or conjugate. This test is simple, easy to perform, does not need host-specific reagents; and in humans, the results match those obtained by the dye test. The method was modified and later named as modified agglutination test (MAT) (Desmonts and Remington 1980; Dubey and Desmonts 1987) by treating sera with 2-mercaptoethanol (to remove non-specific IgM or IgM-like substances) and therefore in MAT, only IgG antibodies are detected. MAT is extensively used for the diagnosis of toxoplasmosis in both animals and humans (Dubey 2010), and hemolysis does not interfere with the test (Dubey 2010). MAT can be modified to differentiate acute from chronic infections (Dubey 2010). The MAT is commercially available (Toxo-screen DA; bioMérieux, France).

Indirect agglutination techniques follow a similar principle. If particles sensitized with *T. gondii* antigens are used in latex or polystyrene, the method is known as LAT (Villard et al. 2012). As a negative point of this technique, the performance for tests based on direct agglutination or indirect hemagglutination was better than those of LAT assays (Villard et al. 2012). In addition, LAT has been suggested not to be very sensitive in livestock sera (Dubey 2010). LAT is also commercially available by several companies.

The indirect hemagglutination test (IHA) is an agglutination test based on the use of erythrocytes (red blood cells) sensitized with *T. gondii* soluble antigens that are agglutinated in the presence of immune serum. It has been mainly used for screening in epidemiological settings and mainly in China (Dubey 2010). However, it is considered an insensitive and non-specific test. In animals, titers lower than 1:128 may be non-specific (Dubey 2010).

Another agglutination-based method is the immunoabsorbent agglutination assay (ISAGA). It was first developed as an IgM ELISA combined with agglutination to avoid the need for a conjugate (Desmonts et al. 1981). It is a fast method based on recombinant antibodies against specific isotypes of immunoglobins. The antibodies are coated into wells, serum samples are added, and then whole tachyzoites. In some variations, whole tachyzoites were substituted with latex beads with soluble antigens (Remington et al. 1983). Results are analyzed for direct agglutination. The ISAGA for IgM has been considering a preferred method to detect of anti-*Toxoplasma* antibodies in infants (Moncada and Montoya 2012; Murat et al. 2012; Pomares and Montoya 2016). IgA and IgE ISAGA have also been described. As a disadvantage, this technique requires large numbers of tachyzoites (Dubey 2010).

Indirect Immunofluorescence Antibody Test (IFAT)

Another classical test commonly used for serological analysis of toxoplasmosis in humans and animals is the indirect IFAT. This test uses whole fixed tachyzoites in glass slides, incubated with serial dilutions of the tested serum. The binding of the test serum to whole parasites is revealed by a fluorescent conjugate against the host species IgG or IgM (Dubey 2010). Some limitations of this method are that it requires conjugate as well as the technical skills to read the slides in a fluorescent microscope. IFAT titers generally correspond to those of the dye test, and IFAT assays have been standardized in some laboratories. However, cross-reactions may occur (Dubey 2010).

Immunoenzymatic Assays: ELISA

Currently, the most used tests for serological analysis in the field of toxoplasmosis are the immunoenzymatic assays or ELISA (Murat et al. 2013a). They are high-throughput methods designed for automated reader devices for simultaneous testing of large numbers of samples and the results are objective (Suzuki et al. 2001). Many of these methods are commercially available and allow wide implementation of toxoplasmosis follow-up in routine laboratories, along with a screening of other pathogens in pregnancy (rubella; cytomegalovirus) (Suzuki et al. 2001; Murat et al. 2013b; Ozgonul and Besirli 2017). They require an ELISA reader to quantify the color reaction.

Toxoplasma gondii tachyzoite lysate antigens (TLA), recombinant parasite antigens, or specific anti-parasite antibodies are coated in a solid phase (wells in a plate or beads), and antibodies will bind to them and then are detected by a conjugate (colored or fluorescent product). A detector antibody, or the antigen-antibody complex, conjugated with an enzyme will produce a colored or fluorescent signal which will be automatically compared to standard values and will give a numerical value. There are several variations in the ELISA method including indirect sandwich ELISA method, double sandwich ELISA, antibody capture (isotype-specific antibodies) ELISA, or competition ELISA (Figure 2). Currently, many manufacturers provide these kits. The kits have been compared in numerous studies (Gay-Andrieu et al. 2009; Maudry et al. 2009; Murat et al. 2013b) with only slight differences in sensitivity, specificity, or antibody kinetics among the different manufacturers (Murat et al. 2013b). ELISAs have been set up for IgG and IgM detection to determine acute or chronic infections and can be used as a kinetic measure of antibodies. They are the most currently used tests in the routine screening of pregnant women, although results may need additional confirmatory methods (Dard et al. 2016) and have also been used in many animal studies.

Avidity ELISA Tests

The major limitation of serological tests is the failure to clarify whether infections are acute (recently acquired) or chronic disease (distant infections with the parasite); this is very important in cases acquired during pregnancy or congenitally. To establish whether infections are acute or chronic, in addition to kinetics analysis of antibodies, the assessment of the avidity of the antibodies (IgG and IgM) may be helpful (Gay-Andrieu et al. 2009; Murat et al. 2013a; Pomares and Montoya 2016; Garnaud et al. 2020). The affinity of specific IgG antibodies is initially low after primary antigenic challenge and increases during subsequent weeks and months by antigen-driven B cell selection (Montoya 2002; Murat et al. 2013a).

Modified Agglutination Test (MAT)

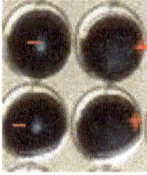

ELISA (Enzyme-Linked Immunosorbent Assay)

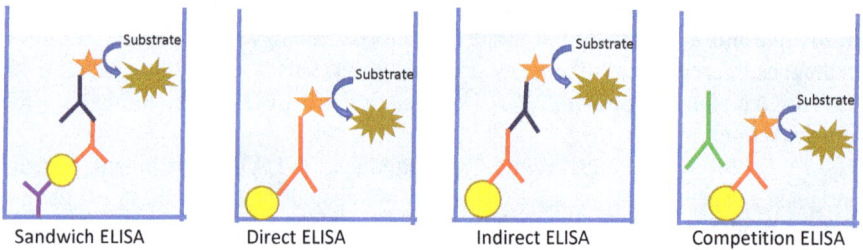

Sandwich ELISA   Direct ELISA   Indirect ELISA   Competition ELISA

Key:

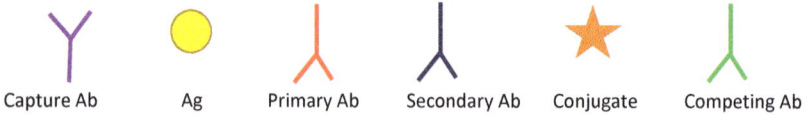

Capture Ab   Ag   Primary Ab   Secondary Ab   Conjugate   Competing Ab

IFAT (Indirect Fluorescent Antibody Test)

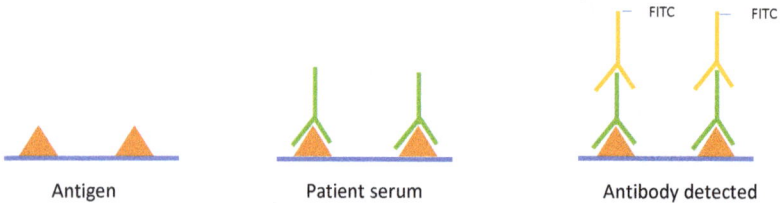

Antigen   Patient serum   Antibody detected

Figure 2. Schematic representation of some main serological methods for detection of *Toxoplasma gondii* antibodies.

The strength of antibody binding can be evaluated by introducing a wash step using a buffer to dissociate the antibody-antigen complex (protein-denaturing reagents such as urea are mostly used), which removes low-avidity antibodies associated with recently acquired infections (Robert-Gangneux and Darde 2012). The resulting titer of detectable IgG is used to calculate a ratio of titers obtained from urea-treated and urea-untreated samples. Then, the avidity result is estimated by the ratios of antibody titration curves of urea-treated and untreated samples (Montoya 2002).

The IgG avidity test was first described by Hedman et al. (1989) and has been proposed to identify latent infections. A high index excludes recent toxoplasmosis. In contrast, an intermediate or low index only suggests recent infection, but some people with latent *T. gondii* infection show IgG with low or intermediate avidity (Villard et al. 2013). In a comparison of four commercialized avidity tests, a positive predictive value of 100% was observed for confirmation of latent toxoplasmosis, but the test did not show enough reliability for the diagnosis of acute infection (Candolfi et al. 2007). Therefore, although early studies suggested that a low or a very low value could confirm a recent infection, it is currently accepted that the only validated use of these tests is the exclusion of recent infection (i.e., in the last four months for most tests) when the value is found above a defined cut-off (Murat et al. 2013a).

Currently, there are several commercial automated or semiautomated immunoassays for IgG- and IgM-*T. gondii* antibody detection, including concurrently with a diagnosis of other pathogens. These commercial techniques have high sensitivity and specificity, reproducibility, and fast and precise measurement of IgG and IgM antibody levels (Sickinger et al. 2008; Gay-Andrieu et al. 2009; Prusa et al. 2010; Murat et al. 2012; Prusa et al. 2012; Murat et al. 2013b; Fricker-Hidalgo et al. 2017; Boquel et al. 2018; Fallahi et al. 2018; Genco et al. 2019). In a comparison study, Vidas *Toxo* IgG avidity had the best performance for the diagnosis of latent toxoplasmosis (Villard et al. 2013). In a more recent comparison of nine commercial immunoassays (Advia Centaur, Architect, AxSYM, Elecsys, Enzygnost, Liaison, Platelia, VIDAS, and VIDIA assays), using panels of serum samples from routine testing of patients with acute or chronic toxoplasmosis, IgG sensitivities ranged from 97.1% to 100%, and IgG specificities ranged from 99.5% to 100%. IgM sensitivities ranged from 65% to 97.9%, and IgM specificities ranged from 92.6% to 100% (Villard et al. 2016a). Another automated analyzer commercially available platform is the BioPlex 2200 (Bio-Rad Laboratories, Hercules, CA) for the detection of anti-*Toxoplasma*, -rubella, and -cytomegalovirus antibodies in the same assay. The BioPlex 2200 ToRC IgG/IgM kit was compared to Platelia IgG/IgM enzyme-linked immunosorbent assays (ELISAs) (Bio-Rad Laboratories) and the Toxo-Screen direct agglutination assay (bioMérieux, Lyon, France). The BioPlex 2200 IgG assay was a sensitive (97.8%) and specific (91.3%) method for IgG detection and showed high specificity (97.4%) of IgM detection (Guigue et al. 2014).

Chemiluminescent Immunoassays (CLIA)

Chemiluminescent immunoassays are also automated techniques that use a similar immunological method to EIA but with alternative detection systems (chemiluminescence or electrochemiluminescence) (Petersen et al. 2005; Prusa et al. 2010; Sickinger et al. 2008).

The most modern systems, combining chemiluminescence (CL) and specific immunoreactions, enable the determination of the concentration of the analyte in relation to the intensity of emitted light, triggered by a chemical reaction. The reagents required for the reaction producing light may be coupled to antibodies or antigens. Chemiluminescent compounds employed have been luminol and isoluminol derivatives, and acridinium chemiluminogenic salts. For *T. gondii* detection, a CLIA was set up using an optimized acridinium label (AL) utilizing SAG2-GRA1-ROP1L chimeric antigens (Ferra et al. 2015a). This antigenic preparation was proved to be very useful in the detection of specific anti-*T. gondii* antibodies in sera of humans and different groups of livestock animals (horses, pigs, and sheep) (Ferra et al. 2015a; 2015b). In a later study by Holec-Gąsior et al. (2018), the same CLIA attached to a secondary antibody (IgG-AL) and SAG2-GRA1-ROP1L chimeric antigen for *T. gondii* was evaluated and compared in the detection of specific antibodies to conventional ELISA in 88 seropositive sera. According to the authors, the new CLIA assay proved to be more sensitive and had better differentiation of sera of patients with *T. gondii* infection from sera of healthy individuals (Holec-Gąsior et al. 2018).

Immunochromatography Tests (ICT) and Rapid Detection
Serological Tests (RDT)

Rapid diagnostic tests (RDTs), especially with point-of-care (POC) features, are promising diagnostic methods in clinical microbiology laboratories, especially in areas with minimal laboratory facilities (Khan and Noordin 2020). Several immunochromatography tests (ICTs) have been developed as RDTs for *T. gondii* using different antigens and formulations, such as colloidal gold-labeled antigens or antibodies as a tracer, and cellulose membrane as solid support. These are easy, fast tests that do not require specialized equipment. The antigens or antibodies are dropped on a pad on the nitrocellulose membrane, which slowly infiltrate by capillary action, reaching the conjugate area, then the antigen-antibody complexes will provide a color reaction (Liu et al. 2015). These rapid and easy-to-use kits have high accuracy and could be a suitable serodiagnostic tool for toxoplasmosis (Kim et al. 2017).

The first likely ICT test for *T. gondii* was developed using SAG2 as an antigen to detect antibodies in cats (Huang et al. 2004). The method was later developed to detect *T. gondii* anti-excretory/secretory circulating antigens IgG antibodies (Wang et al. 2011). Acute infections as early as 2–4 days post-infection were detected with a high agreement with ELISA in sensitivity and specificity (Wang et al. 2011). Also, in cats, using the recombinant SAG1 (rSAG1) antigen, Chong et al. (2011) developed an RDT which showed 0.88 kappa value compared with a commercialized ELISA kit, and an overall sensitivity and specificity of 100% (23/23) and 99.4% (158/159), respectively (Chong et al. 2011).

Another ICT also based on the major surface antigen (SAG1) linked with dense granule protein 2 (GRA2) (GST-GRA2-SAG1A loaded RDT or TgRDT) showed an overall specificity and sensitivity of 100% and 97.1% by comparison with ELISA using *T. gondii* whole cell lysates as the antigen (Song et al. 2013). The authors analyzed seroprevalence levels in residents in several areas of Korea using this method (Kim et al. 2017).

GRA7 and GRA14 were also used in a rapid ICT assay, which was compared to an indirect ELISA and SAG2 as a reference control in infected mice and to LAT in pigs (Terkawi et al. 2013). Results in pig sera examined with the ICT compared favorably with those of LAT and indirect ELISA for GRA7, with kappa values of 0.66 and 0.70 to 0.79, respectively. The authors concluded that the ICT based on GRA7 was a promising diagnostic tool for routine and mass screening of samples.

The *Toxoplasma* ICT IgG-IgM or ICT-IgG-IgM*bk* (LDBIO Diagnostic, Lyon, France) is a commercial rapid point-of-care (POC) test, which simultaneously detects *Toxoplasma*-specific IgG/IgM antibodies. At one end of the cassette, red latex particles conjugated with *T. gondii* antigen and blue latex particles conjugated with goat anti-rabbit antibody are adsorbed on fiberglass support. After dispensing the serum sample, if specific IgG/IgM is present, a red line will appear at the test line while a blue line will appear at the control line (Chapey et al. 2017). These tests could be used advantageously for the screening for toxoplasmosis in pregnant women (Chapey et al. 2017). The *Toxoplasma* ICT IgG-IgM was compared with the Architect automated chemiluminescence test. Diagnostic sensitivity and specificity were 97% and 96%, respectively. However, the test scored eight false-positive IgG and yielded negative results in three sera displaying unspecific IgM in the Architect test. Therefore, the LDBIO test appeared to be a more reliable first-line test, although the false-positive results for IgG deserve further investigation (Chapey et al. 2017). The LDBIO test was also compared in another study (Begeman et al. 2017). The authors found that the LDBIO test was highly sensitive (100%) and specific (100%) for distinguishing IgG/IgM-positive from negative sera and could be used as a reliable cost-saving POC test (Begeman et al. 2017).

Gomez et al. (2018) compared three commercial POC, ICT tests (*Toxo* IgG/IgM Rapid Test (Biopanda), OnSite *Toxo* IgG/IgM Combo-Rapid-*test* that detect IgG and IgM separately, and the *Toxoplasma* ICT-IgG-IgM*bk* (LDBIO) that detects either or both immunoglobulin IgG/IgM in combination). For the detection of *Toxoplasma* IgG, sensitivity was 100% for all three POC kits and specificity was also comparable (96.3%, 97.5%, and 98.8%, respectively). The POC kits did not exhibit cross-reactivity for false-positive *Toxoplasma*-IgM sera. However, diagnostic accuracy was significantly higher for the LDBIO-POC kit (Biopanda, 62.2%, OnSite, 28%, and LDBIO combined IgG/IgM, 100%). The LDBIO-POC test exhibited 100% sensitivity for the combined detection of IgG/IgM in acute and chronic *Toxoplasma* infection. Biopanda and Onsite POC tests exhibited poor sensitivity for *Toxoplasma*-IgM detection (Gomez et al. 2018).

The most recent ICT assay established was based on two monoclonal antibodies: TgSAG3-3A7 and TgSAG3-4D5 from the conserved protein of TgSAG3 that can be expressed in all the infective stages of *T. gondii*. Gold particles were prepared and conjugated with TgSAG3-3A7 Mab and TgSAG3-4D5 monoclonal antibodies were used as the capture antibody (Luo et al. 2018). Using porcine serum samples and compared to a commercial ELISA kit, the relative sensitivity and specificity of this ICT was 100% and 99.65%, respectively.

In cats and dogs, a new test combining the principles of immunochromatography and fluid dynamics (DFICT) was proposed (Jiang et al. 2015). The DFICT is performed by applying a 100 µL aliquot of liquid gold-SPA conjugate to the reagent hole and a 5 µL aliquot of a serum sample to the sample hole. The results were observable within 5 minutes by the naked eye. The lowest detectable limit of the assay was determined as the highest dilution (1:320) of positive serum. No cross-reaction of the antibodies with other related canine or feline pathogens was observed. The DFICT can be stored for 12 months at 4°C or 6 months at room temperature with no loss of sensitivity or specificity. A high degree of consistency was observed between the DFICT and the standard ELISA kit, supporting the reliability of the novel test strip. The introduction of a liquid gold nanoparticle conjugate reagent provides this method with several attractive characteristics, such as ease of manufacture, low sample volume requirements, high selectivity, and high efficiency. This method opens a novel pathway for rapid diagnostic screening and field analysis (Jiang et al. 2015). In addition to the method indicated in the above section, Wang et al. (2011) developed a fluid-phase immune assay for circulating antigens (see section 2.2.3).

Luciferase Immunoprecipitation System (LIPS) and Luciferase-Linked Antibody Capture Assay (LACA)

The luciferase immunoprecipitation system (LIPS) is a relatively simple, highly sensitive, and rapid quantitative immunoassay (Aye et al. 2018). The major advantages of this assay over ELISA are a wider dynamic range, shorter overall assay time, and less sample volume. Recombinant *Toxoplasma* antigens (dense granule antigens GRA6, GRA7, and GRA8 and bradyzoite antigen BAG1) were fused with nanoluciferase (Nluc, a small luciferase enzyme) and analyzed in sera from experimental mice infected with *T. gondii* and a WHO standard anti-*Toxoplasma* human immunoglobulin (TOXM) (Aye et al. 2018). The detection limits were estimated to be 3.9, 2, 1, and 1 IU/mL for rGRA6, rGRA7, rGRA8, and rBAG1, respectively. The LIPS assay for toxoplasmosis could detect antibodies against *T. gondii* in the mouse and human sera with a reasonably high sensitivity. A luciferase-LACA was recently established for the detection of *T. gondii* in chickens (Duong et al. 2020).

Piezoelectric Immunoagglutination Assay and Methods Using Biosensors

An agglutination test based on antigen-coated gold nanoparticles in the presence of antibodies detected by a piezoelectric biosensor device was developed for *T. gondii* using rabbit serum and blood (Wang et al. 2004). The antigen-coated nanoparticles (10 nm in diameter), in the presence of the corresponding antibody, causes a frequency change that is monitored by a piezoelectric device. Using a piezoelectric crystal, the immobilization of antibodies or antigen on the crystal is unnecessary. This is also an ICT that was sensitive to a dilution ratio of anti-*T. gondii* antibody as low as 1:5,500 and was found to be comparable to ELISA, as shown by a high (0.965) correlation (Wang et al. 2004).

A biosensor for luminescence was the basis for the development of a nanoscience plasmonic gold chip (pGOLD) with fluorescence enhancement in the near-infrared region for simultaneous detection of IgG, IgM, and IgA antibodies against *T. gondii* in a small volume of serum or whole blood (only ~ 1 µL) (Li et al. 2016). IgG antibody detection sensitivity, specificity, positive predictive value (PPV), and negative predictive value (NPV) were all 100%. IgM antibody detection achieved 97.6% sensitivity and 96.9% specificity with a 90.9% PPV and a 99.2% NPV. It requires very small blood volumes and could be implemented for screening of *T. gondii* infection during gestation. The approximate time for the detection of IgG, IgM, and IgA antibodies with a plasmonic gold chip is about 2 h for 10 US dollars per patient (Li et al. 2016) with about 1 µL serum sample. The pGOLD was used for the detection of IgG/IgM in sera samples from seroconverted, chronically infected, non-infected, and newborns (Pomares et al. 2017). Results were compared with commercial tests for the detection of IgG and IgM. The IgG and IgM test results on the platform agreed, respectively, at 95% and 93% with the commercial kits. When compared with the overall clinical interpretation of the serological profile, the agreement reached 99.5% and 97.7% for IgG and IgM, respectively. The multiplexed IgG/IgM test on pGOLD platform was considered a strong candidate for its use in the massive screening programs for toxoplasmosis during pregnancy (Pomares et al. 2017). The advantage of plasmonic biosensors is their potential to be miniaturized and multiplexed.

Laser-induced fluorescence (LIF) is another type of optical detection approach. A microfluidic immunosensor based on LIF has been developed for the quantitative detection of IgG antibodies against *T. gondii* in humans. Zinc oxide nanoparticles (ZnO-NPs) were covered with chitosan and then used to conjugate *T. gondii* antigens into the central microfluidic channel (Medawar-Aguilar et al. 2019). Serum samples containing anti-*T. gondii* IgG antibodies were injected into the immunosensor where they interacted immunologically with *T. gondii* antigens. Bound antibodies then were quantified by the addition of anti-IgG antibodies labeled with alkaline phosphatase (ALP). ALP enzymatically converts the non-fluorescent 4-methylumbelliferyl phosphate (4-MUP) to soluble fluorescent methylumbelliferone that is measured using excitation at 355 nm and emission at 440 nm. The relative fluorescent response of methylumbelliferone is proportional to the concentration of anti-*T. gondii* IgG antibodies (Medawar-Aguilar et al. 2019). Results acquired by the LIF immunosensor agreed with those obtained by ELISA, suggesting that the designed sensor represents a promising tool for the quantitative determination of anti-*T. gondii* IgG antibodies of clinical samples (Medawar-Aguilar et al. 2019).

Toxoplasma Antigens and Serotyping

Most ELISA and some other serological tests use *T. gondii* lysate antigen (TLA) for the specific anti-*T. gondii* antibodies diagnosis. Alternative sources of antigens are synthetic recombinant antigens or recombinant chimeric antigens of *T. gondii* obtained using genetic engineering and molecular biology (Drapała et al. 2015; Ferra et al. 2015a; 2015b; Ferra et al. 2020). Several *T. gondii* antigens with strong immunogenicity in mouse and human antibody profiles have been described such as GRA2, GRA3, GRA4, GRA5, GRA6, GRA7, GRA8, GRA14, MIC1, MIC2, and MAG1 (Döşkaya et al. 2018) with GRA6 and GRA7 being the antigens more widely used (Arranz-Solis et al. 2019).

Beghetto et al. (2006) evaluated the diagnostic utility of 6 antigenic regions of *T. gondii*, *MIC2*, *MIC3*, *M2AP*, *GRA3*, *GRA7*, and *SAG1* gene products, assembled in recombinant chimeric antigens by genetic engineering. They analyzed the new chimeric antigen in serum samples from adults with acquired *T. gondii* infection and from infants born to mothers with primary toxoplasmosis contracted during pregnancy, some of whom were congenitally infected. In their study, IgG and IgM Rec-ELISAs with individual chimeric antigens have performance characteristics comparable to those of the corresponding commercial assays. In fact, the IgM-capture assays based on chimeric antigens improved the ability to diagnose congenital toxoplasmosis postnatally compared with the ability to diagnose congenital toxoplasmosis using standard assays (Beghetto et al. 2006).

Serotyping based on the specific antibody-antigen recognition using polymorphic peptides was described as a typing method for *Toxoplasma* strains in several studies in humans (Kong et al. 2003) and in animals (Sousa et al. 2010; Maksimov et al. 2013; Ferra et al. 2015a; Holec-Gąsior et al. 2019). In an early study, Kong et al. (2003) developed an enzyme-linked immunosorbent assay for typing strains using SAG2A, GRA3, GRA6, and GRA7 as *T. gondii* antigens in mice and humans. In eight patients who had dye test titers > 64 and for whom the infecting strain type was known, the peptides correctly distinguished Type II from non-type II infections, and the ELISA analysis of a second group of 10 infected pregnant women from whom the parasite strain had not been isolated, gave a clear prediction of the strain type causing infection (Kong et al. 2003).

Different clonal types of *T. gondii* are thought to be associated with distinct clinical manifestations of infections. Serotyping is used since many host antibodies are raised against immunodominant parasite proteins that are highly polymorphic between strains (Arranz-Solis et al. 2019). Serotyping has been suggested as an inexpensive method for determining the strain type in *Toxoplasma* infections in humans and animals and correlating the genotype with disease outcome. According to several authors, serotyping may allow the determination of the clonal type of *T. gondii* infecting humans (Liu et al. 2015; Dard et al. 2016; Arranz-Solis et al. 2019) and to extend typing studies to larger populations which include infected but non-diseased individuals (Maksimov et al. 2012a). Maksimov et al. (2012a, b) established a peptide-microarray test for *T. gondii* serotyping. Human sera showed reactions against synthetic peptides with sequences specific for clonal Type II peptides. Type I and type III peptides were recognized by 42% (n = 73) or 16% (n = 28) of the human sera, respectively. On the other hand, a proportion of the sera (n = 22; 13%) showed no reaction with type-specific peptides. Individuals with acute toxoplasmosis reacted with a statistically significantly higher number of peptides as compared to individuals with latent *T. gondii* infection or seropositive patients (Maksimov et al. 2012a).

Serotyping was also used in patients with ocular toxoplasmosis (OT). The serotype in uveitis patients with OT were compared with non-OT seropositive patients with noninfectious autoimmune posterior uveitis in Germany (Shobab et al. 2013). A novel, nonreactive (NR) serotype was detected more frequently in serum samples of OT patients (50/114, 44%) than in non-OT patients (4/56, 7%). *Toxoplasma* NR and Type II serotypes were predominant in German OT patients (Shobab et al. 2013).

However, there seems to be cross-reactivity among the classical strains (Types I, II, and III) using the previous versions of the classical antigens; the serological assays could only reliably distinguish type 2 from non-type 2 infections (Arranz-Solis et al. 2019). Additional antigens ROP6, MIC12, SRS29A, and SRS13 were evaluated through IgM/IgG kinetics with well-categorized human sera for acute toxoplasmosis. The antigens were reactive with IgM and IgG antibodies and showed strong immunogenicity; these antigens could be potential antigens for vaccines or diagnostic serological kits (Döşkaya et al. 2018). Additional antigenic peptides have been selected, including several from rhoptry, dense granules, and surface proteins to improve *Toxoplasma* serotyping (Arranz-Solis et al. 2019). A redesigned version of the published GRA7 typing peptide performed better and specifically distinguished type III from non-type III infections in sera from mice, rabbits, and humans (Arranz-Solis et al. 2019). Xicoténcatl-García et al. (2019) also recently designed new peptides of GRA6, GRA7, and SAG1 proteins with more SNPs among the three clonal strains than those previously designed (Xicoténcatl-García et al. 2019). These authors screened the antigens and observed that in Mexico, serotype I was the most frequent (38%); additionally, this serotype was significantly more frequent among mothers who transmitted the infection to their offspring than among those who did not (53% vs. 8%, p = 0.04). They suggest that serotyping using the improved GRA6 peptide triad could be useful to serotype *T. gondii* in humans.

Serotyping still needs refinement in samples of animal origin (Sousa et al. 2010; Holec-Gąsior et al. 2019). In the first attempt to apply the serotyping technology to animal reference material, serotyping was used in naturally infected meat-producing animals. An ELISA test based on peptides derived from GRA6 and GRA7 antigens was used (Sousa et al. 2010). However, only three serum samples from 11 chickens and two from 15 pigs had serotyping results in agreement with genotyping (Sousa et al. 2010). In a study, cats mounted a clonal type-specific antibody response against *T. gondii,* and serotyping revealed for most seropositive field sera patterns resembling those observed after clonal Type II-*T. gondii* infection (Maksimov et al. 2013). In small ruminants, four tetravalent chimeric proteins (AMA1N-SAG2-GRA1-ROP1, AMA1C-SAG2-GRA1-ROP1, AMA1-SAG2-GRA1-ROP1, and SAG2-GRA1-ROP1-GRA2) were compared with an indirect IgG ELISA-based on a TLA in sera. All chimeric proteins were characterized by high specificity (between 96.4% to 100%), whereas the sensitivity of the IgG ELISA was variable (between 78.5% and 96.8%). The highest sensitivity was observed in the IgG ELISA test based on the AMA1-SAG2-GRA1-ROP1. This chimeric protein could be a promising serodiagnostic tool for *T. gondii* infection in small ruminants (Holec-Gąsior et al. 2019). Protein microarrays could be useful for the analysis of antibodies in meat juice and serum for high sample throughput for slaughtering animals (Loreck et al. 2020).

Serology in Saliva or Oral Fluids

Collecting blood by phlebotomy is the general approach used for serological diagnostic analysis; however, it is invasive and has several drawbacks (pain, vasovagal reactions, and anxiety). Saliva has been suggested as an alternative diagnostic source of antibodies against *T. gondii* (Chahed Bel-Ochi et al. 2013; Sampaio et al. 2014; 2019; Macre et al. 2019; Li et al. 2019).

Antibody detection in saliva using different methods has been compared to serological methods: rSAG1 based-ELISA (Chahed Bel-Ochi et al. 2013); Dot-ELISA and protein A IgG capture immunoassay (Sampaio et al. 2014); Sporozoite-specific embryogenesis-related protein (TgERP) as antigen-ELISA (Mangiavacchi et al. 2016); indirect ELISA using a crude extract of the RH strain (Cañedo-Solares et al. 2018); pGOLD a multiplexed serology assay for detection of *T. gondii* IgG and IgM, rubella and CMV IgG antibodies in serum, whole blood, and saliva using novel plasmonic gold (pGOLD chips) (Li et al. 2019); protein A solid phase capture assay for IgG reactive to biotinylated purified proteins (Sampaio et al. 2019); (Macre et al. 2019) Dot-ELISA (as in Sampaio et al. 2014).

The studies using saliva have been performed mainly in children in endemic and low-income areas where different aspects, such as the effect of age and the way of *T. gondii* transmission, were included (Mangiavacchi et al. 2016). However, the method was disappointing as a diagnostic tool because of the low concordance of results in saliva and serum samples in several studies (Sampaio et al. 2014; Cañedo-Solares et al. 2018). Some of those authors suggested that saliva might be reflecting a local immune response against this protozoan (Cañedo-Solares et al. 2018). In a recent study, using a multiplexed serology assay for the detection of *T. gondii* IgG and IgM, together with the analysis of rubella and CMV IgG antibodies in serum, whole blood, and saliva using novel plasmonic gold (pGOLD) chips (Li et al. 2019), good sensitivity and specificity was observed (100% for multiplex *T. gondii*, CMV, and rubella IgG and 100% and 95.4% for IgM, respectively), when compared to commercial test results in serum (Li et al. 2019).

Another study analyzed *T. gondii* antibodies in the saliva of experimentally and naturally exposed pigs to *T. gondii* by immunoblot (IB) for IgG and IgA antibodies against *T. gondii*-SAG1 antigen (Campero et al. 2020). The results showed that IB was not a robust matrix to assess the serological status for *T. gondii* in saliva from individual animals compared to serology; it also showed inconsistent results for some sows (Campero et al. 2020).

Immunoblotting

Immunoblotting or Western blotting is also serological tests used for immunological profiles for *T. gondii*. Currently, there are ready-to use nitrocellulose strips obtained from electro-transfer of an electrophoresis gel containing bands of *T. gondii* antigens. Like an indirect sandwich ELISA, the strips are sequentially incubated with tested serum, anti-isotype and enzyme-antibody conjugate and finally, a substrate that precipitates a colored band. IB has been used to study the immune response in newborns. Comparative IgG and IgM IB were associated with classical serological analysis (Rilling et al. 2003) but IB was more sensitive than ELISA in neonates and seemed to provide earlier detection of IgG in seroconversions when low IgG levels were observed. IB (*Toxo* II IgG) diagnosed toxoplasmosis seroconversion before two routinely used immunoassays (IA) (Platelia and Elecsys *Toxo* IgG) and an indirect IFAT (IB) in 92.3% of cases (36/39) and thus allowed for earlier therapeutic intervention (Jost et al. 2011). For these reasons it is becoming the reference confirmatory method, tending to replace the DT in this usage because of similar performance and greater practicability (Sickinger et al. 2008; Murat et al. 2013).

Indirect Methods by Immunological Tests for T-Cell Based
Responses for Toxoplasma gondii

The interferon-gamma release assay (IGRA) is a less commonly used test. It is an immunological T-cell-based test. It can be performed in whole blood. The method is still at the research stage and requires special training in cell culture for its development, but it could be helpful in the confirmation of highly suspected cases (Ciardelli et al. 2008). The IGRA method has been used for the early detection of *T. gondii* congenital infection (reviewed by Mahmoudi et al. 2017). Guglietta et al. (2007) reported increased levels of IFN-gamma (20- to 40-fold), in the presence of GST-GRA1, in subjects with acquired or congenital infection (Guglietta et al. 2007). There was no statistical difference in T-cell activation between individuals with acquired and congenital infections. However, considerable differences in subgroups of children with congenital toxoplasmosis (cases less than 4 years old and cases more than 4 years old) were found. In addition, the GRA1 stimulation index (SI) displayed significant differences when healthy adults with acquired infection and children with congenital infection less than 4 years old were compared. Recombinant protein, GST-cl16.2, was the most reactive protein inducing PBMC proliferation in 54% of samples from subjects with acquired infection and 70% of samples from patients with congenital toxoplasmosis (Guglietta et al. 2007). Ciardelli et al. (2008) compared routine tests (ELISA, ISAGA, and Western blot to evaluate IgM, IgA, and IgG antibodies to *T. gondii*) to IGRA. While routine tests detected only 52% of congenitally infected newborns at birth, the immunologic tests accurately distinguished infected from uninfected newborns and showed strong lymphocyte activation after *in vitro* culture with *T. gondii* antigens, even during the first days of life (Ciardelli et al. 2008). IFN-gamma production was found to be significantly higher in infected infants than in uninfected cases ($P < 0.001$) and the sensitivity and specificity of IGRA in these patients were 90.3% and 85.7%, respectively (Ciardelli et al. 2008).

Direct Methods for Detection of Parasite Antigens, Parasite DNA, or the Parasite Itself in Tissues and Fluids

Histologic and Microscopical Demonstration of Parasite in Tissues
and Other Samples

Direct detection of tachyzoites and/or tissue cysts of *T. gondii* in host tissues collected by biopsy or at necropsy can be demonstrated by histological techniques. A rapid diagnosis may be made by microscopic examination of tissue sections with lesions, usually after staining (e.g., Giemsa, Toluene, Diff-Quick stains, among others). Well-preserved *T. gondii* are crescent-shaped. In sections, the tachyzoites usually appear round to oval (Dubey 2010). Electron microscopy can aid in the diagnosis. *T. gondii* tachyzoites are always located in parasitophorous vacuoles. Tissue cysts are usually spherical and lack septa, and the cyst wall can be stained with a silver stain. The bradyzoites are strongly positive on periodic acid Schiff (PAS) staining. The diagnosis of *Toxoplasma* in exfoliative cytology specimens can be challenging since organisms are not well visualized on ThinPrep or Pap-stained material and in these materials, Wright-Giemsa staining can be particularly helpful (Monaco et al. 2012). Demonstration of tachyzoites in tissue sections or smears of body fluid (e.g., CSF or amniotic or BAL fluids) establishes the diagnosis of the acute infection, but it is infrequently performed (Murat et al. 2013a).

Immunohistochemical staining of parasites with fluorescent or other types of labeled *T. gondii* antisera can aid in diagnosis (Dubey 2010). The immunoperoxidase technique (IHQ) has proven both more sensitive and specific than stained tissue sections and can be used in unfixed or formalin-fixed paraffin-embedded tissue sections. The IHQ has been used to distinguish tachyzoites from bradyzoites in tissues using antigen BAG1 but will react with other protozoa (Dubey 2010). In humans, the test is mainly performed in some cases of pneumonia in immunosuppressed patients by detection of the parasite in bronchoalveolar lavage fluid (Monaco et al. 2012). However, it is often difficult to demonstrate *T. gondii* stages in conventionally stained tissue sections, and even with IHQ, sensitivity is much lower than PCR. In general, polyclonal antibodies are more useful than monoclonal antibodies for IHQ (Dubey 2010).

Isolation of T. gondii in Cell Culture and Bioassays in Animal Models

Direct demonstration of the organism by isolation of the parasite, which can be performed by inoculation in animal models (primarily mice) or inoculation in tissue cell cultures of virtually any human tissue or body fluid can be used for successful diagnose of *T. gondii* infection. As with detection in tissues, isolation of *T. gondii* from blood or body fluids establishes that the infection is acute (Murat et al. 2013a; Khan and Grigg 2017).

Vero cell lines were optimized for *T. gondii* cultivation and provided maximum yields and viability at 85% confluence in DEM medium, with an inoculum, of 1×10^7 tachyzoites after three days of culture (Saadatnia et al. 2010). The *in vitro* methods using cell culture cells, however, have practically been abandoned today (Murat et al. 2013a). Isolation in animal models *in vivo*, mainly mice by intraperitoneal infection, is more frequently performed than cell culture (Dubey 2010; Wallon et al. 2013). Mice can be inoculated subcutaneously or intraperitoneally to isolate live parasites from either serological or PCR-positive samples, heparinized blood, or cerebrospinal fluid of acutely or chronically infected tissues (Dubey 2010). Infected tissues need to be homogenized before infecting mice. Pepsin or trypsin treatment of chronically infected tissues has also been suggested in order to increase the probability of parasite DNA isolation (Dubey 2010). These methods can also be used as confirmation for infection and non-diagnostic purposes (strain amplification and storage and identification of strains in epidemiological studies).

Detection of Antigens

Detection of circulating antigens was suggested by some authors for diagnosis of active infections. Hassle et al. (1988) used a three-layer ELISA and the ELISA results were reexamined by IB. The antigenemia correlated with IgG and IgM antibody titers with clinical symptoms and with pathological findings in acute infections in AIDS patients. More recent studies reported the development of a rapid ICT for the rapid detection of *T. gondii* circulating antigens in the blood of animals during the acute stage of toxoplasmosis (Wang et al. 2011). In the infected animals, circulating antigens in sera could be detected as early as the second-day post-infection (PI) and in all animals by the fourth day. This test could be a powerful supplement to the current diagnostic methods (Wang et al. 2011).

Detection of Nucleic Acids by Molecular Methods

The detection of nucleic acid (mainly DNA from *T. gondii* in infected fluids or tissues) by molecular methods is becoming one of the most useful methods for the direct detection of *T. gondii*. There are several types of molecular methods for the diagnosis of toxoplasmosis described, such as conventional PCR, nested PCR, semi-nested PCR, quantitative real-time PCR (qPCR), and more recently, loop-mediated isothermal amplification (LAMP) and Recombinase Polymerase Amplification (RPA). Droplet digital PCR (ddPCR) is a new technique considered to resist inhibitors and to provide accurate quantitative information without the need for preparation of standard curves, but to our knowledge, has not been used for *T. gondii* analysis to date.

Conventional PCR and Quantitative PCR (qPCR)

PCR is an excellent method for the diagnosis and confirmation of *T. gondii* in infected tissues in animals and humans (Montoya 2002). In addition, genomic sequencing and typing methods based on previous PCRs are important for the characterization of *T. gondii* strains. PCR is also needed to distinguish DNA from *Toxoplasma* oocysts from other morphologically identical oocysts; also present in cat feces, such as *Hammondia hammondi*, which are not pathogenic (Schares et al. 2008).

Conventional PCR and qPCR, both have high sensitivity and specificity. Historically, Burg et al. (1989) first reported the detection of *T. gondii* DNA from a single tachyzoite by amplification of the B1 gene by PCR. Although other target genes have been used by PCR [P30, ITS1, 18S rRNA (this last one consider by some authors as an excellent marker for the identification of *T. gondii* infection (Robertson et al. 2019))], the repetitive regions of the 35-copy B1 gene (Burg et al. 1989), and 300-copy 529-bp repetitive element (529-bp RE) of *T. gondii* (Homan et al. 2000; Reischl et al. 2003) are still the most frequently used in humans and animals (Su and Dubey 2009). The 529-bp RE is 10–100 folds more sensitive than the B1 marker, and because of its high sensitivity has become a preferred marker for the detection of *T. gondii* in human and animal tissues (Edvinsson et al. 2006; Su and Dubey 2009; Darwich et al. 2012; Murat et al. 2013a; Mcleod et al. 2020). Bier et al. (2019) compared two real-time PCRs (qPCRs; Tg-qPCR1, Tg-qPCR2) and one conventional endpoint PCR (cPCR), all targeting the 529 repeated elements in DNA of the three clonal *T. gondii* types prevailing in Europe and North America. The qPCR efficiencies for all three clonal types ranged between 93.8 and 94.4% (Tg-qPCR1) and 94.3–95.6% (Tg-qPCR2). The cPCR using primer pair TOX5/Tox-8 was considered unsuitable for the detection of *T. gondii* DNA in pork as unspecific amplification of porcine DNA was observed.

In general, monoplex and multiplex PCR has been shown to be able to specifically identify *T. gondii* in many different tissue biopsies and fluids, including brain and cerebrospinal fluid (toxoplasmic encephalitis); in bronchoalveolar lavage (BAL) fluid (in patients with AIDS); in peripheral blood (in disseminated infections); in vitreous and aqueous fluids (in ocular toxoplasmosis); in fetal blood and amniotic fluid (in fetal toxoplasmosis) or in biopsies from immunocompromised patients undergoing hematopoietic stem cell transplantation (reviewed Montoya 2002; Weiss and Dubey 2009; Belfort et al. 2017).

Quantitative real-time PCR techniques (qPCR), in addition to high sensitivity and specificity, allow quantification of the parasite burdens in the infected samples, if standards are included in the reaction and allow fast detection with a low risk of laboratory contamination with amplicons. There are several technologies used for qPCR. Fluorescent resonance energy transfer (FRET) and hydrolysis probes, both seem to provide enhanced specificity compared to DNA-binding dyes such as SYBR Green (Murat et al. 2013a). An internal amplification control is advised to ensure the detection of PCR inhibition. As conventional PCR, qPCR for *T. gondii* is also mainly based on the B1 gene, 18 S rRNA gene, and 529-bp DNA sequences.

A commercialized qPCR is now available targeting the *T. gondii* 529-bp RE (Bio-Evolution *Toxoplasma gondii* assay). The commercial kit was evaluated using different specimens (including blood, bronchoalveolar fluids, and cerebrospinal fluids) from immunocompromised patients (Ammar et al. 2019). The concordance rate was 99.3% when comparing 529-bp RE laboratory-developed PCR methods and the commercial kit. The sensitivity and specificity of the commercial kit were calculated at 98.8% and 100%, respectively (Ammar et al. 2019).

For congenital infections, PCR on amniotic fluid is an accurate method for the diagnosis with high sensitivity and has shown a high specificity (about 100%), although its sensitivity varies with the date of infection. Sensitivities in first, second, and trimester were 33–75%, 80–97%, and 68–88%, respectively (Fallahi et al. 2018). For diagnosis in animals, the same PCR gene targets have been used (See animal studies section).

Loop-Mediated Isothermal Amplification (LAMP)

Isothermal techniques, such as LAMP are cost-effective potential alternatives to PCR, especially in developing countries. LAMP is a relatively novel and simple molecular technique for monitoring microbial pathogens. LAMP allows the amplification of small traces of the nucleic acid of interest at a constant temperature under isothermal conditions (60–65°C) using two or more sets of primers and a Bst DNA polymerase (large fragment) with both high strand displacement and replication activities that facilitates replication of the nucleic acid. The amplification product is detected at the end of the reaction by gel electrophoresis or in a real-time fashion by employing fluorophores, such as SYBR green that enable the detection of the amplicon at an interval of time (Tefera et al. 2018). LAMP is increasingly gaining interest. The technology requires less equipment and is more available than PCR (Lin et al. 2012; Fallahi et al. 2014; Fallahi et al. 2015); however, the contamination rate of LAMP seems higher than that of PCR (Fallahi et al. 2015).

The procedure of the LAMP assay is very simple, as the reaction would be carried out in a single tube under isothermal conditions and the result would be read out within 1 hour (as early as 35 minutes with loop primers) (Sun et al. 2017). Due to its simplicity, sensitivity, and specificity, LAMP is suggested as an appropriate method for routine diagnosis of active toxoplasmosis in humans, including the diagnoses of primary toxoplasmosis among high-risk pregnant women (Lau et al. 2010; El Aal et al. 2018). This method is also expected to play an important role in the monitoring of *T. gondii* contamination in various food products (Zhuo et al. 2015).

LAMP assays have been designed for several genes targets such as SAG1, SAG2, and B1 genes (Lau et al. 2010), B1 (Fallahi et al. 2014; El Aal et al. 2018; Valian et al. 2019), 529 bp RE (Fallahi et al. 2014; Sun et al. 2017; Valian et al. 2019), or ITS-1 (Zhuo et al. 2015). A commercial LAMP assay [Iam TOXO Q-LAMP or Iam TOXO (DiaSorin®)] is based on the 529-bp RE.

Historically, three LAMP assays based on the SAG1, SAG2, and B1 genes of *T. gondii* were developed by Lau et al. (2010). When the sensitivities and specificities of the LAMP assays were compared to those in a conventional nested PCR, the LAMPs were highly sensitive and had a detection limit of 0.1 tachyzoites, with no cross-reactivity to other parasites. The methods were used on blood samples from patients with active toxoplasmosis (n = 40), negative controls (n = 40), and patients with other parasitic infections (n = 15). The SAG2-based LAMP (SAG2-LAMP) had greater sensitivity (87.5%) than the SAG1-LAMP (80%), B1-LAMP (80%), and nested PCR (62.5%) and all assays and nested PCR were 100% specific. A LAMP and nested-PCR targeting the 529 bp RE and the B1 gene were compared to each other for the detection of *T. gondii* DNA in blood samples of children with leukemia (Fallahi et al. 2014). Another LAMP designed based on the conserved sequence of 529 bp repetitive fragment of *T. gondii* could detect a single tachyzoite or 10 copies of recombinant plasmid (Sun et al. 2017).

As mentioned above, a LAMP technique for *T. gondii* is commercially available [Iam TOXO Q-LAMP (DiaSorin®)] named later as Iam TOXO (Khan and Noordin 2020) based on the 529 bp RE. The method was evaluated and compared to a reference laboratory-developed method using qPCR in amniotic fluid (AF) and plasma samples and then in a cohort of AF, placental tissues, and blood (Varlet-Marie et al. 2017). Although the LAMP assay was less sensitive than the laboratory-developed method at very low parasite concentrations (0.1 *T. gondii* genome equivalents/mL), the two methods yielded identical results qualitatively and, in some instances, quantitatively, particularly for AF samples (Varlet-Marie et al. 2017). LAMP techniques have also been used in the peripheral blood of women who experienced spontaneous abortion (El Aal et al. 2018) and compared to real-time PCR. Both techniques obtained positive results in eight samples confirming primary toxoplasmosis.

The LAMP technique has also been applied for the detection of *T. gondii* in pork (Zhuo et al. 2015) as well as ready-to-eat-salads and was found to be more sensitive than conventional PCR (Lalle et al. 2018) (see more information on detection in fresh produce). In pork, when the assay targeted the ITS-1 sequence, the detection limit of the LAMP assay was 0.9 fg *T. gondii* genomic DNA, a sensitivity that was 10-fold higher than that of a conventional PCR assay. When both the LAMP assay and conventional PCR were applied to detect *T. gondii* genomic DNA in diaphragm samples from pig farms in Zhejiang Province, China, LAMP was more sensitive than conventional PCR (13.56% and 9.32%) (Zhuo et al. 2015).

Recently, *T. gondii* LAMP assays targeting the 529-bp RE and the B1 gene were designed using uracil DNA glycosylase in blood samples (Valian et al. 2019). When those techniques were compared with the qPCR method, a better detection rate was observed for toxoplasmosis among at-risk people using the 529 bp RE sequence compared to the B1 gene (Valian et al. 2019). Among 110 studied cases, 39 (35.45%) and 36 (32.7%) were positive for *T. gondii* DNA with the 529 bp RE-LAMP and B1-LAMP, respectively. However, there were false-negative results when compared with the real-time PCR method (Valian et al. 2019). Therefore, although the UDG-LAMP seemed a highly sensitive, accurate, and reliable method with no false-positive results for the diagnosis of *T. gondii* infection in blood specimens, some cases may be missed.

Recombinase Polymerase Amplification (RPA)

Recombinase polymerase amplification (RPA) is another nucleic isothermal acid amplification technology. By adding a reverse transcriptase enzyme to an RPA reaction, the technology can be used to detect RNA as well as DNA, without the need for a separate step to produce cDNA. The amplification product can be visualized using a lateral flow (LF) strip by adding a specific probe into the RPA reaction solution. RPA uses a combination of enzymes such as recombinase, polymerase, and single-stranded DNA-binding (SSB) protein in the cycle of nucleic acid amplification. The amplification is fast and is designed to work at temperatures between 37°C and 42°C. Besides, the specificity, sensitivity, portability, and the possibility of both endpoint and real-time detection make RPA an attractive alternative to other molecular methods. LF-RPA is a complex assay that has a higher efficiency and is more suitable to be used in field detection than conventional PCR (Wu et al. 2017).

The method was recently developed (B1-LF-RPA) for the detection and monitoring of *T. gondii* in the environment from soil and water samples based on the B1 gene of *T. gondii* (Wu et al. 2017). Applications of this method, in addition to the environment, could be used for diagnosis of *T. gondii* infections in humans and animals, epidemiological surveys of *T. gondii,* fundamental research of *T. gondii* in the laboratory and for the detection of the parasite in fresh produce (Wu et al. 2017).

Genotyping

In humans and animals, the clinical presentation of *T. gondii* infection varies widely depending on the strain of the parasite, the host-species susceptibility to the parasite, and other factors such as the immune status of the host, the infective dose or the parasite life-cycle stage ingested (Su and Dubey 2020).

There are two main multilocus techniques used for genotyping of *T. gondii.* The first genetic typing of *T. gondii* strains was performed using six multilocus PCR-restriction fragment length polymorphism (RFLP) (PCR-RFLP) markers as described by Howe and Sibley (1995). These markers were complemented to a multiplex multilocus PCR-RFLP method using 10 markers (Su et al. 2010). Currently, the 10 PCR-RFLP markers in a nested PCR include SAG1, SAG2, SAG3, BTUB, GRA6, c22-8, c29-2, L358, PK1, and Apico. This method has been used in numerous studies (Su et al. 2012; Shwab et al. 2014; Jiang et al. 2018; Su and Dubey 2020).

In addition, a second multilocus method was developed based on microsatellite (MS) analysis (Ajzenberg et al. 2010). The two genotyping techniques have different strain designations. Genotyping results obtained with the current multilocus MS or RFLP techniques seem to match results obtained by highly resolutive techniques, such as multilocus sequence typing (MLST) (Su et al. 2012) or whole genome sequencing (WGS) (Lorenzi et al. 2016). Most of these studies reported genotyping results of five to 15 markers from those two different techniques.

Other genotyping techniques described for *T. gondii,* but not commonly used in clinical samples, are multilocus sequence typing (MLST) and random amplified polymorphic DNA-PCR (RAPD-PCR). The MLST method, based on DNA polymorphisms, including SNPs (single nucleotide polymorphisms), deletions, and insertions has the highest resolution among all typing methods but only when enough genomic DNA is available (Su et al. 2010) which is not usual in clinical samples (Liu et al. 2015). RAPD-PCR can identify DNA polymorphisms without predetermined genetic data, using single short arbitrary primers under low stringency conditions. This PCR was used for the genetic differentiation of *T. gondii* from closely related organisms and to identify virulence markers. On the other hand, the band profiles in this technique are very difficult to reproduce and DNA must be highly pure, so cannot be used for clinical samples (Liu et al. 2015). Vilares et al. (2020) used a multilocus amplicon-based sequencing strategy targeting genome-dispersed polymorphic loci in samples associated with human infection and they concluded that a discrete loci panel has the potential to improve the molecular epidemiology of *T. gondii* toward better monitoring of circulating genotypes with clinical importance.

A supplementary test to conventional genotyping is High-Resolution Melting (HRM). It is a post-PCR method to analyze genetic variation based on the melting temperatures related to their sequences that is more informative than MS (Liu et al. 2015).

Although there are only a few lineages of *T. gondii*, the population structure of the parasite is complicated with several haplogroups worldwide (Su et al. 2012; Ajzenberg 2015). Even though there is a strong correlation between *T. gondii* strain genetic diversity with geographic origin, there is no correlation with host specificity or disease outcome (Ajzenberg 2015). Thus, to date, genetic markers used for genotyping *T. gondii* strains are very good to reveal the association of genotypes with geographical locations but there is no allele or allelic combination that can predict the course and severity of strains in human toxoplasmosis (Ajzenberg 2015; Darde et al. 2020).

Next-Generation Sequencing (NGS) and Comparative Genomics

Next-generation sequencing (NGS) opens the possibility of finding new markers for new techniques for the detection or improving the ones available. NGS can screen a single DNA sample and detect pathogen DNA from thousands of host DNA sequence reads, making it a versatile and informative tool for the investigation of pathogens in diseased animals (Cooper et al. 2016). ToxoDB, a free online resource that provides access to genomic and functional genomic data, was created by Kissinger et al. (2003). Currently, ToxoDB contains 32 fully sequenced and annotated genomes of *T. gondii* (Harb and Roos 2020). It also contains numerous functional genomic datasets including microarray, RNAseq, proteomics, ChIP-seq, and phenotypic data. ToxoDB has served as a model for other pathogens, resulting in its expansion into EuPathDB on the major eukaryotic pathogens (Boothroyd 2020).

WGS has been performed on 65 *T. gondii* strains, including Type I, Type II, Type III, and recombinant strains (reviewed by Lau et al. 2016; Lorenzi et al. 2016). Recently, proteomic and transcriptomic analyses of early and late-chronic *T. gondii* infection identified a novel stage-specific transcript, the bradyzoite-specific isoform of sporoAMA1; this shows novel and stage-specific transcripts (Garfoot et al. 2019). New markers could be used to improve *T. gondii* detection. The utilization of diverse omics-based methods can also help to identify promising drug targets (Cowell and Winzeler 2019). NGS studies have also been performed for comparative genomics (Reid et al. 2012; Lorenzi et al. 2016) and have also been used for confirmation of toxoplasmosis cases in animals, such as the case of a Risso's dolphin (*Grampus griseus*) (Cooper et al. 2016).

Imaging Techniques and Bioluminescence Imaging

When clinical signs suggest involvement of the CNS and/or spinal cord, tests should include computed tomography (CT) or magnetic resonance imaging (MRI) of the brain and/or spinal cord. Masamed et al. (2009) reviewed the previously described CT and T1-weighted (W) MRI target signs seen in toxoplasmosis and included a new imaging sign, the T2W/FLAIR (fluid attenuated inversion recovery) target as an aid to the diagnosis of cerebral toxoplasmosis (Masamed et al. 2009). Neuroimaging studies of the brain are also used in the support of therapy in cerebral toxoplasmosis (Dard et al. 2016; Rostami et al. 2018) and in ocular toxoplasmosis after swept-source optical coherence tomography angiography (SS-OCTA) (de Oliveira Dias et al. 2020).

Parasite growth in experimental infections in mice was followed using Bioluminescence imaging (Saeij et al. 2005). The parasite strains used in the intraperitoneal infection were engineered to stably express luciferase as a light-emitting protein and the progress of the infection was visualized noninvasively following the injection of a substrate for the luciferase enzyme. This method had the potential to study the dissemination and growth of different strains for the course of infection ad reactivation induced by immunosuppressants (Saeij et al. 2005).

Potential Future Diagnostic Methods (Not Developed to Date or Not Conventionally Used to Date)

Matrix-Assisted Laser Desorption Ionization Time-of-Flight Mass Spectrometry (MALDI-TOF MS) for Protein Profiles of Parasites

MALDI-TOF MS is a high-throughput analytical technique based on the detection of mass spectral fingerprints of proteins. It has many applications in different fields of science, including clinical and veterinary parasitology for the detection and identification of parasites. MALDI-TOF MS has been used for the characterization of the protein profile of parasites such as *Cryptosporidium* spp., *Giardia* spp., and *Entamoeba* spp.; to our knowledge, it has not been designed for *T. gondii* to date. Mass spectrometry (gas chromatography-mass spectrometry and liquid chromatography-mass spectrometry) were recently used to study the metabolite profiling (metabolomic) of *T. gondii* tachyzoites in cultures of human foreskin fibroblasts (King et al. 2020).

Aptamers

The latest progress made in aptamer use for parasite diagnosis confirmed that DNA and RNA aptamers represent attractive alternative molecules in the search for new tools to detect parasitic infections that affect human health worldwide (Ospina-Villa et al. 2018). A recent study reported an enzyme-linked aptamer assay (ELAA) for the detection of newly developed aptamers against *Toxoplasma* ROP18 protein in human serum (Varga-Montes et al. 2019). The authors found a significant association between ELAA test positive for human serum samples and severe congenital toxoplasmosis (p = 0.006) and considered that the development and testing of aptamers-based assays opens a window for low-cost and rapid tests looking for *T. gondii* biomarkers for human toxoplasmosis

Lab-on-Chip Methods

Currently, microfluidic chips are not ready for testing for *T. gondii* as they are still expensive due to manufacturing procedures, non-scalability, and the requirement of a microscopic syringe pump for operation. A device is still not available commercially for *Toxoplasma* detection.

Other Isothermal Reactions

There are several isothermal amplification techniques, which differ in the specifics of primer design and reaction mechanism such as nucleic acid sequence-based amplification (NASBA) (used to amplify RNA sequences), helicase-dependent amplification (HDA) (for DNA amplification), and nicking enzyme amplification reaction (NEAR) (also for DNA amplification) that to date have not been used for *T. gondii* detection.

Detection of Oocysts in Cats and in the Environment

Autofluorescence

Autofluorescence is a useful characteristic in the microscopic detection of *T. gondii* oocysts (Lindquist et al. 2003). *Toxoplasma gondii* autofluorescence is of enough intensity and duration to allow the identification of these oocysts from complex microscopic sample backgrounds. Oocysts glow pale blue when illuminated with an ultraviolet (UV) light source and viewed with the correct UV excitation and emission filter set (Lindquist et al. 2003). The intense autofluorescence of *T. gondii* oocysts follows a distinctive pattern with the sporocyst walls being readily visible; this pattern is distinct from the pattern of autofluorescence seen in *Cyclospora cayetanensis* (Lindquist et al. 2003) and could be used in detection in water and other complex mixtures.

Detection of T. gondii Oocysts in Cats

Oocysts are very difficult to find in cats. At any given time only 1% of cats will be shedding oocysts (Dubey and Beattie 1988) and they will need to be detected in the feces of infected cats by concentration methods (e.g., flotation in high density sucrose solution) since there may be too few present to be detected by direct smear (Ruiz and Frenkel 1980a). Schares et al. (2008) examined microscopically 24,106 fecal samples from cats from Germany and other European countries and found oocysts with a morphology like that of *H. hammondi* and *T. gondii* in only 74 samples (0.31%). Therefore, for epidemiological surveys, the detection of *T. gondii* oocysts in cat feces is impractical and not very informative (Dubey 2004). In one study, oocysts were detected microscopically in only 12.7% and an additional 87.3% of naturally infected cats by mouse assay (Ruiz and Frenkel 1980a). For definitive identification, *T. gondii* oocysts should be sporulated and then bioassayed in mice to distinguish them from other related coccidians since *T. gondii* oocysts cannot be distinguished by direct microscopic examination from the oocysts of at least four other coccidians, *H. hammondi*, *H. heydorni*, *Neospora caninum*, and *Besnoitia* species (Dubey 2004).

Sroka et al. (2018) compared different saturated solutions for the concentration of oocysts using centrifugal flotation in saccharose, $MgSO_4$, $ZnSO_4$, and $NaNO_3$ in water samples spiked with *T. gondii* tachyzoites and oocysts. They found the highest efficiency in oocysts detection using sodium nitrate solution ($NaNO_3$) and saccharose. They used $NaNO_3$ flotation followed by DNA extraction with the removal of inhibitors to compare real-time PCR and nested PCR and found the best results for the detection of *T. gondii* as observed by real-time PCR targeting the B1 gene compared to nested PCR. They did not include an analysis of other genes.

In a study in Southern Thailand, Chemoh et al. (2016) observed 19.3% positive samples of 254 feline fecal specimens to the presence of coccidian oocysts. When samples were analyzed by 529 bp PCR and ITS-1 PCR for *T. gondii*, only 0.8% of samples were positive by the first method, and only 6.67% positive by the second; none of the positive samples by PCR were microscopically positive. Veronesi et al. (2017) analyzed fecal samples from owned cats in Italy; they observed that biomolecular approaches were more sensitive than microscopic detection (16 samples PCR positive versus only two samples positive by microscopy). When they compared two amplification protocols (B1 and the 529-bp RE) for the molecular diagnosis of *Toxoplasma* infection in fecal samples from 78 owned cats, after sucrose flotation, the two stool samples microscopically positive for *T. gondii*-like oocysts also tested positive by both B1 and 529-bp RE-PCRs. The amplification sets targeting B1 and 529-bp RE showed substantially different yields, but while 529-bp RE was a standard conventional PCR, B1 was a nested PCR.

The determination of serological prevalence could be more helpful to show contact with *T. gondii* in cats than searching for oocysts in fecal samples (Dubey 2004). In an epidemiological survey for *T. gondii* on pig farms, oocysts were detected in only 5 of 274 (1.8%) samples of cat feces, 2 of 491 (0.4%) samples of feed and 1 of 79 (1.3%) samples of soil, but 267 of 391 (68.3%) cats had antibodies to *T. gondii* (Dubey et al. 1995). Seropositive cats will have already shed *T. gondii* oocysts (Dubey and Frenkel 1972). Serologic surveys found that approximately up to 50% of cats surveyed in the United States have antibodies to *T. gondii*, and most of these cats probably ceased shedding oocysts (Dubey 2004; Dubey 2010).

Detection of *T. gondii* Oocysts in the Environment

Oocysts are a major source of infection for humans and animals. The sporulated oocysts that can infect definitive and IH are very resistant to environmental conditions and can persist in the environment for long periods (Tenter et al. 2000; Dubey 2010; Galvani et al. 2019). Equally, oocysts are highly resistant to the various chemical inactivation processes commonly used by water supply systems (Galvani et al. 2019). Currently, there are no commercial reagents available to detect *T. gondii* oocysts in the environment. Hohweyer et al. (2016), developed an immunomagnetic separation assay (IMS *Toxo*), using a specific purified monoclonal antibody. This IMS *Toxo* coupled with microscopic and qPCR analyses was evaluated in raspberries and basil (Hohweyer et al. 2016). Due to the intrinsic characteristics of the matrixes, few oocysts are expected to be detected in environmental samples, including soil, water, and fresh produce.

Oocysts Detection in Water

Detection of *T. gondii* oocysts in water is more difficult than that of other coccidian oocysts and there are no standardized methods (Dubey et al. 2020e). The concentration of oocysts by centrifugation, filtration through small pore filters, elution of oocysts from filters, immunomagnetic separation, and fluorescence-activated cell sorting have been suggested based on experiences with the detection of the related coccidians, *Cryptosporidium,* and *Giardia* species (Dubey 2004). As indicated above, *T. gondii* oocysts have a specific pattern of autofluorescence that may be useful in identification (Lindquist et al. 2003). However, since *T. gondii* oocysts cannot be distinguished by direct microscopic examination from at least four other coccidians, additional methods will be needed to confirm that oocysts are from *T. gondii*. PCR will help in that differentiation in any of the methods available [conventional PCR, qPCR, LAMP, and RPA (Liu et al. 2015; Wu et al. 2017; Galvani et al. 2019)]. Villena et al. (2004) found *T. gondii* DNA in 10 of 125 environmental water samples but none were positive by bioassay in mice.

Wells et al. (2015) compared the molecular detection of *T. gondii* in water samples from Scotland using the 529-bp RE qPCR and ITS1 nested PCR. *T. gondii* DNA was detected in 8.8% of samples using 529-bp RE qPCR, and of those only 50% were positive by ITS1 nested PCR, less than 4.4% of total samples. Another technique developed for the detection of DNA from *T. gondii* oocysts used in soil and water was an RPA method in combination with an LF strip for the detection of DNA of *T. gondii* oocysts (Wu et al. 2017). DNA of *T. gondii* oocysts was amplified by a pair of specific primers based on the *T. gondii* B1 gene over 15 minutes at a constant temperature ranging from 30°C to 45°C. The amplification product was visualized by the LF strip within 5 minutes using the specific probe added to the RPA reaction system. The sensitivity of the established assay was 10 times higher than that of nested PCR with a lower detection limit of 0.1 oocyst per reaction, and there was no cross-reactivity with other closely related protozoan species (Wu et al. 2017). When sample detection (50) was compared using LF-RPA assay with nested PCR based on the B1 gene sequence, both agreed showing 5 out of the 50 environmental samples were positive. The B1-LF-RPA method was also proven to be sufficiently tolerant of existing inhibitors in the environment (Wu et al. 2017).

A real-time PCR technique (qPCR) performed after concentration by filtration in Envirocheck® HV capsules of volumes of 20 L used a 62-base-pair fragment of the B1 gene as the target sequence in a recent study in Brazil (Galvani et al. 2019). Characteristics of the samples and climatic conditions produced very different mean recoveries in samples from the rainy season [3.2% (SD ± 3.2)] and in the dry period [62.0% (SD ± 6.2) (Galvani et al. 2019)].

Recently, a rapid detection method for infectious oocysts by cell culture of their sporocysts combined with qPCR (sporocyst-CC-qPCR) was assessed (Rousseau et al. 2019b). This sporocyst-based CC-qPCR appeared to be a good alternative to the mouse bioassay for monitoring infectious *T. gondii* oocysts directly in water and using biosentinel mussel species (blue and zebra mussels). The method was able to detect fewer than 10 infectious oocysts in water within four days (one day of contact and three days of cell culture), compared to four weeks by mouse bioassay and as low as ten infective oocysts in experimentally contaminated mussels (Rousseau et al. 2019b).

Previous studies detected oocyst DNA in mussels, which as filter-feeders can accumulate and concentrate *T. gondii* oocysts, using conventional PCR (Arkush et al. 2003; Villena et al. 2004), qPCR (Coupe et al. 2019), or LAMP (Durand et al. 2020). Arkush et al. (2003) reported the detection of DNA in tissues of mussels that were experimentally contaminated with *T. gondii* oocysts. They found *T. gondii* DNA up to 18 days post-exposure of mussels; however, viable oocysts were detected only for 3 days of exposure (Arkush et al. 2003). Detection of *T. gondii* in different bivalves has also been achieved by qPCR targeting the B1 gene and the 529 bp (reviewed by Edvinsson et al. 2020). Recently, Durand et al. (2020), using a LAMP in experimentally spiked mussels, detected 5 oocysts/g in tissue and 5 oocyst/mL in hemolymph, which could make this method a promising alternative to qPCR. Coupe et al. (2019) analyzed detection methods in filter-feeding shellfish in green-lipped mussel (*Perna canaliculus*) hemolymph using oocyst spiking experiments and suggested that the 529-bp RE qPCR assay may be preferable for future mussel studies, but direct sequencing is required for definitive confirmation of *T. gondii* DNA detection. The lowest limit of detection was 5 oocysts using 529-bp RE qPCR assays, with a good correlation between oocyst concentrations and Cq values, and an acceptable efficiency. Both qPCR assays were sensitive to *T. gondii*, but cross-reacted with *Sarcocystis* spp. and the 529-bp RE primers also cross-reacted with *N. caninum* DNA. In a study in mussels farmed or sold at retail outlets in Italy, Tedde et al. (2019) investigated the occurrence and seasonality of zoonotic protozoans, including *T. gondii,* and reported for the first-time *T. gondii* found in *M. galloprovincialis* in Italy and *M. edulis* in Europe. Based on experimental exposure of several protozoans to zebra mussels (*Dreissena polymorpha*), showed that the percentage of *T. gondii*-positive mussels reflected the contamination level in the freshwater and could be used for biomonitoring of aquatic ecosystems (Geba et al. 2019).

Oocysts Detection in Soil

Although *T. gondii* oocysts have been isolated from soil, there is not a simple reliable method for large-scale epidemiological use. The method used by Wu et al. (2017) in water was also applied to soil samples. Because feral chickens on small farms feed from the ground, finding *T. gondii* in chickens may be a better indicator of infection in the environment (Dubey 2004). In a study of feral chickens, *T. gondii* was isolated from 54% of 50 chickens by bioassay in mice (Ruiz and Frenkel 1980b). Serological surveys and isolation of viable *T. gondii* in free-ranging chickens were used to assess environmental contamination with oocysts (Dubey 2010).

A recent study in China analyzed the presence and genotype distribution of *T. gondii* DNA in soil samples (Cong et al. 2020). Soil samples collected from farms and parks had the highest prevalence. Using PCR assays for 529-bp RE and ITS-1 gene sequences were more sensitive than the B1 gene-based assay. Positive PCR products were genotyped using multilocus PCR-RFLP, and *Toxo*DB #9 was the predominant genotype found in the contaminated soil samples from six geographic regions in that country, which is also one of the most prevalent genotypes in China (ToxoDB#9) (Cong et al. 2020). Detection of as low as one oocyst/g in soil was achieved by a protocol established by Escotte-Binet et al. (2019).

Oocysts Detection in Fresh Produce

An association between *T. gondii* infection and the consumption of unwashed raw fruits and vegetables contaminated with oocysts has been reported and the increasing habit to eat pre-washed ready-to-eat salads poses a new potential risk for consumers (Lalle et al. 2018). Most methods for detection of *T. gondii*, as with other parasites, in fresh produce are largely based on the direct identification of the parasitic stages by microscopy (which is difficult since few oocysts are expected) or detection of the nucleic acids of the parasites by molecular techniques, which is the most sensitive method. For the recovery of parasites in fresh produce, the wash step and washing solutions used are crucial to detach the parasitic forms. A low number of parasitic forms will be expected to be found and there are PCR inhibitors (Tefera et al. 2018; Almeria et al. 2019; Berrouch et al. 2020).

Molecular detection of *T. gondii* in vegetables and fruits by molecular techniques has been achieved by conventional PCR targeting the B1 gene or the 529 bp RE with positive results in lettuce, chicory, rocket, and parsley from production sites and stores, in both organic and nonorganic samples in South America (Marchioro et al. 2016); *T. gondii* DNA was amplified with the primers B22-B23 in strawberries and crisphead lettuce samples contaminated by dripping and when DNA extraction was carried out after freeze-thaw cycles or ultrasound in Brazil (de Souza et al. 2016).

Quantitative PCR has been used in several studies and multiplex qPCR methods have been developed (Lalonde and Gajadhar 2011; Hohweyer et al. 2016; Temesgen et al. 2019; Shapiro et al. 2019). The detection limit of a qPCR targeting the B1 locus in artificially infected radish (amount of sample tested not reported) was 100 oocysts (Lass et al. 2012). Lalonde and Gajadhar (2011) developed a very sensitive qPCR assay using melting curve analysis (MCA) to detect, differentiate, and identify DNA from *Cryptosporidium parvum*, *T. gondii*, *C. cayetanensis*, *Eimeria bovis*, *Eimeria acervulina*, *Cystoisospora suis*, and *Sarcocystis cruzi* using qPCR with SYBR Green detection and this qPCR assay consistently detected DNA from as few as 10 *T. gondii* oocysts. The method was optimized and validated on leafy green vegetables and berry fruits (Lalonde and Gajadhar 2016a). As few as 3 oocysts per gram of fruit or 5 oocysts per gram of herbs or green onions could reliably be detected using the optimized method (Lalonde and Gajadhar 2016a). The method was used in a survey in Canada on imported leafy green vegetables and *T. gondii* was identified in three samples of baby spinach (origin USA or USA and Mexico) and was the first finding of *T. gondii* in leafy greens in North America (Lalonde and Gajadhar 2016b). Another multiplex qPCR was developed for the detection of *Echinococcus multilocularis*, *T. gondii*, and *C. cayetanensis* on berries (Temesgen et al. 2019). The limit of detection was estimated to be 10 oocysts for *Toxoplasma* per 30 g of raspberries or blueberries (Temesgen et al. 2019). An additional, multiplex assay was evaluated in spiked spinach for simultaneous detection of *Cryptosporidium*, *Giardia*, *C. cayetanensis*, and *T. gondii* followed by parasite differentiation via either a nested-specific PCR or a restriction fragment length polymorphism (RFLP) assay. The lowest limits of detection using the nested mPCR assay were 1–10 (oo)cysts/g spinach (in 10 g samples processed), and this method proved more sensitive than qPCR for parasite detection (Shapiro et al. 2019).

Hohweyer et al. (2016) developed a new immunomagnetic separation assay (IMS *Toxo*) using a specific purified monoclonal antibody. This IMS *Toxo* coupled with microscopic and qPCR targeting 529 bp RE was evaluated in raspberries and basil. The limit of detection on a simple matrix was 5 oocysts and, in both matrixes (raspberries and basil) 33 oocysts/g, and recovery rates were between 0.2 and 35% (Hohweyer et al. 2016).

A LAMP assay, targeting the 529-bp RE locus, could detect 25 oocysts/50 g in ready-to-eat baby lettuce *T. gondii* oocysts (Lalle et al. 2018). The qPCR method is faster than LAMP because the latter requires visualization of the amplification products. However, the comparable sensitivity of the two assays and the cheapest equipment required for LAMP makes this a valuable molecular test to be performed, also in a resource-limited setting (Lalle et al. 2018).

Recently, Lass et al. (2019), using a real-time PCR assay targeting the B1 gene and multilocus genotyping, detected 10 (3.6%) positive samples of 279 fresh vegetable samples [lettuce (*Lactuca sativa*) (5/71: 7.0%), spinach (*Spinacia oleracea*) (2/50: 4.0%), Bok choy (*Brassica rapa* subsp. *chinensis*) (1/34: 2.9%), rape (*Brassica napus*) (1/22: 4.5), red cabbage (*Brassica oleracea*) (1/8: 1.2%)] tested in open markets in the Qinghai province in China. Of those, eight were *T. gondii* type I and the remaining two *T. gondii* Type II. Based on quantitative real-time PCR (qPCR) oocysts per sample ranged between less than one and 27,000 oocysts, with the majority not exceeding a few oocysts per sample.

The molecular methods mentioned above cannot differentiate viable oocysts from dead/noninfective ones in any environmental sample. Incubation at 22°C with propidium monoazide (PMA) coupled to qPCR targeting the 529-bp RE was used to discriminate dead and viable oocysts (Rousseau et al. 2019a). The principle is that PMA binding to DNA would inhibit PCR amplification in dead but not viable oocysts. Untreated and heat-killed oocysts incubated with PMA were differentiated by this qPCR. However, the reduction of viability by heating at high temperatures was slight and qPCR was not suppressed by heat, underestimating the efficacy of this treatment (Rousseau et al. 2019a).

Methodologies for the Study of T. gondii as a Model Organism

Toxoplasma is also considered as a model organism for cellular, biochemical, molecular, and genetic studies for other clinically important apicomplexan organisms such as malaria parasites (*Plasmodium* spp.) due to their easy growth in culture and availability of a broad array of genetics tools for the genetic manipulation of the *T. gondii* genome (Roos et al. 1992; Piro et al. 2020). The molecular tools regularly used to manipulate this parasite include both forward and reverse genetics, including transfection, transformation, and gene knockout mutagenesis (reviewed by Sidik et al. 2014; Wang et al. 2016; Piro et al. 2020).

Some of those studies could be helpful in the identification of novel targets of therapeutic, diagnostic, and immunoprophylactic interests (Ma et al. 2019). These authors analyzed the potential antigenicity of *T. gondii* ME49 ES proteins using an Abundance of Antigenic Regions (AAR) values at RNA and microarray levels for those purposes (Ma et al. 2019). In a recent review of these methodologies, Boothroyd (2020) compiled the research methods and developments in the study of *T. gondii* during the last 30 years, including studies of parasite organelles, genes, immune response, parasite proteins, and interactions with the host. Calarco et al. (2020) reviewed sequence variants in clinically important protozoan parasites, including *T. gondii*. In addition, Piro et al. (2020) developed a simple and fast method to screen single clones of *T. gondii* directly from the 96-well plates without previous parasite expansion or time-consuming genomic extraction. This approach would permit screening at an earlier point than previously for the assessment of gene functions (Piro et al. 2020).

Detection in Clinical Situations in Humans

Diagnosis in Immunocompetent Patients

Most cases of *T. gondii* infection in adults and children are asymptomatic; however, with respect to primary infections, around 10%–20% of otherwise immunocompetent individuals will present some symptoms (Moncada and Montoya 2012). Lymphadenopathy is the most common manifestation with less common symptoms being chorioretinitis, myocarditis, and/or polymyositis among others (Moncada and Montoya 2012).

On immunocompetent patients, initial serological testing should include analysis for the presence of IgG and IgM anti-*T. gondii* antibodies with a second specimen analyzed 3–4 weeks apart (in parallel) (Moncada and Montoya 2012). Negative results in both tests virtually rule out toxoplasmosis. A single high titer of any immunoglobulin is insufficient to make the diagnosis. Acute infection is supported by seroconversion of IgG and IgM antibodies or a greater than the four-fold rise in IgG antibody titer in sera run in parallel at both times (Moncada and Montoya 2012).

In lymphadenitis, the histological analysis could complement serology. In cases of myocarditis and polymyositis in immunocompetent patients, endomyocardial biopsy and biopsy of skeletal muscle have been successfully used to establish *T. gondii* as the etiologic agent. On the other hand, parasite isolation studies and PCR have rarely proven useful for diagnosis in immunocompetent patients (Moncada and Montoya 2012).

Immunosuppressed Patients

Immunosuppressed patients might include those with cancer, HIV infection, long term-treatments with corticosteroids, hematologic malignancies (lymphomas), and transplant recipients (hematopoietic stem cells or solid organ transplant) (including heart, lung, liver, or kidney).

Toxoplasma gondii infection is a severe problem in cancer patients; integrated measures should be conducted to prevent and control *T. gondii* infection in those patients (Cong et al. 2015). In patients with neoplasia, such as lymphoma, leukemia, or multiple myeloma, a high percentage of seropositivity was detected (Yazar et al. 2004) and the authors recommended periodical parasitological surveys of those patients. Similarly, in a case-control study of 900 cancer patients and 900 control individuals, the prevalence of anti-*T. gondii* IgG in cancer patients by ELISA (35.56%) was significantly higher than that in controls (17.44%). The highest *T. gondii* seroprevalence was detected in lung cancer patients (60.94%), followed by cervical cancer patients (50%), brain cancer patients (42.31%), and endometrial cancer patients (41.67%). Exposure to contaminated soil and consumption of raw/undercooked contaminated meat was significantly associated with *T. gondii* infection in cancer patients. Three *T. gondii* genotypes (ToxoDB#9, ToxoDB#10, and Type I variant) were identified (Cong et al. 2015). Huang et al. (2016) and Gharavi et al. (2017) also suggested that *T. gondii* infection might be a main risk factor for leukemia patients, but further studies are needed to confirm this conclusion.

In suspected immunosuppressed patients infected by *T. gondii*, the initial assessment should include a serological test for specific antibodies. Those with a positive result are at risk of reactivation of the infection (Moncada and Montoya 2012). The reactivation of *T. gondii* in immunosuppressed patients can lead to more widespread forms and increased mortality (Villard et al. 2016b) and is the most common cause of toxoplasmosis in immunosuppressed patients, with exception of those with a heart transplant in which the main risk of developing disease is to acquire the infection from a seropositive donor (Murat et al. 2013a). In previous seronegative patients, seroconversion indicates primary infection, while in previously seropositive patients increases in IgG and/or IgM may mean reactivation. A negative result, however, does not completely exclude infection because immunosuppressed patients may not have IgM or IgG antibodies, even in the event of a reactivation, while having latent infections. Therefore, serological results are to be considered with caution, and additional methods of detection used as confirmation are strongly recommended. In cases of negative serology, direct evidence of the parasite by PCR is recommended in blood and body fluids (bronchoalveolar lavage fluid, tissue biopsies, bone marrow), (cerebrospinal fluid (CSF), or vitreous and aqueous humor), according to symptoms and accessibility of the lesion (Murat et al. 2013a; Ozgonul and Bersirli 2017), preferably by qPCR. In disseminated infections analysis of blood, CSF, and/or bronchoalveolar lavage PCRs are recommended (Dard et al. 2016). Biopsies and histological examination of available tissues using *T. gondii* immunohistochemistry analysis and qPCR are also recommended (Dard et al. 2016).

When clinical signs suggest the involvement of the CNS and/or spinal cord, tests should include CT or magnetic resonance imaging (MRI) of the brain and/or spinal cord. This clinical and radiologic response is also used in the support of therapy in cerebral toxoplasmosis. In immunosuppressed patients, toxoplasmosis encephalitis is the main lesion (Kaplan et al. 2009), and other organs commonly involved in immunocompromised patients with toxoplasmosis are the lungs, eyes, and heart. Fatal toxoplasmosis cases with no encephalitic symptoms have been associated with fulminant myocarditis and pneumonitis (Eza and Lucas 2006). In those cases, toxoplasmosis was only diagnosed on autopsy (Eza and Lucas 2006).

In cases of encephalitis, PCR can be performed in brain biopsies (Kaplan et al. 2009), but this is a very invasive procedure, and PCR in CSF shows poor sensitivity (Kaplan et al. 2009). If a brain biopsy is not feasible, a lumbar puncture should be considered if it is safe to perform. In myopericarditis, PCR in the blood can help in the diagnosis of acute toxoplasmosis as an alternative to serology (Leveque et al. 2019).

In solid transplants the treatment administered to prevent organ rejection causes profound immunosuppression, thus the patient is exposed to the additional risk of reactivation of tissue cysts contained in the transplanted organ (Murat et al. 2013a). Chemotherapy should be prescribed to the seronegative recipient if the donor was seropositive. Serological status of donor and recipient should be performed at least for cardiac donors (Fischer et al. 2009).

Ocular Toxoplasmosis

Toxoplasma gondii infection can have ocular manifestations known as ocular toxoplasmosis. In ocular toxoplasmosis, diagnosis can be reached by classic ophthalmic examination (funduscopic examination) because ocular lesions are often distinctive consisting of chorioretinal scars (Montoya 2002; Murat et al. 2013a; Ozgonul and Besirli 2017; Khan and Khan 2018; Greigert et al. 2019). A toxoplasmosis scar can be associated with severe visual field loss when it occurs close to the optic disk (Ozgonul and Besirli 2017). Congenitally infected newborns, who are asymptomatic at birth, are at high risk of developing ocular lesions in childhood and adolescence (Gilbert and Stanford 2000).

Atypical forms, however, have misleading symptoms that may require confirmation of the diagnosis by different combinations of biological methods (Greigert et al. 2019). Currently, the detection of *Toxoplasma*-specific antibodies or DNA of the parasite in ocular specimens is the main basis of the diagnosis (Maenz et al. 2014; Gomez-Marin and de-la-Torre 2020). Serological tests including serum anti-*Toxoplasma* titers of IgM and IgG may be needed to support the diagnosis. Seropositivity for *T. gondii* infection indicates previous systemic exposure to the parasite but does not confirm the diagnosis of ocular toxoplasmosis. Seronegativity will exclude *T. gondii* as the origin of the disease, but positive serological evidence of antibodies is not predictive of active ocular toxoplasmosis (Gomez-Marin and de-la-Torre 2020).

The detection of parasite DNA is better performed in aqueous humor sampling by anterior chamber paracentesis. Real-time PCR assays in aqueous humor samples showed relative success (Simon et al. 2004; Choi et al. 2020). Of the 23 clinically toxoplasmosis suspect patients, 22 showed serological evidence of exposure to *Toxoplasma*; one had a serological profile indicative of active infection. The analysis of paired aqueous humor and serum samples revealed an intraocular antibody production in 9 of 23 cases (39.1%). The quantitative real-time PCR revealed positive and high parasite numbers and high *Toxoplasma*/human genome ratios in three cases. Furthermore, PCR was the only positive confirmatory test in two cases (11.1%) (Simon et al. 2004).

The Goldmann-Witmer coefficient (GWC), has been used for intraocular antibody production analysis. This coefficient compares the *Toxoplasma*-specific antibodies in ocular fluids and serum, as does a calculation based on the ratio of total IgG antibodies in ocular aqueous humor divided by that of peripheral blood. Although a ratio > 2 should indicate intraocular antibody production, this may also occur in healthy controls, and therefore a ratio of at least three is often used to confirm the diagnosis (De Groot-Mijnes et al. 2006). Another similar coefficient is the Candolfi coefficient (Greigert et al. 2019). Comparative IB has been evaluated as an alternative to calculating immune load (Robert-Gangneux et al. 2004). Rothova et al. (2008) compared the efficiency of PCR to GWC in the aqueous humor of patients with toxoplasmic chorioretinitis. Rothova et al. (2008) reported a GWC sensitivity of 57% in immunocompromised patients, whereas the sensitivity was 93% in immunocompetent patients, showing that GWC was a more sensitive test. PCR was negative in 84% of toxoplasmic chorioretinitis patients, in contrast to

7% for GWC. In another study, the combination of GWC with PCR significantly improved the diagnostic sensitivity from 81 to 93% (Fekkar et al. 2008). Sugita et al. (2011) established a two-2-step PCR protocol to improve OT diagnoses by qPCR. By using this method, it was possible to detect a very small amount of DNA in small amounts of ocular samples with a sensitivity of 85%. Belfort et al. (2017) used qPCR 529 bp RE in peripheral blood in patients with different forms of uveitis. Patients with acute toxoplasmosis (Belfort et al. 2017) could not detect *T. gondii* DNA in peripheral blood.

In a recent cross-sectional study of patients in Iran with suspected active ocular toxoplasmosis (Arshadi et al. 2019), the clinical manifestations, serological analysis (IgG and IgM analyzed by CIT and ELISA, and IgG avidity test), and molecular detection (B1 gene) were highly correlated in the diagnosis of ocular toxoplasmosis. Ocular toxoplasmosis showed no significant correlation with gender, age, behavior, occupation, or education (Arshadi et al. 2019).

A recent study also observed significant differences in the clinical characteristics of ocular toxoplasmosis according to serum IgM status. IgM$^+$ patients were older, less likely to report pain, and had lower levels of intraocular inflammation, but were more likely to have macular involvement. Age was found to be correlated with larger and more peripheral lesions (Ajamil-Rodanes et al. 2020).

In summary, accurate identification of the disease is difficult by a single method and many authors support the implementation of combined strategies to increase the possibility of adequate diagnosis.

Congenital Toxoplasmosis in Pregnant Women and Follow-Up in Newborns

Pregnant Women

Toxoplasma is vertically transmitted to the fetus through the placenta. If primary infection occurs in a pregnant woman during pregnancy, a risk of abnormalities in the fetus and even abortion can occur. Establishing primary *T. gondii* infection early in the first trimester of pregnancy is of critical importance for medical intervention to minimize transmission and damage to the fetus (Montoya 2002; Murat et al. 2013a; Khan and Khan 2018).

The diagnosis of a *T. gondii* acute infection in a pregnant woman can be made by detecting antibodies if it is known that the woman was previously seronegative. A second sample should be collected 2–4 weeks after the first sample. The presence of IgG in absence of IgM before conception or at the beginning of pregnancy (first trimester) assures fetal protection against the parasite and excludes acute infection in the last 6 months (Ozgonul and Besirli 2017). In immunocompetent women presence of IgG before pregnancy indicates a low risk for transplacental transmission (Ozgonul and Besirli 2017). On the other hand, seronegative women (negative IgG and IgM) are not immunized against the parasite and vertical transmission to the fetus in pregnancy. They are advised to avoid undercooked meat consumption or to have contact with cat feces to avoid *T. gondii* infection during pregnancy and serological follow-up should be done one month after delivery (Ozgonul and Besirli 2017). Prenatal treatment has been associated with a decrease in the transmission rate and an improvement in children's clinical outcomes (Wallon et al. 2013). According to the French national program, prenatal screening for *Toxoplasma* infection and treatment allows outstanding cost savings (Prusa et al. 2017).

When a case of positivity to both IgG and IgM occurs, this may be due to nascent toxoplasmosis seroconversion, non-specific IgM reaction, or residual IgM (Boquel et al. 2018). An IgG avidity test is recommended since it can provide information regarding the time of exposure. An avidity test with a high index in the first trimester indicates that the infection was acquired before conception because high-avidity IgG antibodies take 3–4 months to appear and exclude recent infections acquired during pregnancy. However, the serum sample needs to be tested before four months of pregnancy. The avidity test performed later is not useful (Murat et al. 2013a). Low-avidity IgG antibodies should not be used to confirm the diagnosis of a recent infection due to the persistence of these antibodies for many months after the acute infection (Ozgonul and Besirli 2017). In cases of IgG results by ELISA with very low antibody titers (equivocal zone) confirmatory tests (DT or IB for IgG and ISAGA for IgM) are recommended (Dard et al. 2016).

In cases of seroconversion during pregnancy, detection of IgM (by ELISA or a confirmatory test such as ISAGA) with or without IgG in a previously negative patient implies infection of less than one month (if monthly sampled). Further sampling needs to be done until IgG antibodies are detected and in these patients, treatment should be proposed to the mother until delivery (Dard et al. 2016; Ozgonul and Besirli 2017).

Diagnosis in Newborn or Congenitally Infected Babies

Prenatal diagnosis of congenital toxoplasmosis in women can be based on ultrasonography and amniocentesis (Prusa et al. 2015; Fallahi et al. 2018) since intracranial calcification, microcephaly, hydrocephalus, ascites, hepato-splenomegaly, or severe growth restriction of the fetus are detectable by ultrasound scanning (Lopes et al. 2007; Paquet et al. 2013). To decline the risk of maternal to fetal transmission, it is suggested that monthly ultrasonography of the fetus be applied throughout gestation (Moncada and Montoya 2012). Other techniques that can help in the diagnosis of fetal toxoplasmosis are CT and magnetic resonance imaging (MRI) but these are also not specific. Ultrasonography is used in prenatal diagnosis and CT can detect lesions in infants and MRI is considered more suitable for the evaluation of the extent of damage (Liu et al. 2015). If the fetal ultrasound is abnormal, then amniocentesis and collection of AF should be performed and direct detection of the parasite by qPCR and/or mouse inoculation is the elected test to diagnose the infection. If there is a positive PCR or if mice become seropositive, there is fetal infection (Year et al. 2009; Wallon et al. 2010).

Vera et al. (2020) reported congenital toxoplasmosis in a four-week-old male neonate with a history of intermittent hypothermia. The infant presented with an acute onset of bilateral lower extremity paralysis and areflexia, and eosinophilic encephalomyelitis with spinal cord hemorrhage (Vera et al. 2020). The infant had IgG by the dye test, IgA ELISA, and IgM ISAGA. His mother acquired the infection during gestation as evidenced by a maternal IgG dye test result, IgM, IgA, and IgE by ELISA, and low IgG avidity. At the six-month follow-up, the infant had marginal improvement in his retinal lesions and paraplegia of the lower extremities. An MRI demonstrated encephalomalacia with possible cortical laminar necrosis and spinal cord atrophy in the areas of previous hemorrhage (Vera et al. 2020).

The presence or absence of IgM antibodies, titers of IgG antibodies, and antibody kinetics may help in the determination of the recency of infection (Montoya 2002; Weiss and Dubey 2009; Murat et al. 2013a). Detection of maternal anti-*Toxoplasma* IgG by 12 months of age is the gold standard, and the combination of IgA and IgM antibodies results are also suitable for serological diagnosis of toxoplasmosis in newborns (Moncada and Montoya 2012). IgA ELISA and IgM ISAGA are the preferred methods to detect of anti-*Toxoplasma* antibodies in infants (Moncada and Montoya 2012; Pomares and Montoya 2016; Dard et al. 2016). For discrimination between the infection of newborns and maternal contamination, IgA testing should be repeated about 10 days after birth. In addition, IB can separate maternal antibodies from fetal and/or infant antibodies (Pinon et al. 2001). A PCR assay in combination with serological tests will help in the definitive diagnosis of congenital toxoplasmosis in infants. Direct isolation of the parasite or amplification of the parasite-specific DNA using PCR in CSF fluid, peripheral blood, and urine can be useful for early diagnosis of congenital toxoplasmosis (Olariu et al. 2014).

If the diagnosis is performed at delivery, collection of samples from the placenta and/or cord blood is recommended. Testing will be then performed as in AF by qPCR and/or mouse inoculation (Robert-Gangneux et al. 2011). During postnatal follow-up, neonatal serology is carried out at birth on blood samples and umbilical cord for neosynthesized antibodies IgM, IgG, and IgA between 3–10 days of life. Congenital toxoplasmosis is characterized by specific IgG after the first year of life. IgG crosses the placenta whereas IgM and IgA do not, but they may contaminate a newborn's blood during labor (Pinon et al. 2001). Children should be tested for IgG and IgM until one year old or until IgG levels become undetectable. The decrease in specific IgG levels down to a negative level before 1 year of age is strong evidence of the absence of congenital infection (Murat et al. 2013a). Because of false-negative results associated with fetal diagnosis, all children of mothers with acute toxoplasmosis must be tested for the possibility of congenital infection (Fallahi et al. 2018). Serological and clinical examinations are the most common for the detection of congenital toxoplasmosis in newborns. Obtaining the clinical history, physical examination, and pediatric neurologic and ophthalmologic examination is mandatory for newborns who are suspected to have congenital toxoplasmosis (Moncada and Montoya 2012).

The Executive Council of the Obstetricians and Gynecologists of Canada recommended that amniocentesis and PCR should be offered if the maternal primary infection is diagnosed, or if serology cannot confirm or exclude acute infection or in presence of abnormal ultrasound findings (Paquet et al. 2013). A recent study also recommended a combination of serological and PCR methods (Yamada et al. 2019). Congenital *T. gondii* infection screening using IgG avidity and multiplex-nested PCR methods for pregnant women with a positive test for *T. gondii* antibody plus a positive or equivocal test for *T. gondii* IgM was useful for detecting a high-risk pregnancy and diagnosing congenital *T. gondii* infection.

Detection in Animals and Meat for Human Consumption

The consumption of raw or uncooked meat from infected animals (mostly ovine, caprine, and pork) is considered one of the major transmission routes for *T. gondii* infection in humans (Dubey et al. 2005; Hill et al. 2006). In addition, the parasite causes important economic losses linked to reproductive disorders in susceptible animal species and, therefore, is of animal health concern in farm animals. Small ruminants and pigs are very susceptible to *T. gondii* infection; many epidemiological studies, including the study of risk factors associated with seroprevalence against the parasite, have been performed in these species, as well as in many other species, including wildlife species, worldwide (Dubey 2010; Stelzer et al. 2019).

Like in humans, in animals, a great variety of serological assays are available for *T. gondii* antibody detection such as IFAT, agglutination test (MAT), ELISA, or IB, among others. Of those, MAT and ELISA are probably the most commonly used with some commercially available tests (*Toxo*-Spot IF, bioMérieux, France) such as ELISA kits (PrioCHECK® *Toxoplasma* Ab SR, Prionics Schlieren-Zurich, Switzerland; Safepath Laboratory, Carlsbad, CA). The IFAT is a well-established technique for detecting anti-*T. gondii* antibodies in different animal species but require conjugate and are not automated. The use of some of the assays in animals is more limited since species-specific secondary antibodies and conjugates are often not available for many species, and for that reason, many studies, particularly in wildlife species, rely on competitive ELISA techniques (cELISA) and agglutination tests (MAT) that do not need species-specific secondary antibodies. For competitive ELISAs, the principle of competition makes this test theoretically possible to be used in any other species; validation data are not yet available for many species, however (Almeria 2015). The specificity, sensitivity, and cut-off value of serological tests has not been evaluated in many animal species. Therefore, confirmation of the results by several tests should be implemented, including molecular techniques. Molecular techniques include conventional and nested PCR and qPCR as in humans. LAMP technology has also been applied for the rapid detection of *T. gondii* in pork (Zhuo et al. 2015) (see LAMP section).

If abortion rates in a herd/flock are high, diagnostics should also include testing for *T. gondii*, and an initial step in the diagnosis should include the serological status of the animals in the herd to identify high-risk herds or animals. In aborted dams and fetuses, a positive serological result in fluids from an aborted fetus in ruminants would confirm infection by the parasite, since antibodies do not cross the placenta. Confirmation by PCR in fetal tissues would be advised in other species, but presence of parasite DNA will need to be related to the general status of *T. gondii* infection in the whole herd, and elimination of other causes of abortion in the same herd. The absence of antibody in the aborted dam excludes toxoplasmosis but a positive result does not prove etiology.

A recent European research project focusing on detection of *T. gondii* in farm animals (Opsteegh et al. 2016), concluded that with the currently available serological methods for pigs, poultry, and small ruminants serological screening can be used to identify high-risk herds or animals. However, a negative result in an indirect test cannot be used to declare that the meat is safe. On the other hand, in cattle and horses, MAT-based detection of antibodies, and possibly serological screening in general, are not recommended as an indicator of the presence of viable *T. gondii,* and direct detection methods are preferred. In fact, the detection of the presence of the parasite in the tissues of a seropositive animal bioassay is still the reference method to confirm parasite viability. However, bioassays are expensive, time-consuming and have some ethical problems, so cell culture methods are currently preferred (Opsteege et al. 2020; Rousseau et al. 2019b). Other alternatives such as various highly sensitive and specific PCR methods have mainly proved not to have sufficient sensitivity due to the limited amount of sample to be tested, low tissue cyst density, and random distribution of tissue cysts, particularly when low numbers of *T. gondii* tissue cysts are in those tissues (Hill et al. 2016).

Detection of the parasite in meat from species for human consumption is important. Some alternative methods using cardiac fluid for toxoplasmosis surveys in meat have been evaluated (Halos et al. 2010; Villena et al. 2012). To improve detection in meat samples a direct detection and genotyping of *T. gondii* using magnetic capture (MC) and PCR for detection and quantification of *T. gondii* was developed (MC-qPCR) (Opsteegh et al. 2010). The method involved the preparation of crude DNA extract from 100 g samples of meat, MC of *T. gondii* DNA, and quantitative real-time PCR targeting the *T. gondii* 529-bp RE. The detection limit of this assay was approximately 230 tachyzoites per 100 g of meat sample. Importantly, the results obtained with the PCR method were comparable to the bioassay results from experimentally infected pigs, and to serological findings in sheep (Opsteegh et al. 2010). This PCR method could be an alternative to bioassay for the detection and genotyping of *T. gondii*, and to quantify the organism in meat samples from various sources. The MC-qPCR method was evaluated, among others, in experimentally infected goats (Juránková et al. 2013), serrano cured pork ham (Gomez-Samblas et al. 2015), chickens in field conditions (Schares et al. 2018), experimentally infected calves (Burrells et al. 2018) and in calves and adult cattle in natural conditions (Opsteegh et al. 2019).

Serological results of chicken sera by ELISA, IFAT, and MAT showed good performance in identifying chickens that were positive using either a mouse bioassay, MC-qPCR, or on acidic pepsin digests (PD-qPCR), showing diagnostic sensitivities of 87.5%, 87.5%, and 65.2%, respectively, and diagnostic specificities of 86.2%, 82.8%, and 100%, respectively (Schares et al. 2018). However, a combination of methods should be performed in cattle at least since one individual test will not provide an answer as to whether a calf harbors *T. gondii* tissue cysts in experimental infections (Burrells et al. 2018) and in samples collected from naturally infected calves and adult cattle in several countries (Opsteegh et al. 2019). When a selection of individual tissues, previously used in the mouse bioassay, was examined by MC-qPCR, parasite DNA could only be detected from two animals, despite all calves showing seroconversion after infection (Burrells et al. 2018). A study showed a lack of concordance among the bioassay and MC-qPCR in veal calves and adult cattle collected in Italy, the Netherlands, Romania, and the United Kingdom. Some cattle that tested positive in the bioassay tested negative by MC-qPCR and vice-versa. The methods used were not reliable indicators of the presence of *T. gondii* parasites or DNA in cattle (Opsteegh et al. 2019).

Gisbert Algaba et al. (2017) improved the MC-PCR by adding the co-capture of cellular r18S as a means of tracking the extraction and as a non-competitive PCR inhibition control, by making the release of the target DNA from the streptavidin-coated paramagnetic beads UV-dependent (Gisbert Algaba et al. 2017). The modified MC-PCR could be an alternative to the mouse bioassay for the screening of various types of tissues and meat with the additional advantage of being quantitative. The authors improved efficiency by reducing incubation times and by reducing the cost through a comparison of reagents (99% limit of detection: 65.4 tachyzoites per 100 g of meat), and once optimized the method was subjected to an ISO 17025 validation with pork as the main matrix (Gisbert Algaba et al. 2017). In organic pigs, a positive result was obtained by MC-qPCR for *T. gondii* in 14 out of the 92 hearts sampled; parasites were isolated by mouse bioassay, from 9 of these 14 samples, demonstrating the presence of viable *T. gondii* in animals intended for human consumption (Gisbert Algaba et al. 2020).

Some recent studies have been performed for detecting the presence of the parasite in animal species for human consumption. A recent study confirmed small ruminants' meat as a possible source of *T. gondii* infection for consumers eating raw or undercooked meat, particularly in those countries where the consumption of sheep and goats' meat products is a traditional gastronomic habit (Gazzonis et al. 2020). In that study, meat juices from small ruminants slaughtered or commercialized in Italy were analyzed by a commercial ELISA, and the muscles of positive samples were analyzed by PCR and sequencing. *T. gondii* DNA was detected in 15 sheep and three goats and shown by sequencing of the B1 gene to be Type II. The presence of the parasite was also observed in backyard pigs intended for familial consumption in Romania (Paştiu et al. 2019). The animals were serologically analyzed by IFAT while heart samples were analyzed by PCR targeting the 529-bp repeat region. The *T. gondii* isolates were genotyped by the analysis of 15 MS markers. In addition, heart samples from IFAT-positive animals were bioassayed in mice and observed for the viable parasite in the seropositive animals (Paştiu et al. 2019). Tissue samples from wild boars from southern Italy revealed the high prevalence and parasite load (Santoro et al. 2019), while in a national survey of 750 randomly selected samples of fresh, unfrozen lamb and 750 samples of fresh and unfrozen pork from retail meat stores in the USA, and low prevalence of viable *T. gondii* infection (two positive lamb samples and one positive pork sample) was observed using the mouse bioassay (Dubey et al. 2020f).

Rabbits are hunted and consumed in some countries. A recent study analyzing brain and heart samples from 470 slaughtered domestic rabbits in Central China showed the occurrence rate of *T. gondii* DNA was 2.8% by nested PCR. The frequency of infection was not related to sex, breed, or region. One of the samples was identified as ToxoDB genotype #9 (Qian et al. 2019). New genotypes and mixed infections in feral cats and atypical new genotypes of *T. gondii* and mixed infections in stray dogs were found in Chiapas, Mexico, by quantitative real-time PCR (qPCR) and endpoint PCR and PCR-RFLP genotyping (Valenzuela-Moreno et al. 2020). In China, meat from sheep and goats collected from rural markets (16.04%) had a significantly higher *T. gondii* prevalence than those collected from supermarkets (6.84%) (p < 0.001) and sheep and goats raised in backyards were more likely to be infected by *T. gondii* compared with those raised on farms (p < 0.001) (Ai et al. 2019).

A recent study tested cervical lymph node samples of horses from northern China for the presence of the *T. gondii* B1 gene by semi-nested PCR. The B1-positive samples were genotyped at nine nuclear loci using PCR-RFLP. 6.1% of 231 samples were *T. gondii* positive. Only two were successfully genotyped at all loci; five were successfully genotyped at five to eight loci, and all typed samples belonged to *Toxo*DB genotype number 9 (Ren et al. 2019). In retail raw meat in Poland, including cured bacon, raw or smoked sausages, ham, and minced meat, Sroka et al. (2019) digested the samples using a pepsin solution and performed nested and real-time PCR (B1 gene). In the selected B1-positive samples, multiplex PCR was performed using several genetic markers. The percentages of positive results for meat products—sausages, smoked meat products, ham, and minced meat—ranged from 4.5% to 5.8% and the differences between them were not significant. We would like to emphasize that detection of DNA does not equate with infectivity.

Summary and Future Needs

There have been several developments and advancements in *T. gondii* detection methods in the last decade. New multiplex testing for prenatal care, the design of several RDTs, and the use of new molecular techniques such as LAMP are paradigms in the diagnosis and control of the parasite. Future needs should include careful serologic screening during gestation to diagnose primary infection in the pregnant woman, nontoxic medicines to eliminate encysted bradyzoites and tachyzoites, and a vaccine to prevent the infection in humans (McLeod et al. 2020). Correct diagnosis of toxoplasmosis is necessary for the control and prevention of this important disease of public and animal health importance.

References

Abo-Al-Ela, H.G. 2019. Toxoplasmosis and psychiatric and neurological disorders: a step toward understanding parasite pathogenesis. ACS Chem. Neurosci. Jul 3. doi: 10.1021/acschemneuro.9b00245.

Ajamil-Rodanes, S., J. Luis, R. Bourkiza, B. Girling, A. Rees, C. Cosgrove, C. Pavesio et al. 2020. Ocular toxoplasmosis: phenotype differences between toxoplasma IgM positive and IgM negative patients in a large cohort. Br. J. Ophthalmol. Apr 28: bjophthalmol-2019-315522. doi: 10.1136/bjophthalmol-2019-315522.

Ajzenberg, D., F. Collinet, A. Mercier, P. Vignoles and M.L. Dardé. 2010. Genotyping of *Toxoplasma gondii* isolates with 15 microsatellite markers in a single multiplex PCR assay. J. Clin. Microbiol. 48: 4641–4645. doi: 10.1128/JCM.01152-10.

Ajzenberg, D. 2015. 20 years of *Toxoplasma gondii* genotyping. Future Microbiol. 10: 689–691.

Ai, K., C.Q. Huang, J.J. Guo, H. Cong, S.Y. He, C.X. Zhou and W. Cong. 2019. Molecular detection of *Toxoplasma gondii* in the slaughter sheep and goats from Shandong Province, Eastern China. Vector Borne Zoonotic Dis. Sep 23. doi: 10.1089/vbz.2019.2488.

Ammar, N.A., H. Year, J. Bigot, F. Botterel, C. Hennequin and J. Guitard. 2019. Multicentric evaluation of the bio-evolution *Toxoplasma gondii* assay for the detection of *Toxoplasma* DNA in immunocompromised patients. J. Clin. Microbiol. Dec 4. pii: JCM.01231-19. doi: 10.1128/JCM.01231-19.

Arkush, K.D., M.A. Mille, C.M. Leutenegger, I.A. Gardner, A.E. Packham, A.R. Heckeroth, A.M. Tenter, B.C. Barr and P.A. Conrad. 2003. Molecular and bioassay-based detection of *Toxoplasma gondii* oocyst uptake by mussels (*Mytilus galloprovincialis*). Int. J. Parasitol. 33: 1087–1097.

Arranz-Solís, D., C. Cordeiro, L.H. Young, M.L. Dardé, A.G. Commodaro, M.E. Grigg and J.P.J. Saeij. 2019. Serotyping of *Toxoplasma gondii* infection using peptide membrane arrays. Front. Cell. Infect. Microbiol. 9: 408. doi: 10.3389/fcimb.2019.00408.

Arshadi, M., L. Akhlaghi, A.R. Meamar, L. Alizadeh Ghavidel, K. Nasiri, M. Mahami-Oskouei, F. Mousavi, Z. Rampisheh, M. Khanmohammadi and E. Razmjou. 2019. Sero-molecular detection, multi-locus genotyping, and clinical manifestations of ocular toxoplasmosis in patients in northwest Iran. Trans. R. Soc. Trop. Med. Hyg. 113: 195–202. doi: 10.1093/trstmh/try137.

Aye, K.M., E. Nagayasu, M. Baba, A. Yoshida, Y. Takashima and H. Maruyama. 2018. Evaluation of LIPS (luciferase immunoprecipitation system) for serodiagnosis of toxoplasmosis. J. Immunol. Methods 462: 91–100.

Batz, M.B., S. Hoffmann and J.G. Jr. Morris. 2012. Ranking the disease burden of 14 pathogens in food sources in the United States using attribution data from outbreak investigations and expert elicitation. J. Food Prot. 75: 1278–1291.

Begeman, I.J., J. Lykins, Y. Zhou, B.S. Lai, P. Levigne, K. El Bissati, K. Boyer, S. Withers, F. Clouser, A.G. Noble, P. Rabiah, C.N. Swishe, P.T. Heydemann, D.G. Contopoulos-Ioannidis, J.G. Montoya, Y. Maldonado, R. Ramirez, C. Press, E. Stillwaggon, F. Peyron and R. McLeod. 2017. Point-of-care testing for *Toxoplasma gondii* IgG/IgM using *Toxoplasma* ICT IgG-IgM test with sera from the United States and implications for developing countries. PLoS Negl. Trop. Dis. 11: e0005670. doi: 10.1371/journal.pntd.0005670.

Beghetto, E., A. Spadoni, L. Bruno, W. Buffolano and N. Gargano. 2006. Chimeric antigens of *Toxoplasma gondii*: toward standardization of toxoplasmosis serodiagnosis using recombinant products. J. Clin. Microbiol. 44: 2133–2140.

Belfort, R.N., J. Isenberg, B.F. Fernandes, S. Di Cesare, R. Jr. Belfort and M.N. Jr. Burnier. 2017. Evaluating the presence of *Toxoplasma gondii* in peripheral blood of patients with diverse forms of uveitis. Int. Ophthalmol. 37: 19–23. doi:10.1007/s10792-016-0221-8.

Berrouch, S., S. Escotte-Binet, R. Harrak, A. Huguenin, P. Flori, L. Favennec, I. Villena et al. 2020. Detection methods and prevalence of transmission stages of *Toxoplasma gondii*, *Giardia duodenalis* and *Cryptosporidium* spp. in fresh vegetables: a review. Parasitol. 147: 516–532. doi: 10.1017/S0031182020000086.

Bier, N.S., G. Schares, A. Johne, A. Martin, K. Nöckler and A. Mayer-Scholl. 2019. A. Performance of three molecular methods for detection of *Toxoplasma gondii* in pork. Food Waterborne Parasitol. 14: e00038. doi:10.1016/j.fawpar.2019.e00038.

Boquel, F., L. Monpierre, S. Imbert, F. Touafek, R. Courtin, R. Piarroux and L. Paris. 2019. Interpretation of very low avidity indices acquired with the Liaison XL *Toxo* IgG avidity assay in dating toxoplasmosis infection. Eur. J. Clin. Microbiol. Infect. Dis. 38: 253–257. doi: 10.1007/s10096-018-3421-5.

Bowen, L.N., B. Smith, D. Reich, M. Quezado and A. Nath. 2016. HIV-associated opportunistic CNS infections: pathophysiology, diagnosis and treatment. Nat. Rev. Neurol. 12: 662–674. doi: 10.1038/nrneurol.2016.149.

Burg, J.L., C.M. Grover, P. Pouletty and J.C. Boothroyd. 1989. Direct and sensitive detection of a pathogenic protozoan, *Toxoplasma gondii*, by polymerase chain reaction. J. Clin. Microbiol. 27: 1787–1792.

Burrells, A., A. Taroda, M. Opsteegh, G. Schares, J. Benavides, C. Dam-Deisz, P.M. Bartley, F. Chianini, I. Villena, J. van der Giessen, E.A. Innes and F. Katzer. 2018. Detection and dissemination of

Toxoplasma gondii in experimentally infected calves, a single test does not tell the whole story. Parasit. Vectors 11: 45. doi: 10.1186/s13071-018-2632-z.

Buxton, D., S.W. Maley, S.E. Wright, S. Rodger, P. Bartley and E.A. Innes. 2007. *Toxoplasma gondii* and ovine toxoplasmosis: new aspects of an old story. Vet. Parasitol. 149: 25–28.

Calarco, L., J. Barratt and J. Ellis. 2020. Detecting sequence variants in clinically important protozoan parasites. Int. J. Parasitol. 50: 1–18. doi: 10.1016/j.ijpara.2019.10.004.

Campero, L.M., F. Schott, B. Gottstein, P. Deplazes, X. Sidler and W. Basso. 2019. Detection of antibodies to *Toxoplasma gondii* in oral fluid from pigs. Int. J. Parasitol. Dec 20. pii: S0020-7519(19)30302-9. doi: 10.1016/j.ijpaDojakra.2019.11.002.

Candolfi, E., R. Pastor, R. Huber, D. Filisetti and O. Villard. 2007. IgG avidity assay firms up the diagnosis of acute toxoplasmosis on the first serum sample in immunocompetent pregnant women. Diagn. Microbiol. Infect. Dis. 58: 83–88.

Cañedo-Solares, I., F. Gómez-Chávez, H. Luna-Pastén, L.B. Ortiz-Alegría, Y. Flores-García, R. Figueroa-Damián, C.A. Macedo-Romero and D. Correa. 2018. What do anti-*Toxoplasma gondii* IgA and IgG subclasses in human saliva indicate? Parasite Immunol. 40: e12526. doi: 10.1111/pim.12526.

Chapey, E., M. Wallon and F. Peyron. 2017. Evaluation of the LDBIO point of care test for the combined detection of toxoplasmic IgG and IgM. Clin. Chim. Acta 464: 200–201. doi: 10.1016/j.cca.2016.10.023.

Choi, W., H.G. Kang, E.Y. Choi, S.S. Kim, C.Y. Kim, H.J. Koh, S.C. Lee et al. 2020. Clinical utility of aqueous humor polymerase chain reaction and serologic testing for suspected infectious uveitis: a single-center retrospective study in South Korea. BMC Ophthalmol. 20: 242. doi: 10.1186/s12886-020-01513-x.

Chong, C.K., W. Jeong, H.Y. Kim, D.J. An, H.Y. Jeoung, J.E. Ryu, A.R. Ko, Y.J. Kim, S.J. Hong, Z. Yang and H.W. Nam. 2011. Development and clinical evaluation of a rapid serodiagnostic test for toxoplasmosis of cats using recombinant SAG1 antigen. Korean J. Parasitol. 49: 207–212

Ciardelli, L., V. Meroni, M.A. Avanzini, L. Bollani, C. Tinelli, F. Garofoli et al. 2008. Early and accurate diagnosis of congenital toxoplasmosis. Pediatr. Infect. Dis. J. 27: 125e129.

Cong, W., G.H. Liu, Q.F. Meng, W. Dong, S.Y. Qin, F.K. Zhang et al. 2015. *Toxoplasma gondii* infection in cancer patients: prevalence, risk factors, genotypes and association with clinical diagnosis. Cancer Lett. 359: 307–313. doi: 10.1016/j.canlet.2015.01.036.

Cong, W., N.Z. Zhang, R.S. Hu, F.C. Zou, Y. Zou, W.Y. Zhong et al. 2020. Prevalence, risk factors and genotype distribution of *Toxoplasma gondii* DNA in soil in China. Ecotoxicol Environ. Saf. 189: 109999. doi: 10.1016/j.ecoenv.2019.109999.

Cook, A.J, R.E. Gilbert, W. Buffolano, J. Zufferey, E. Petersen, P.A. Jenum et al. 2000. Sources of *Toxoplasma* infection in pregnant women: European multicentre case-control study. European Research Network on Congenital Toxoplasmosis. BMJ 321: 142–147.

Cooper, M.K., D.N. Phalen, S.L. Donahoe, K. Rose and J. Šlapeta. 2016. The utility of diversity profiling using Illumina 18S rRNA gene amplicon deep sequencing to detect and discriminate *Toxoplasma gondii* among the cyst-forming coccidia. Vet. Parasitol. 216: 38–45. doi: 10.1016/j.vetpar.2015.12.011.

Coupe, A., L. Howe, K. Shapiro and W.D. Roe. 2019. Comparison of PCR assays to detect *Toxoplasma gondii* oocysts in green-lipped *mussels* (*Perna canaliculus*). Parasitol. Res. 118: 2389–2398. doi: 10.1007/s00436-019-06357-z.

Cowell, A.N. and E.A. Winzeler. 2019. Advances in omics-based methods to identify novel targets for malaria and other parasitic protozoan infections. Genome Med. 11: 63. doi: 10.1186/s13073-019-0673-3.

Dard, C., H. Fricker-Hidalgo, M.P. Brenier-Pinchart and H. Pelloux. 2016. Relevance of and new developments in serology for toxoplasmosis. Trends Parasitol. 32: 492–506.

Darde, M.L., A. Mercier, C. Su, A. Khan and M. Grigg. Molecular epidemiology and population structure of *Toxoplasma gondii*. pp. 64–76. *In*: Weiss, L.M. and K. Kim (eds.). *Toxoplasma gondii*-The Model Apicomplexan-Perspectives and Methods. Third edition, Academic Press.

Darwich, L., O. Cabezón, I. Echeverria, M. Pabón, I. Marco, R. Molina-López et al. 2012. Presence of *Toxoplasma gondii* and *Neospora caninum* DNA in the brain of wild birds. Vet. Parasitol. 183: 377–381. doi: 10.1016/j.vetpar.2011.07.024.

De Groot-Mijnes, J.D., A. Rothova, A.M. Van Loon, M. Schuller, N.H. Ten Dam-Van Loon, J.H. De Boer et al. 2006. Polymerase chain reaction and Goldmann-Witmer coefficient analysis are complimentary for the diagnosis of infectious uveitis. Am. J. Ophthalmol. 141: 313–318.

Demar, M., D. Ajzenberg, D. Maubon, F. Djossou, D. Panchoe, W. Punwasi et al. 2007. Fatal outbreak of human toxoplasmosis along the Maroni River: epidemiological, clinical, and parasitological aspects. Clin. Infect. Dis. 45: e88–e95.

de Oliveira Dias, J.R., C. Campelo, E.A. Novais, G.C. de Andrade, P. Marinho, Y.F. Zamora, L.F. Peixoto et al. 2020. New findings useful for clinical practice using swept-source optical coherence tomography angiography in the follow-up of active ocular toxoplasmosis. Int. J. Retina Vitreous 6: 30. doi: 10.1186/s40942-020-00231-2.

Desmonts, G. and J.S. Remington. 1980. Direct agglutination test for diagnosis of *Toxoplasma* infection: method for increasing sensitivity and specificity. J. Clin. Microbiol. 11: 562–528.

Desmonts, G., Y. Naot and J.S. Remington. 1981. Immunoglobulin M-immunosorbent agglutination assay for diagnosis of infectious diseases: diagnosis of acute congenital and acquired *Toxoplasma* infections. J. Clin. Microbiol. 14: 486–491.

de Souza, C.Z., K. Rafael, A.P. Sanders, B.T. Tiyo, A.A. Marchioro, C.M. Colli et al. 2016. An alternative method to recover *Toxoplasma gondii* from greenery and fruits. Int. J. Environ. Health Res. 26: 600–605.

Döşkaya, M., L. Liang, A. Jain, H. Can, S. Gülçe İz, P.L. Felgner et al. 2018. Discovery of new *Toxoplasma gondii* antigenic proteins using a high throughput protein microarray approach screening serum of murine model infected orally with oocysts and tissue cysts. Parasit. Vectors 11: 393. doi: 10.1186/s13071-018-2934-1.

Dubey, J.P. am J.K. Frenkel. 1972. Cyst-induced toxoplasmosis in cats. J. Protozool. 19: 155–177.

Dubey, J.P. and G. Desmonts. 1987. Serological responses of equids fed *Toxoplasma gondii* oocysts. Equine Vet. J. 19: 337–339.

Dubey, J.P. and C.P. Beattie. 1988 Toxoplasmosis of Animals and Humans. CRC Press, Boca Raton, FL, pp. 1–220.

Dubey, J.P., R.M. Weigel, A.M. Siegel, P. Thulliez, U.D. Kitron, M.A. Mitchell et al. 1995. Sources and Reservoirs of *Toxoplasma gondii* Infection on 47 Swine Farms in Illinois. J. Parasitol. 81: 723–729.

Dubey, J.P. 2004. Toxoplasmosis—a waterborne zoonosis. Vet. Parasitol. 126: 57–72.

Dubey, J.P., D.E. Hill, J.L. Jones, A.W. Hightower, E. Kirkland, J.M. Roberts et al. 2005. Prevalence of viable *Toxoplasma gondii* in beef, chicken, and pork from retail meat stores in the United States: risk assessment to consumers. J. Parasitol. 91: 1082–1093.

Dubey, J.P. 2009. Toxoplasmosis in sheep—the last 20 years. Vet. Parasitol. 163: 1–14. doi: 10.1016/j.vetpar.2009.02.026.

Dubey, J.P. 2010. Toxoplasmosis of Animals and Humans. Second Edition. CRC Press. Boca-Raton, FL. 313p.

Dubey, J.P. 2020. The history and life cycle of *Toxoplasma gondii*. pp. 1–19. *In*: Weiss, L.M. and K. Kim (eds.). *Toxoplasma gondii*—The Model Apicomplexan-Perspectives and Methods. Third edition, Academic Press.

Dubey, J.P., F.H.A. Murata, C.K. Cerqueira-Cézar, O.C.H. Kwok and C. Su. 2020a. Economic and public health importance of *Toxoplasma gondii* infections in sheep: the last decade. Vet. Parasitol. (in press).

Dubey, J.P., F.H.A. Murata, C.K. Cerqueira-Cézar and O.C.H. Kwok. 2020b. Public health and economic importance of *Toxoplasma gondii* infections in goats: The last decade. Res. Vet. Sci. 132: 292–307. doi: 10.1016/j.rvsc.2020.06.014.

Dubey, J.P., F.H.A. Murata, C.K. Cerqueira-Cézar, O.C.H. Kwok and Y.R. Yang. 2020c. Public health significance of *Toxoplasma gondii* infections in cattle: 2009–2020. J. Parasitol (in press).

Dubey, J.P., C.K. Cerqueira-Cézar, F.H.A. Murata, O.C.H. Kwok, Y.R. Yang and C. Su. 2020d. All about toxoplasmosis in cats: the last decade. Vet. Parasitol. 283: 109145. doi: 10.1016/j.vetpar.2020.109145.

Dubey, J.P., H.F.J. Pena, C.K. Cerqueira-Cézar, F.H.A. Murata, O.C.H. Kwok, Y.R. Yang, S.M. Gennari and C. Su. 2020e. Epidemiologic significance of *Toxoplasma gondii* infections in chickens (*Gallus domesticus*): the past decade. Parasitology Jul 14: 1–27. doi:10.1017/S0031182020001134.

Dubey, J.P., D.E. Hill, V. Fournet, D. Hawkins-Cooper, C.K. Cerqueira-Cézar, F.H.A. Murata et al. 2020f. Low prevalence of viable *Toxoplasma gondii* in fresh, unfrozen, American pasture-raised pork and lamb from retail meat stores in the United States. Food Control. 109: 106961.

entreprise.

Duong, H.D., C. Appiah-Kwarteng, Y. Takashima, K.M. Aye, E. Nagayasu and A. Yoshida. 2020. A novel luciferase-linked antibody capture assay (LACA) for the diagnosis of *Toxoplasma gondii* infection in chickens. Parasitol. Int. 77: 102125. doi: 10.1016/j.parint.2020.102125.

Durand, L., S. La Carbona, A. Geffard, A. Possenti, J.P. Dubey and M. Lalle. 2020. Comparative evaluation of loop-mediated isothermal amplification (LAMP) vs qPCR for detection of *Toxoplasma gondii* oocysts DNA in mussels. Exp. Parasitol. 208: 107809. doi: 10.1016/j.exppara.2019.107809.

Edvinsson, B., M. Lappalainen and B. Evengard and Toxoplasmosis ESGf. 2006. Real-time PCR targeting a 529-bp repeat element for diagnosis of toxoplasmosis. Clin. Microbiol. Infect. 12: 131–136

El Aal, A.A.A., R.R. Nahnoush, M.A. Elmallawany, W.S. El-Sherbiny, M.S. Badr and G.M. Nasr. 2018. Isothermal PCR for feasible molecular diagnosis of primary toxoplasmosis in women recently experienced spontaneous abortion. Open Access Maced. J. Med. Sci. 6: 982–987.

Escotte-Binet, S., A.M. Da Silva, B. Cancès, D. Aubert, J.P. Dubey, S. La Carbona, I. Villena et al. 2019. A rapid and sensitive method to detect *Toxoplasma gondii* oocysts in soil samples. Vet. Parasitol. 274: 108904. doi: 10.1016/j.vetpar.2019.07.012.

Eza, D.E. and S.B. Lucas. 2006. Fulminant toxoplasmosis causing fatal pneumonitis and myocarditis. HIV Med. 7: 415–420.

Fallahi, S., S.J. Seyyed Tabaei, Y. Pournia, N. Zebardast and B. Kazemi B. 2014. Comparison of loop-mediated isothermal amplification (LAMP) and nested-PCR assay targeting the RE and B1 gene for detection of *Toxoplasma gondii* in blood samples of children with leukaemia. Diagn. Microbiol. Infect. Dis. 79: 347–354.

Fallahi, S., Z.A. Mazar, M. Ghasemian and A. Haghighi. 2015. Challenging loop-mediated isothermal amplification (LAMP) technique for molecular detection of *Toxoplasma gondii*. Asian Pac. J. Trop. Med. 8: 366–72. doi: 10.1016/S1995-7645(14)60345-X.

Fallahi, S., A. Rostami, M. Nourollahpour Shiadeh, H. Behniafar and S. Paktinat. 2017. An updated literature review on maternal-fetal and reproductive disorders of *Toxoplasma gondii* infection. J. Gynecol. Obstet. Hum. Reprod. 47: 133–140. doi: 10.1016/j.jogoh.2017.12.003.

FAO/WHO [Food and Agriculture Organization of the United Nations/World Health Organization]. 2014. Multicriteria-based ranking for risk management of food-borne parasites. Microbiological Risk Assessment Series No. 23. Rome. 302pp

Fekkar, A., B. Bodaghi, F. Touafek, P. Le Hoang, D. Mazier and L. Paris. 2008. Comparison of immunoblotting, calculation of the Goldmann-Witmer coefficient, and real-time PCR using aqueous humor samples for diagnosis of ocular toxoplasmosis. J. Clin. Microbiol. 46: 1965–1967.

Fernández-Aguilar, X., D. Ajzenberg, O. Cabezón, A. Martínez-López, L. Darwich, J.P. Dubey et al. 2013. Fatal toxoplasmosis associated with an atypical *Toxoplasma gondii* strain in a Bennett's wallaby (*Macropus rufogriseus*) in Spain. Vet. Parasitol. 196: 523–527. doi: 10.1016/j.vetpar.2013.03.001.

Ferra, B., L. Holec-Gąsior and J. Kur. 2015a. Serodiagnosis of *Toxoplasma gondii* infection in farm animals, horses, swine, and sheep) by enzyme-linked immunosorbent assay using chimeric antigens. Parasitol. Int. 64: 288–294.

Ferra, B., L. Holec-Gąsior and J. Kur. 2015b. A new *Toxoplasma gondii* chimeric antigen containing fragments of SAG2, GRA1, and ROP1 proteins—impact of immunodominant sequences size on its diagnostic usefulness. Parasitol. Res. 114: 3291–3299.

Ferra, B., L. Holec-Gąsior and W. Grąźlewska. 2020. *Toxoplasma gondii* recombinant antigens in the serodiagnosis of toxoplasmosis in domestic and farm animals. Animals 10: E1245. doi: 10.3390/ani10081245.

Flegr, J., J. Prandota, M. Sovickova and Z.F. Israili. 2014. Toxoplasmosis—a global threat: Correlation of latent toxoplasmosis with specific disease burden in a set of 88 countries. PLoS ONE 9: e90203.

Fricker-Hidalgo, H., B. Cimon, C. Chemla, M.L. Darde, L. Delhaes, C. L'ollivier et al. 2013. *Toxoplasma* seroconversion with negative or transient immunoglobulin M in pregnant women: myth or reality? A French multicenter retrospective study. J. Clin. Microbiol. 51: 2103–2011. doi: 10.1128/JCM.00169-13. Epub 2013 Apr 24.

Fricker-Hidalgo, H., C. L'Ollivier, C. Bosson, S. Imbert, S. Bailly, C. Dard et al. 2017. Interpretation of the Elecsys Toxo IgG avidity results for very low and very high index: study on 741 sera with a determined date of toxoplasmosis. Eur. J. Clin. Microbiol. Infect. Dis. 36: 847–852. doi: 10.1007/s10096-016-2870-y.

Fuglewicz, A.J., P. Piotrowski and A. Stodolak. 2017. Relationship between toxoplasmosis and schizophrenia: A review. Adv. Clin. Exp. Med. 26: 1033–1038, DOI: 10.17219/acem/61435.

Galvani, A.T., A.P.G. Christ, J.A. Padula, M.R.F. Barbosa, R.S. de Araújo, M.I.Z. Sato et al. 2019. Real-time PCR detection of *Toxoplasma gondii* in surface water samples in São Paulo, Brazil. Parasitol. Res. 118: 631–640. doi: 10.1007/s00436-018-6185-z.

Garfoot, A.L., G.M. Wilson, J.J. Coon and L.J. Knoll. 2019. Proteomic and transcriptomic analyses of early and late-chronic *Toxoplasma gondii* infection shows novel and stage specific transcripts. BMC Genomics 20: 859. doi:10.1186/s12864-019-6213-0.

Garnaud, C., H. Fricker-Hidalgo, B. Evengård, M.J. Álvarez-Martínez, E. Petersen, L.M. Kortbeek et al. 2020. *Toxoplasma gondii*-specific IgG avidity testing in pregnant women. Clin. Microbiol. Infect. 22: garS1198-743X(20)30220-2. doi: 10.1016/j.cmi.2020.04.014.

Gay-Andrieu, F., H. Fricker-Hidalgo, E. Sickinger, A. Espern, M.P. Brenier-Pinchart, H.B. Braun et al. 2009. Comparative evaluation of the ARCHITECT Toxo IgG, IgM, and IgG avidity assays for anti-*Toxoplasma* antibodies detection in pregnant women sera. Diagn. Microbiol. Infect. Dis. 65: 279–287.

Gazzonis, A.L., S.A. Zanzani, L. Villa and M.T. Manfredi. 2020. *Toxoplasma gondii* infection in meat-producing small ruminants: Meat juice serology and genotyping. Parasitol. Int. Jan 18: 102060. doi: 10.1016/j.parint.2020.102060.

Geba, E., D. Aubert, L. Durand, S. Escotte, S. La Carbona, C. Cazeaux et al. 2020. Use of the bivalve *Dreissena polymorpha* as a biomonitoring tool to reflect the protozoan load in freshwater bodies. Water Res. 170: 115297. doi: 10.1016/j.watres.2019.115297.

Genco, F., A. Sarasini, M. Parea, M. Prestia, L. Scudeller and V. Meroni. 2019. Comparison of the LIAISON®XL and ARCHITECT IgG, IgM, and IgG avidity assays for the diagnosis of *Toxoplasma*, cytomegalovirus, and rubella virus infections. New Microbiol. 42: 88–93.

Gilbert, R.E. and M.R. Stanford. 2000. Is ocular toxoplasmosis caused by prenatal or postnatal infection? Br. J. Ophthalmol. 84(2): 224–226.

Gisbert Algaba, I., M. Geerts, M. Jennes, W. Coucke, M. Opsteegh, E. Cox et al. 2017. A more sensitive, efficient and ISO 17025 validated Magnetic Capture real time PCR method for the detection of archetypal *Toxoplasma gondii* strains in meat. Int. J. Parasitol. 47: 875–884. doi: 10.1016/j.ijpara.2017.05.005.

Gisbert Algaba, I., B. Verhaegen, J.B. Murat, W. Coucke, A. Mercier, E. Cox et al. 2020. Molecular study of *Toxoplasma gondii* isolates originating from humans and organic pigs in Belgium. Foodborne Pathog. Dis. Jan 6. doi: 10.1089/fpd.2019.2675.

Gomez-Marin, J.E. and A. de-la Torre. 2020. Ocular disease due to *Toxoplasma gondii*. pp. 257–268. *In*: Weiss, L.M. and K. Kim (eds.). *Toxoplasma gondii*—The Model Apicomplexan-Perspectives and Methods. Third edition, Academic Press.

Gomez-Samblas, M., S. Vílchez, J.C. Racero, M.V. Fuentes and A. Osuna. 2015. Quantification and viability assays of *Toxoplasma gondii* in commercial "Serrano" ham samples using magnetic capture real-time qPCR and bioassay techniques. Food Microbiol. 46: 107–113. doi: 10.1016/j.fm.2014.07.003.

Greigert, V., E. Di Foggia, D. Filisetti, O. Villard, A.W. Pfaff, A. Sauer et al. 2019. When biology supports clinical diagnosis: review of techniques to diagnose ocular toxoplasmosis. Br. J. Ophthalmol. 103: 1008–1012. doi:10.1136/bjophthalmol-2019-313884.

Guglietta, S., E. Beghetto, A. Spadoni, W. Buffolano, P. Del Porto and N. Gargano. 2007. Age-dependent impairment of functional helper T cell responses to immunodominant epitopes of *Toxoplasma gondii* antigens in congenitally infected individuals. Microbes Infect. 9: 127–133.

Guigue, N., J. Menotti, S. Hamane, F. Derouin and Y.J. Garin. 2014. Performance of the BioPlex 2200 flow immunoassay in critical cases of serodiagnosis of toxoplasmosis. Clin. Vaccine Immunol. 21: 496–500. doi: 10.1128/CVI.00624-13.

Halos, L., A. Thébault, D. Aubert, M. Thomas, C. Perret, R. Geers, A. Alliot et al. 2010. An innovative survey underlining the significant level of contamination by *Toxoplasma gondii* of ovine meat consumed in France. Int. J. Parasitol. 40: 193–200. doi: 10.1016/j.ijpara.2009.06.009.

Havelaar, A.H., J.A. Haagsma, M.J.J. Mangen, J.M. Kemmeren, L.P.B. Verhoef, S.M.C. Vijgen et al. 2012. Disease burden of foodborne pathogens in the Netherlands. Int. J. Food Microbiol. 156: 231–238.

Hedman, K., M. Lappalainen, I. Seppaia and O. Makela. 1989. Recent primary *Toxoplasma* infection indicated by a low avidity of specific IgG. J. Infect. Dis. 159: 736–740.

Hill, D.E., S. Chirukandoth, J.P. Dubey, J.K. Lunney and H.R. Gamble. 2006. Comparison of detection methods for *Toxoplasma gondii* in naturally and experimentally infected swine. Vet. Parasitol. 141: 9–17.

Hill, D.E. and J.P. Dubey. 2016. *Toxoplasma gondii* as a parasite in food: Analysis and control. Microbiol. Spect. 4.

Hoffmann, S., M.B. Batz and J.G. Jr. Morris. 2012. Annual cost of illness and quality-adjusted life year losses in the United States due to 14 foodborne pathogens. J. Food Protect. 75: 1292–1302.

Hohweyer, J., C. Cazeaux, E. Travaillé, E. Languet, A. Dumètre, D. Aubert et al. 2016. Simultaneous detection of the protozoan parasites *Toxoplasma*, *Cryptosporidium* and *Giardia* in food matrices and their persistence on basil leaves. Food Microbiol. 57: 36–44. doi: 10.1016/j.fm.2016.01.002.

Holec-Gąsior, L., B. Ferra and W. Grążlewska. 2019. *Toxoplasma gondii* tetravalent chimeric proteins as novel antigens for detection of specific immunoglobulin G in sera of small ruminants. 9. Animals (Basel). 9: pii: E1146. doi: 10.3390/ani9121146.

Homan, W.L., M. Vercammen, J. De Braekeleer and H. Verschueren. 2000. Identification of a 200- to 300-fold repetitive 529 bp DNA fragment in *Toxoplasma gondii*, and its use for diagnostic and quantitative PCR. Int. J. Parasitol. 30: 69–75.

Howe, D.K. and L.D. Sibley. 1995. *Toxoplasma gondii* comprises three clonal lineages: correlation of parasite genotype with human disease. J. Infect. Dis. 172: 1561–1566.

Huang, X., X. Xuan, H. Hirata, N. Yokoyama, L. Xu, N. Suzuki et al. 2004. Rapid immunochromatographic test using recombinant SAG2 for detection of antibodies against *Toxoplasma gondii* in cats. J. Clin. Microbiol. 42: 351–353.

Huang, Y., Y. Huang, A. Chang, J. Wang, X. Zeng and J. Wu. 2016. Is *Toxoplasma gondii* infection a risk factor for leukemia? an evidence-based meta-analysis. Med. Sci. Monit. 22: 1547–1552.

Jiang, W., Y. Liu, Y. Chen, Q. Yang, P. Chun, K. Yao et al. 2015. A novel dynamic flow immunochromatographic test (DFICT) using gold nanoparticles for the serological detection of *Toxoplasma gondii* infection in dogs and cats. Biosens. Bioelectron. 72: 133–139. doi:10.1016/j.bios.2015.04.035.

Jiang, T., E.K. Shwab, R.M. Martin, R.W. Gerhold, B.M. Rosenthal, J.P. Dubey et al. 2018. A partition of *Toxoplasma gondii* genotypes across spatial gradients and among host species, and decreased parasite diversity towards areas of human settlement in North America. Int. J. Parasitol. 48: 611–619. doi: 10.1016/j.ijpara.2018.01.008.

Jones, J.L. and J.P. Dubey. 2010. Waterborne toxoplasmosis—recent developments. Exp. Parasitol. 124: 10–25. doi: 10.1016/j.exppara.2009.03.013.

Jones, J.L. and J.P. Dubey. 2012. Foodborne toxoplasmosis. Clin. Infect. Dis. 55: 845–851. doi: 10.1093/cid/cis508.

Jost, C., F. Touafek, A. Fekkar, R. Courtin, M. Ribeiro and D. Mazier. 2011. Utility of immunoblotting for early diagnosis of toxoplasmosis seroconversion in pregnant women. Clin. Vaccine Immunol. 18: 1908–1912.

Juránková, J., M. Opsteegh, H. Neumayerová, K. Kovařčík, A. Frencová, V. Baláž et al. 2013. Quantification of *Toxoplasma gondii* in tissue samples of experimentally infected goats by magnetic capture and real-time PCR. Vet. Parasitol. 193: 95–99. doi: 10.1016/j.vetpar.2012.11.016.

Khan, A., J.S. Shaik, M. Behnke, Q. Wang, J.P. Dubey, H.A. Lorenzi et al. 2014. NextGen sequencing reveals short double crossovers contribute disproportionately to genetic diversity in *Toxoplasma gondii*. B.M.C. Genomics 15: 1168. doi: 10.1186/1471-2164-15-1168.

Khan, A. and M.E. Grigg. 2017. *Toxoplasma gondii*: Laboratory maintenance and growth. Curr. Protoc. Microbiol. 44: 20C.1.1-20C.1.17. doi: 10.1002/cpmc.26.

Khan, K. and W. Khan. 2018. Congenital toxoplasmosis: An overview of the neurological and ocular manifestations. Parasitol. Int. 67: 715–721. doi: 10.1016/j.parint.2018.07.004.

Khan, A.H. and R. Noordin. 2020. Serological and molecular rapid diagnostic tests for *Toxoplasma* infection in humans and animals. Eur. J. Clin. Microbiol. Infect. Dis. 39: 19–30. doi: 10.1007/s10096-019-03680-2.

Kim, Y.H., J.H. Lee, S.K. Ahn, T.S. Kim, S.J. Hong, C.K. Chong et al. 2017. Seroprevalence of Toxoplasmosis with ELISA and Rapid Diagnostic Test among Residents in Gyodong-do, Inchon city, Korea: A Four-Year Follow-up. Korean J. Parasitol. 55: 247–254. doi: 10.3347/kjp.2017.55.3.247.

King, E.F.B., S.A. Cobbold, A.D. Uboldi C.J. Tonkin and M.J. McConville. 2020. Metabolomic analysis of *Toxoplasma gondii* tachyzoites. Methods Mol. Biol. 2071: 435–452. doi: 10.1007/978-1-4939-9857-9_22.

Kissinger, J.C., B. Gajria, L. Li, I.T. Paulsen and D.S Roos. 2003. ToxoDB: accessing the *Toxoplasma gondii* genome. Nucleic Acids Res. 31: 234–236.

Kong, J.T., M.E. Grigg, L. Uyetake, S. Parmley and J.C. Boothroyd. 2003. Serotyping of *Toxoplasma gondii* infections in humans using synthetic peptides. J. Infect. Dis. 187: 1484–1495.

Lalle, M., A. Possenti, J.P. Dubey and E. Pozio. 2018. Loop-mediated isothermal amplification-lateral-flow dipstick (LAMP-LFD) to detect *Toxoplasma gondii* oocyst in ready-to-eat salad. Food Microbiol. 70: 137–142.

Lalonde, L.F. and A.A. Gajadhar. 2011. Detection and differentiation of coccidian oocysts by real-time PCR and melting curve analysis. J. Parasitol. 97: 725–730. doi: 10.1645/GE-2706.1.

Lalonde, L.F. and A.A. Gajadhar. 2016a. Optimization and validation of methods for isolation and real-time PCR identification of protozoan oocysts on leafy green vegetables and berry fruits. Food and Waterborne Parasitol. 2: 1–7.

Lalonde, L.F. and A.A. Gajadhar. 2016b. Detection of *Cyclospora cayetanensis*, *Cryptosporidium* spp., and *Toxoplasma gondii* on imported leafy green vegetables in Canadian survey. Food and Waterborne Parasitol. 2: 8–14.

Lass, A., H. Pietkiewicz. B. Szostakowska and P. Myjak. 2012. The first detection of *Toxoplasma gondii* DNA in environmental fruits and vegetables samples. Eur. J. Clin. Microbiol. Infect. Dis. 31: 1101–1108. doi: 10.1007/s10096-011-1414-8.

Lass, A., L. Ma, I. Kontogeorgos, X. Zhang, X. Li and P. Karanis. 2019. First molecular detection of *Toxoplasma gondii* in vegetable samples in China using qualitative, quantitative real-time PCR and multilocus genotyping. Sci. Rep. 9: 17581. doi: 10.1038/s41598-019-54073-6.

Lau, Y.L., P. Meganathan, P. Sonaimuthu, G. Thiruvengadam, V. Nissapatorn and Y. Chen. 2010. Specific, sensitive, and rapid diagnosis of active toxoplasmosis by a loop-mediated isothermal amplification method using blood samples from patients. J. Clin. Microbiol. 48: 3698–3702.

Lau, Y.L., W.C. Lee, R. Gudimella, G. Zhang, X.T. Ching, R. Razali et al. 2016. Deciphering the Draft Genome of *Toxoplasma gondii* RH strain. PLoS One 11: e0157901. doi: 10.1371/journal.pone.0157901.

Leveque, M.F., D. Chiffré, C. Galtier, S. Albaba, C. Ravel, L. Lachaud et al. 2019. Molecular diagnosis of toxoplasmosis at the onset of symptomatic primary infection: A straightforward alternative to serological examinations. Int. J. Infect. Dis. 79: 131–133. doi: 10.1016/j.ijid.2018.11.368.

Li, X., C. Pomares, G. Gonfrier, B. Koh, S. Zhu, M. Gong et al. 2016. Multiplexed anti-*Toxoplasma* IgG, IgM, and IgA assay on plasmonic gold chips: towards making mass screening possible with dye test precision. J. Clin. Microbiol. 54: 1726–1733. doi: 10.1128/JCM.03371-15.

Li, X., C. Pomares, F. Peyron, C.J. Press, R. Ramirez, G. Geraldine et al. 2019. Plasmonic gold chips for the diagnosis of *Toxoplasma gondii*, CMV, and rubella infections using saliva with serum detection precision. Eur. J. Clin. Microbiol. Infect. Dis. 38: 883–890. doi:10.1007/s10096-019-03487-1.

Lin, Z., Y. Zhang, H. Zhang, Y. Zhou, J. Cao and J. Zhou. 2012. Comparison of loop-mediated isothermal amplification (LAMP) and real-time PCR method targeting a 529-bp repeat element for diagnosis of toxoplasmosis. Vet. Parasitol. 185: 296–300.

Lindquist, H.D.A., W.J. Bennett, J.D. Hester, M.W. Ware, J.P. Dubey and W.V. Everson. 2003. Autofluorescence of *Toxoplasma gondii* and related coccidian oocysts. J. Parasitol. 89: 865–867.

Liu, Q., Z.D. Wang, S.Y. Huang and X.Q. Zhu. 2015. Diagnosis of toxoplasmosis and typing of *Toxoplasma gondii*. Parasit. Vectors 8: 292. doi: 10.1186/s13071-015-0902-6.

Lopes, F.M., D.D. Gonçalves, R. Mitsuka-Breganó, R.L. Freire and I.T. Navarro. 2007. *Toxoplasma gondii* infection in pregnancy. Braz. J. Infect. Dis. 11: 496–506.

Loreck, K., S. Mitrenga, R. Heinze, R. Ehricht, C. Engemann, C. Lueken, M. Ploetz et al. 2020. Use of meat juice and blood serum with a miniaturised protein microarray assay to develop a multi-parameter IgG screening test with high sample throughput potential for slaughtering pigs. BMC Vet. Res. 16: 106. doi: 10.1186/s12917-020-02308-4.

Lorenzi, H., A. Khan, M.S. Behnke, S. Namasivayam, L.S. Swapna, M. Hadjithomas et al. 2016. Local admixture of amplified and diversified secreted pathogenesis determinants shapes mosaic *Toxoplasma gondii* genomes. Nat. Commun. 7: e10147.

Luo, J., H. Sun, X. Zhao, S. Wang, X. Zhuo, Y. Yang et al. 2018. Development of an immunochromatographic test based on monoclonal antibodies against surface antigen 3 (TgSAG3) for rapid detection of *Toxoplasma gondii*. Vet. Parasitol. 252: 52–57. doi: 10.1016/j.vetpar.2018.01.015.

Ma, Z., A.M. Mutashar Alhameed, A.C. Kaminga, B. Lu, X. Li, J. Zhang et al. 2019. Bioinformatics of excretory/secretory proteins of *Toxoplasma gondii* strain ME49. Microb. Pathog. 140: 103951. doi: 10.1016/j.micpath.2019.103951.

Macre, M.S., L.R. Meireles, B.F.C. Sampaio and H.F. Andrade Júnior. 2019. Saliva collection and detection of anti-*T. gondii* antibodies of low-income school-age children as a learning strategy on hygiene, prevention and transmission of toxoplasmosis. Rev. Inst. Med. Trop. São Paulo 61: e48. doi: 10.1590/S1678-9946201961048.

Maenz, M., D. Schlüter, O. Liesenfeld, G. Schares, U. Gross and U. Pleyer. 2014. Ocular toxoplasmosis past, present and new aspects of an old disease. Prog. Retin Eye Res. 39: 77–106. doi: 10.1016/j.preteyeres.2013.12.005.

Mahmoudi, S., S. Mamishi, X. Suo and H. Keshavarz. 2017. Early detection of *Toxoplasma gondii* infection by using a interferon gamma release assay: A review. Exp. Parasitol. 172: 39–43. doi: 10.1016/j.exppara.2016.12.008.

Maksimov, P., J. Zerweck, A. Maksimov, A. Hotop, U. Gross, U. Pleyer et al. 2012. Peptide microarray analysis of *in silico*-predicted epitopes for serological diagnosis of *Toxoplasma gondii* infection in humans. Clin. Vaccine Immunol. 19: 865–874. doi: 10.1128/CVI.00119-12.

Maksimov, P., J. Zerweck, V. Maksimov, A. Hotop, U. Gross, K. Spekker et al. 2012. Analysis of clonal type-specific antibody reactions in *Toxoplasma gondii* seropositive humans from Germany by peptide-microarray. PLoS One 7: e34212. doi: 10.1371/journal.pone.0034212.

Maksimov, P., J. Zerweck, J.P. Dubey, N. Pantchev, C.F. Frey and A. Maksimov. 2013. Serotyping of *Toxoplasma gondii* in cats (*Felis domesticus*) reveals predominance of type II infections in Germany. PLoS One 8: e80213. doi: 10.1371/journal.pone.0080213.

Mangiavacchi, B.M., F.P. Vieira, L.M. Bahia-Oliveira and D. Hill. 2016. Salivary IgA against sporozoite-specific embryogenesis-related protein TgERP) in the study of horizontally transmitted toxoplasmosis via *T. gondii* oocysts in endemic settings. Epidemiol. Infect. 144: 2568–77. doi: 10.1017/S0950268816000960.

Marchioro, A.A., B.T. Tiyo, C.M. Colli, C.Z. de Souza, J.L. Garcia, M.L. Gomes et al. 2016. First detection of *Toxoplasma gondii* DNA in the fresh leafs of vegetables in South America. Vector Borne Zoonotic Dis. 16: 624–626. doi: 10.1089/vbz.2015.1937.

Marcolino, P.T., D.A. Silva, P.G. Leser, M.E. Camargo and J.R. Mineo. 2000. Molecular markers in acute and chronic phases of human toxoplasmosis: determination of immunoglobulin G avidity by Western blotting. Clin. Diagn. Lab. Immunol. 7: 384–389.

Martinez, V.O., F.W. de Mendonça Lima, C.F. de Carvalho and J.A. Menezes-Filho. 2018. *Toxoplasma gondii* infection and behavioral outcomes in humans: a systematic review. Parasitol. Res. 117: 3059–3065. doi: 10.1007/s00436-018-6040-2.

Martinot, M., V. Greigert, C. Farnarier, M.L. Dardé, C. Piperoglou and M. Mohseni-Zadeh. 2019. Spinal cord toxoplasmosis in a young immunocompetent patient. Infection. Dec 9. doi: 10.1007/s15010-019-01380-9.

Masamed, R., A. Meleis, E.W. Lee and G.M. Hathout. 2009. Cerebral toxoplasmosis: case review and description of a new imaging sign. Clinical Radiology 64: 560e563.

Maudry, A., G. Chene, R. Chatelain, H. Patural, B. Bellete, B. Tisseur et al. 2009. Bicentric evaluation of six anti-*Toxoplasma* immunoglobulin G (IgG) automated immunoassays and comparison to the Toxo II IgG Western blot. Clin. Vaccine Immunol. 16: 1322–1326. doi: 10.1128/CVI.00128-09.

Mcleod, R., W. Cohen, S. Dovgin, A. Finkelstein and K.M. Boyer. 2020. Human *Toxoplasma* infection. pp. 143–159. *In*: Weiss, L.M. and K. Kim (eds.). *Toxoplasma gondii*—The Model Apicomplexan-Perspectives and Methods. Third edition, Academic Press.

Medawar-Aguilar, V., C.F. Jofre, M.A. Fernández-Baldo, A. Alonso, S. Angel, J. Raba et al. 2019. Serological diagnosis of toxoplasmosis disease using a fluorescent immunosensor with chitosan-ZnO-nanoparticles. Anal. Biochem. 564-565: 116–122.

Moncada, P.A. and J.G. Montoya. 2012. Toxoplasmosis in the fetus and newborn: an update on prevalence, diagnosis and treatment. Expert. Rev. Anti-infect. Ther. 10: 815–828.

Montoya, J.G. 2002. Laboratory diagnosis of *Toxoplasma gondii* infection and toxoplasmosis. J. Infect. Dis. 18: S73–S82.

Montoya, J.G. and O. Liesenfeld. 2004. Toxoplasmosis. Lancet. 363: 1965–1976.

Murat, J.B., C. L'Ollivier, H. Fricker-Hidalgo, J. Franck, H. Pelloux and R. Piarroux. 2012. Evaluation of the new Elecsys Toxo IgG avidity assay for toxoplasmosis and new insights into the interpretation of avidity results. Clin. Vaccine Immunol. 19: 1838–1843.

Murat, J.B., H.F. Hidalgo, M.P. Brenier-Pinchart and H. Pelloux. 2013a. Human toxoplasmosis: which biological diagnostic tests are best suited to which clinical situations? Expert. Rev. Anti Infect. Ther. 11: 943–956. doi:10.1586/14787210.2013.825441.

Murat, J.B., C. Dard, H. Fricker-Hidalgo, M.L. Dardé, M.P. Brenier-Pinchart and H. Pelloux. 2013b. Comparison of the Vidas system and two recent fully automated assays for diagnosis and follow-up of toxoplasmosis in pregnant women and newborns. Clin. Vaccine Immunol. 20: 1203–1212: 117–120.

Nayeri Chegeni, T., S. Sarvi, M. Moosazadeh, M. Sharif, S.A. Aghayan, A. Amouei et al. 2019. Is *Toxoplasma gondii* a potential risk factor for Alzheimer's disease? A systematic review and meta-analysis. Microb. Pathog. 137: 103751. doi: 10.1016/j.micpath.2019.103751.

Olariu, T.R., J.S. Remington and J.G. Montoya. 2014. Polymerase chain reaction in cerebro-spinal fluid for the diagnosis of congenital toxoplasmosis. Pediatr. Infect. Dis. J. 33: 566–570.

Olariu, T.R., B.G. Blackburn, C. Press, J. Talucod, J.S. Remington and J.G. Montoya. 2019. Role of *Toxoplasma* IgA as part of a reference panel for the diagnosis of acute toxoplasmosis during pregnancy. J. Clin. Microbiol. 57: e01357–18. doi: 10.1128/JCM.01357-18.

One-Reid, A.J., S.J. Vermont, J.A. Cotton, D. Harris, G.A. Hill-Cawthorne, S. Könen-Waisman et al. 2012. Comparative genomics of the apicomplexan parasites *Toxoplasma gondii* and *Neospora caninum*: Coccidia differing in host range and transmission strategy. PLoS Pathog. 8: e1002567. doi: 10.1371/journal.ppat.1002567.

Opsteegh, M., M. Langelaar, H. Sprong, L. den Hartog, S. De Craeye and G. Bokken. 2010. Direct detection and genotyping of *Toxoplasma gondii* in meat samples using magnetic capture and PCR. Int. J. Food Microbiol. 139: 193–201. doi: 10.1016/j.ijfoodmicro.2010.02.027.

Opsteegh, M., G. Schares, R. Blaga and J. van der Giessen. 2016. EFSA Experimental studies on *Toxoplasma gondii* in the main livestock species (GP/EFSA/BIOHAZ/2013/01) Final report. EFSA Supporting publications. Ting Publications 16 February. https://doi.org/10.2903/sp.efsa.2016.EN-995

Opsteegh, M., F. Spano, D. Aubert, A. Balea, A. Burrells, S. Cherchi et al. 2019. The relationship between the presence of antibodies and direct detection of *Toxoplasma gondii* in slaughtered calves and cattle in four European countries. Int. J. Parasitol. 49: 515–522. doi: 10.1016/j.ijpara.2019.01.005.

Opsteegh, M., C. Dam-Deisz, P. de Boer, S. DeCraeye, A. Faré, P. Hengeveld, R. Luiten et al. 2020. Methods to assess the effect of meat processing on viability of *Toxoplasma gondii*: towards replacement of mouse bioassay by *in vitro* testing. Int. J. Parasitol. 50: 357–369. doi: 10.1016/j. ijpara.2020.04.001.

Ozgonul, C. and C.G. Besirli. 2017. Recent developments in the diagnosis and treatment of ocular toxoplasmosis. Ophthalmic Res. 57: 1–12.

Paştiu, A.I., A. Cozma-Petruţ, A. Mercier, A. Balea, L. Galal, V. Mircean et al. 2019. Prevalence and genetic characterization of *Toxoplasma gondii* in naturally infected backyard pigs intended for familial consumption in Romania. Parasit. Vectors. 12: 586. doi: 10.1186/s13071-019-3842-8.

Petersen, E., M.V. Borobio, E. Guy, O. Liesenfeld, V. Meroni, A. Naessens et al. 2005. European multicenter study of the LIAISON automated diagnostic system for determination of *Toxoplasma gondii*-specific immunoglobulin G (IgG) and IgM and the IgG avidity index. J. Clin. Microbiol. 43: 1570–1574.

Peyron, F., M. Wallon, F. Kieffer and G. Graweg. 2016. Toxoplasmosis. pp. 949–1042. *In*: Remington, J.S., J.O. Klein, C.B. Wilson, V. Nizet and Y.A. Maldonado (8th eds). Infectious Diseases of the Fetus and Newborn Infant. Philadelphia, USA: Elsevier Saunders.

Pinon, J.M., H. Dumon, C. Chemla, J. Franck, E. Petersen, M. Lebech et al. 2001. Strategy for diagnosis of congenital toxoplasmosis: evaluation of methods comparing mothers and newborns and standard methods for postnatal detection of immunoglobulin G, M, and A antibodies. J. Clin. Microbiol. 39: 2267–2271.

Piro, F., V.B. Carruthers and M. Di Cristina. 2020. PCR Screening of *Toxoplasma gondii* single clones directly from 96-well plates without DNA purification. *In*: Tonkin, C. (eds.). *Toxoplasma gondii*. Methods in Molecular Biology, vol 2071. Humana, New York, NY doi: 10.1007/978-1-4939-9857-9_6.

Pomares, C. and J.G. Montoya. 2016. Laboratory diagnosis of congenital toxoplasmosis. J. Clin. Microbiol. 54: 2448–2454.

Pomares, C., B. Zhang, S. Arulkumar, G. Gonfrier, P. Marty, S. Zhao et al. 2017. Validation of IgG, IgM multiplex plasmonic gold platform in French clinical cohorts for the serodiagnosis and follow-up of *Toxoplasma gondii* infection. Diagn. Microbiol. Infect. Dis. 87: 213–218.

Prusa, A.R., M. Hayde, L. Unterasinger, A. Pollak and K.R. Herkner. 2010. Evaluation of the Roche Elecsys Toxo IgG and IgM electrochemiluminescence immunoassay for the detection of gestational *Toxoplasma* infection. Diagn. Microbiol. Infect. Dis. 68: 352–357.

Prusa, A.R., M. Hayde, A. Pollak, K.R. Herkner and D.C. Kasper. 2012. Evaluation of the Liaison automated testing system for diagnosis of congenital toxoplasmosis. Clin. Vaccine Immunol. 19: 1859–1863.

Prusa, A.R., D.C. Kasper, A. Pollak, M. Olischar, A. Gleiss and M. Hayde. 2015. Amniocentesis for the detection of congenital toxoplasmosis: results from the nationwide Austrian prenatal screening program. Clin. Microbiol. Infect. 21: 191.e1-8. doi: 10.1016/j.cmi.2014.09.018.

Prusa, A.R., D.C. Kasper, L. Sawers, E. Walter, M. Hayde and E. Stillwaggon. 2017. Congenital toxoplasmosis in Austria: Prenatal screening for prevention is cost-saving. PLoS Negl. Trop. Dis. 11: e0005648. doi: 10.1371/journal.pntd.0005648.

Qian, W., W. Yan, C. Lv, R. Bai and T. Wang. 2019. Occurrence and genetic characterization of *Toxoplasma gondii* and *Neospora caninum* in slaughtered domestic rabbits in Central China. Parasite 26: 36. doi: 10.1051/parasite/2019035.

Ren, W.X., X.X. Zhang, C.Y. Long, Q. Zhao, T. Cheng, J.G. Ma et al. 2019. Molecular detection and genetic characterization of *Toxoplasma gondii* from horses in Three Provinces of China. Vector Borne Zoonotic Dis. 19: 703–707. doi: 10.1089/vbz.2018.2423.

Reischl, U., S. Bretagne, D. Krüger, P. Ernault and J.M. Costa. 2003. Comparison of two DNA targets for the diagnosis of Toxoplasmosis by real-time PCR using fluorescence resonance energy transfer hybridization probes. B.M.C. Infect. Dis. 3: 7.

Remington, J.S., W.M. Eimstad and F.G. Araujo. 1983. Detection of immunoglobulin M antibodies with antigen-tagged latex particles in an immunosorbent assay. J. Clin. Microbiol. 17: 939–941.

Rilling, V., K. Dietz, D. Krczal, F. Knotek and G. Enders. 2003. Evaluation of a commercial IgG/IgM Western blot assay for early postnatal diagnosis of congenital Toxoplasmosis. Eur. Clin. Microbiol. Infect. Dis. 22: 174–180.

Robert-Gangneux, F., P. Binisti, D. Antonetti, A. Brezin, H. Year and J. Dupouy-Camet. 2004. Usefulness of immunoblotting and Goldmann-Witmer coefficient for biological diagnosis of toxoplasmic retinochoroiditis. Eur. J. Clin. Microbiol. Infect. Dis. 23: 34–38.

Robert-Gangneux, F., J.B. Murat, H. Fricker-Hidalgo, M.P. Brenier-Pinchart, J.P. Gangneux and H. Pelloux. 2011. The placenta: a main role in congenital toxoplasmosis? Trends Parasitol. 27: 530–536. doi: 10.1016/j.pt.2011.09.005.

Robert-Gangneux, F. and M.L. Darde. 2012. Epidemiology of and diagnostic strategies for toxoplasmosis. Clin. Microbiol. Rev. 25: 264–296.

Roos, D.S., R.G. Donald, N.S. Morrissette and A.L. Moulton. 1992. Molecular tools for genetic dissection of the protozoan parasite *Toxoplasma gondii*. Methods Cell. Biol. 45: 27–63. doi:10.1016/s0091-679x(08)61845-2.

Rostami, A., P. Karanis and S. Fallahi. 2018. Advances in serological, imaging techniques and molecular diagnosis of *Toxoplasma gondii* infection. Infection 46: 303–315. doi: 10.1007/s15010-017-1111-3.

Rostami, A., S.M. Riahi, H.R. Gamble, Y. Fakhri, M.N. Shiadeh, M. Danesh et al. 2020. Global prevalence of latent toxoplasmosis in pregnant women: a systematic review and meta-analysis. 1. Clin. Microbiol. Infect. Jan 20. pii: S1198-743X(20)30033-1. doi:10.1016/j.cmi.2020.01.008.

Rothova, A., J.H. de Boer, N.H. Ten Dam-van Loon, G. Postma, L. de Visser, S.J. Zuurveen et al. 2008. Usefulness of aqueous humor analysis for the diagnosis of posterior uveitis. Ophthalmol. 115: 306–311.

Rousseau, A., I. Villena, A. Dumètre, S. Escotte-Binet, L. Favennec, J.P. Dubey et al. 2019a. Evaluation of propidium monoazide-based qPCR to detect viable oocysts of *Toxoplasma gondii*. Parasitol. Res. 118: 999–1010. doi: 10.1007/s00436-019-06220-1.

Rousseau, A., S. Escotte-Binet, S. La Carbona, A. Dumètre, S. Chagneau, L. Favennec et al. 2019b. *Toxoplasma gondii* oocyst infectivity assessed using a sporocyst-based cell culture assay combined with quantitative PCR for environmental applications. Appl. Environ. Microbiol. 85: e01189-19. doi: 10.1128/AEM.01189-19.

Ruiz, A. and F.K. Frenkel. 1980a. Intermediate and transport hosts of *Toxoplasma gondii* in Costa Rica. Am. J. Trop. Med. Hyg. 29: 1161–1166.

Ruiz, A. and F.K. Frenkel. 1980b. *Toxoplasma gondii* in Costa Rican cats. Am. J. Trop. Med. Hyg. 29: 1150–1160.

Saadatnia, G., H. Haj Ghani, B.Y. Khoo, A. Maimunah and N. Rahmah. 2010. Optimization of *Toxoplasma gondii* cultivation in VERO cell line. Trop. Biomed. 27: 125–130.

Sabin, A.B. and H.A. Feldman. 1948. Dyes as microchemical indicators of a new immunity phenomenon affecting a protozoan parasite (*Toxoplasma*). Science 108: 660–663.

Saeij, J.P., J.P. Boyle, M.E. Grigg, G. Arrizabalaga and J.C. Boothroyd. 2005. Bioluminescence imaging of *Toxoplasma gondii* infection in living mice reveals dramatic differences between strains. Infect. Immun. 73: 695–702.

Sampaio, B.F., M.S. Macre, L.R. Meireles and H.F. Jr. Andrade. 2014. Saliva as a source of anti-*Toxoplasma gondii* IgG for enzyme immunoassay in human samples. Clin. Microbiol. Infect. 20: O72–74. doi: 10.1111/1469-0691.12295.

Sampaio, B.C.F., J.P. Rodrigues, L.R. Meireles and H.F. Jr. Andrade Junior. 2019. Measles, rubella, mumps and *Toxoplasma gondii* antibodies in saliva of vaccinated students of schools and universities in São Paulo City, Brazil. Braz. J. Infect. Dis. Dec 19. pii: S1413-8670(19)30489-1. doi: 10.1016/j. bjid.2019.11.005.

Santoro, M., M. Viscardi, G. Sgroi, N. D'Alessio, V. Veneziano, R. Pellicano et al. 2019. Real-time PCR detection of *Toxoplasma gondii* in tissue samples of wild boars (*Sus scrofa*) from southern Italy reveals high prevalence and parasite load. Parasit. Vectors 12: 335. doi: 10.1186/s13071-019-3586-5.

Scallan, E., R.M. Hoekstra, F.J. Angulo, R.V. Tauxe, M.A. Widdowson, S.L. Roy et al. 2011. Foodborne illness acquired in the United States—major pathogens. Emerg. Infect. Dis. 17: 7–15. doi: 10.3201/ eid1701.P11101.

Schares, G., M.G. Vrhovec, N. Pantchev, D.C. Herrmann and F.J. Conraths. 2008. Occurrence of *Toxoplasma gondii* and *Hammondia hammondi* oocysts in the faeces of cats from Germany and other European countries. Vet. Parasitol. 152: 34–45. doi: 10.1016/j.vetpar.2007.12.004.

Schares, G., M. Koethe, B. Bangoura, A.C. Geuthner, F. Randau, M. Ludewig et al. 2018. *Toxoplasma gondii* infections in chickens—performance of various antibody detection techniques in serum and meat juice relative to bioassay and DNA detection methods. Int. J. Parasitol. 48: 751–762. doi:10.1016/j.ijpara.2018.03.007.

Shapiro, K., M. Kim, V.B. Rajal, M.J. Arrowood, A. Packham, B. Aguilar and S. Wuertz. 2019. Simultaneous detection of four protozoan parasites on leafy greens using a novel multiplex PCR assay. Food Microbiol. 84: 103252. doi: 10.1016/j.fm.2019.103252.

Shobab, L., U. Pleyer, J. Johnsen, S. Metzner, E.R. James, N. Torun et al. 2013. *Toxoplasma* serotype is associated with development of ocular toxoplasmosis. J. Infect. Dis. 208: 1520–1528. doi: 10.1093/ infdis/jit313.

Shwab, E.K., X.Q. Zhu, D. Majumdar, H.F. Pena, S.M. Gennari, J.P. Dubey et al. 2014. Geographical patterns of *Toxoplasma gondii* genetic diversity revealed by multilocus PCR-RFLP genotyping. Parasitol. 141: 453–461.

Sickinger, E., F. Gay-Andrieu, G. Jonas, J. Schultess, M. Stieler, D. Smith et al. 2008. Performance characteristics of the new ARCHITECT Toxo IgG and Toxo IgG Avidity assays. Diagn. Microbiol. Infect. Dis. 62: 235–244.

Sidik, S.M., C.G. Hackett, F. Tran, N.J. Westwood and S. Lourido. 2014. Efficient genome engineering of *Toxoplasma gondii* using CRISPR/Cas9. PLoS One 9: e100450. doi: 10.1371/journal.pone.0100450.

Simon, A., P. Labalette, I. Ordinaire, E. Fréalle, E. Dei-Cas, D. Camus et al. 2004. Use of fluorescence resonance energy transfer hybridization probes to evaluate quantitative real-time PCR for diagnosis of ocular toxoplasmosis. J. Clin. Microbiol. 42: 3681–3685.

Song, K.J., Z. Yang, C.K. Chong, J.S. Kim, K.C. Lee, T.S Kim et al. 2013. A rapid diagnostic test for toxoplasmosis using recombinant antigenic N-terminal half of SAG1 linked with intrinsically unstructured domain of gra2 protein. Korean J. Parasitol. 51: 503–510. doi: 10.3347/kjp.2013.51.5.503.

Sousa, S., N. Canada, J.M. Correia da Costa and M.L. Dardé. 2010. Serotyping of naturally *Toxoplasma gondii* infected meat-producing animals. Vet. Parasitol. 169: 24–28. doi: 10.1016/j.vetpar.2009.12.025.

Sroka, J., J. Karamon, J. Dutkiewicz, A. Wójcik-Fatla and T. Cencek. 2018. Optimization of flotation, DNA extraction and PCR methods for detection of *Toxoplasma gondii* oocysts in cat faeces. Ann. Agric. Environ. Med. 25: 680–685. doi: 10.26444/aaem/97402.

Sroka, J., E. Bilska-Zając, A. Wójcik-Fatla, V. Zając, J. Dutkiewicz, J. Karamon et al. 2019. Detection and molecular characteristics of *Toxoplasma gondii* DNA in retail raw meat products in Poland. Foodborne Pathog. Dis. 16: 195–204. doi: 10.1089/fpd.2018.2537.

Su, C., E.K. Shwab, P. Zhou, X.Q. Zhu and J.P. Dubey. 2010. Moving towards an integrated approach to molecular detection and identification of *Toxoplasma gondii*. Parasitol. 137: 1–11. doi: 10.1017/S0031182009991065.

Su, C., A. Khan, P. Zhou, D. Majumdar, D. Ajzenberg, M.L. Dardé et al. 2012. Globally diverse *Toxoplasma gondii* isolates comprise six major clades originating from a small number of distinct ancestral lineages. Proc. Natl. Acad. Sci. USA 109: 5844–5849.

Su, C. and J.P Dubey. 2020. Isolation and genotyping of *Toxoplasma gondii* strains. Chapter 3 In Christopher J. Tonkin (ed.). *Toxoplasma gondii*: Methods and Protocols, Springer Science LLC. Humana, New York, NY. Methods Mol. Biol. 2071: 49–80. doi: 10.1007/978-1-4939-9857-9_3.

Sugita, S., M. Ogawa, S. Inoue, N. Shimizu and M. Mochizuki. 2011. Diagnosis of ocular toxoplasmosis by two polymerase chain reaction (PCR) examinations: qualitative multiplex and quantitative real-time. Jpn. J. Ophthalmol. 55: 495–501.

Sun, X.M., Y.S. Ji, X.Y. Liu, M. Xiang, G. He, L. Xie et al. 2017. Improvement and evaluation of loop-mediated isothermal amplification for rapid detection of *Toxoplasma gondii* infection in human blood samples. PLoS One 12: e0169125. doi: 10.1371/journal.pone.0169125.

Sutterland, A.L., A. Kuin, B. Kuiper, T. van Gool, M. Leboyer, G. Fond et al. 2019. Driving us mad: the association of *Toxoplasma gondii* with suicide attempts and traffic accidents—a systematic review and meta-analysis. Psychol. Med. 49: 1608–1623. doi: 10.1017/S0033291719000813.

Suzuki, L.A., R.J. Rocha and C.L. Rossi. 2001. Evaluation of serological markers for the immunodiagnosis of acute acquire toxoplasmosis. J. Med. Microbiol. 50: 62–70.

Tefera, T., K.R. Tysnes, K.S. Utaaker and L.J. Robertson. 2018. Parasite contamination of berries: Risk, occurrence, and approaches for mitigation. Food Waterborne Parasitol. 10: 23–38. doi: 10.1016/j.fawpar.2018.04.002.

Temesgen, T.T., L.J. Robertson and K.R. Tysnes. 2019. A novel multiplex real-time PCR for the detection of *Echinococcus multilocularis*, *Toxoplasma gondii*, and *Cyclospora cayetanensis* on berries. Food Res. Int. 125: 108636. doi:10.1016/j.foodres.2019.108636.

Tenter, A.M., A.R. Heckeroth and L.M. Weiss. 2000. *Toxoplasma gondii*: from animals to humans. Int. J. Parasitol. 30: 1217–1258.

Terkawi, M.A., K. Kameyama, N.H. Rasul, X. Xuan and Y. Nishikawa. 2013. Development of an immunochromatographic assay based on dense granule protein 7 for serological detection of *Toxoplasma gondii* infection. Clin. Vaccine Immunol. 20: 596–601. doi: 10.1128/CVI.00747-12.

Torgerson, P.R., B. Devleesschauwer, N. Praet, N. Speybroeck, A.L. Willingham, F. Kasuga et al. 2015. World Health Organization estimates of the global and regional disease burden of 11 foodborne parasitic diseases, 2010: a data synthesis. PLoS Med. 12: e1001920. doi:10.1371/journal.pmed.1001920.

Valenzuela-Moreno, L.F., C.P. Rico-Torres, C. Cedillo-Peláez, H. Luna-Pastén, S.T. Méndez-Cruz, M.E. Reyes-García et al. 2020. Stray dogs in the tropical state of Chiapas, Mexico, harbour atypical and novel genotypes of *Toxoplasma gondii*. Int. J. Parasitol. 50: 85–90. doi: 10.1016/j.ijpara.2019.12.001.

Valian, H.K., H. Mirhendi, M. Mohebali, S. Shojaee, S. Fallahi, R. Jafari et al. 2019. Comparison of the RE-529 sequence and B1 gene for *Toxoplasma gondii* detection in blood samples of the at-risk seropositive cases using uracil DNA glycosylase supplemented loop-mediated isothermal amplification (UDG-LAMP) assay. Microb. Pathog. 140: 103938. doi: 10.1016/j.micpath.2019.103938.

Varlet-Marie, E., Y. Sterkers, M. Perrotte and P. Bastien. 2018. A new LAMP-based assay for the molecular diagnosis of toxoplasmosis: comparison with a proficient PCR assay. Int. J. Parasitol. 48: 457–462

Vera, C.N., W.M. Linam, J.A. Gadde, D.S. Wolf, K. Walson, J.G. Montoya et al. 2020. Congenital toxoplasmosis presenting as eosinophilic encephalomyelitis with spinal cord hemorrhage. Pediatrics. 145: e20191425. doi:10.1542/peds.2019-1425.

Vilares, A., V. Borges, D. Sampaio, I. Ferreira, S. Martins, L. Vieira et al. 2020. Towards a rapid sequencing-based molecular surveillance and mosaicism investigation of *Toxoplasma gondii*. Parasitol. Res. 119: 587–599. doi:10.1007/s00436-019-06523-3.

Villard, O., B. Cimon, J. Franck, H. Fricker-Hidalgo, N. Godineau, S. Houze et al. 2012. Network from the french national reference center for toxoplasmosis. Evaluation of the usefulness of six commercial agglutination assays for serologic diagnosis of toxoplasmosis. Diagn. Microbiol. Infect. Dis. 73: 231–235. doi:10.1016/j.diagmicrobio.2012.03.014.

Villard, O., L. Breit, B. Cimon, J. Franck, H. Fricker-Hidalgo, N. Godineau et al. 2013. Comparison of four commercially available avidity tests for *Toxoplasma gondii*-specific IgG antibodies. Clin. Vaccine Immunol. 20: 197–204.

Villard, O., B. Cimon, C. L'Ollivier, H. Fricker-Hidalgo, N. Godineau, S. Houze et al. 2016a. Help in the choice of automated or semiautomated immunoassays for serological diagnosis of toxoplasmosis: evaluation of nine immunoassays by the French National Reference Center for Toxoplasmosis. J. Clin. Microbiol. 54: 3034–3042.

Villard, O., B. Cimon, C. L'Ollivier, H. Fricker-Hidalgo, N. Godineau, S. Houze et al. 2016b. Serological diagnosis of *Toxoplasma gondii* infection: Recommendations from the French National Reference Center for Toxoplasmosis. Diagn. Microbiol. Infect. Dis. 84: 22–33. doi:10.1016/j.diagmicrobio.2015.09.009.

Villena, I., D. Aubert, P. Gomis, H. Ferté, J.C. Inglard, H. Denis-Bisiaux et al. 2004. Evaluation of a strategy for *Toxoplasma gondii* oocyst detection in water. Appl. Environ. Microbiol. 70: 4035–4039.

Villena, I., B. Durand, D. Aubert, R. Blaga, R. Geers, M. Thomas, C. Perret et al. 2012. New strategy for the survey of *Toxoplasma gondii* in meat for human consumption. Vet. Parasitol. 183: 203–208.

Wallon, M., J. Franck, P. Thulliez, C. Huissoud, F. Peyron, P. Garcia-Meric et al. 2010. Accuracy of real-time polymerase chain reaction for *Toxoplasma gondii* in amniotic fluid. Obstet. Gynecol. 115: 727–733. doi:10.1097/AOG.0b013e3181d57b09.

Wallon, M., F. Peyron, C. Cornu, S. Vinault, M. Abrahamowicz, C.B. Kopp et al. 2013. Congenital *Toxoplasma* infection: monthly prenatal screening decreases transmission rate and improves clinical outcome at age 3 years. Clin. Infect. Dis. 56: 1223–1231. doi: 10.1093/cid/cit032.

Wang, H., C. Lei, J. Li, Z. Wu, G. Shen and R. Yu. 2004. A piezoelectric immunoagglutination assay for *Toxoplasma gondii* antibodies using gold nanoparticles. Biosens. Bioelectron. 19: 701–709.

Wang, Y.H., X.R. Li, G.X. Wang, H. Yin, X.P. Cai, B.Q. Fu et al. 2011. Development of an immunochromatographic strip for the rapid detection of *Toxoplasma gondii* circulating antigens. Parasitol. Int. 60: 105–107.

Wang, H., T. Wang, Q. Luo, X. Huo, L. Wang, T. Liu et al. 2012. Prevalence and genotypes of *Toxoplasma gondii* in pork from retail meat stores in Eastern China. Int. J. Food Microbiol. 157: 393–397. doi: 10.1016/j.ijfoodmicro.2012.06.011.

Wang, J.L., S.Y. Huang, M.S. Behnke, K. Chen, B. Shen and X.Q. Zhu. 2016. The past, present, and future of genetic manipulation in *Toxoplasma gondii*. Trends Parasitol. 32: 542–553. doi: 10.1016/j.pt.2016.04.013.

Weiss, L.M. and J.P. Dubey. 2009. Toxoplasmosis: a history of clinical observations. Int. J. Parasitol. 39: 895–901.

Wells, B., H. Shaw, G. Innocent, S. Guido, E. Hotchkiss, M. Parigi et al. 2015. Molecular detection of *Toxoplasma gondii* in water samples from Scotland and a comparison between the 529 bp real-time PCR and ITS1 nested PCR. Water Res. 87: 175–181. doi: 10.1016/j.watres.2015.09.015.

Wu, Y.D., M.J. Xu, Q.Q. Wang, C.X. Zhou, M. Wang, X.Q. Zhu et al. 2017. Recombinase polymerase amplification (RPA) combined with lateral flow (LF) strip for detection of *Toxoplasma gondii* in the environment. Vet. Parasitol. 243: 199–203.

Xicoténcatl-García, L., S. Enriquez-Flores and D. Correa. 2019. Testing new peptides From *Toxoplasma gondii* SAG1, GRA6, and GRA7 for serotyping: Better definition using GRA6 in mother/newborns pairs with risk of congenital transmission in Mexico. Front. Cell Infect. Microbiol. 9: 368. doi: 10.3389/fcimb.2019.00368.

Yamada, H., K. Tanimura, M. Deguchi, S. Tairaku, M. Morizane. A. Uchida et al. 2019. A cohort study of maternal screening for congenital *Toxoplasma gondii* infection: 12 years' experience. J. Infect. Chemother. 25: 427–430. doi: 10.1016/j.jiac.2019.01.009.

Yan, C., L.J. Liang, K.Y. Zheng and X.Q. Zhu. 2016. Impact of environmental factors on the emergence, transmission and distribution of *Toxoplasma gondii*. Parasit. Vectors. 9: 137. doi: 10.1186/s13071-016-1432-6.

Yazar, S., O. Yaman, B. Eser, F. Altuntas, F. Kurnaz and I. Sahin. 2004. Investigation of anti-*Toxoplasma gondii* antibodies in patients with neoplasia. J. Med. Microbiol. 53: 1183–1186.

Year, H., D. Filisetti, P. Bastien, T. Ancelle, P. Thulliez and L. Delhaes. 2009. Multicenter comparative evaluation of five commercial methods for *Toxoplasma* DNA extraction from amniotic fluid. J. Clin. Microbiol. 47: 3881–3886. doi:10.1128/JCM.01164-09.

Yücesan, B., C. Babür, N. Koç, F. Sezen, S. Kılıç and Y. Gürüz. 2019. Investigation of anti-*Toxoplasma gondii* antibodies in cats using sabin-feldman dye test in Ankara in 2016. Turkiye Parazitol. Derg. 43: 5–9. doi:10.4274/tpd.galenos.2019.6126.

Zhou, N., H. Fu, Z. Wang, H. Shi, Y. Yu, T. Qu et al. 2019. Seroprevalence and risk factors of *Toxoplasma gondii* infection in children with leukemia in Shandong Province, Eastern China: a case-control prospective study. Peer J. 7: e6604. doi: 10.7717/peerj.6604.

Zhuo, X., B. Huang, L. Luo, H. Yu, B. Yan, Y. Yang et al. 2015. Development and application of loop-mediated isothermal amplification assays based on ITS-1 for rapid detection of *Toxoplasma gondii* in pork. Vet. Parasitol. 208: 246–249. doi: 10.1016/j.vetpar.2015.01.008.

Chapter 8

Biology and Epidemiology of *Cyclospora cayetanensis*

Ynes R. Ortega* and Lordwige Atis

Introduction

Cyclospora cayetanensis is a coccidian parasite that infects humans and causes gastrointestinal illness. Individuals can acquire infection by ingestion of contaminated food or water. In the late 1980s and early 1990s, *Cyclospora* was considered to be a coccidian-like or cyanobacteria-like body. This assumption was based on the autofluorescent properties of its oocyst and morphology. In 1992 and 1994, *Cyclospora* was fully described and named *Cyclospora cayetanensis* (Ortega et al. 1992; 1994). This chapter provides an update on the epidemiology, histopathology, treatment, detection, and control of *C. cayetanensis*.

Biology

Cyclospora oocysts measure 8–10 μm and are excreted in the feces of infected individuals. These oocysts are unsporulated and undifferentiated, with characteristic morula internal appearance. When exposed to 24 ± 2°C, oocysts start to sporulate, resulting in the formation of two sporocysts, each containing two sporozoites (Figure 1). The sporulation process generally takes about 7–15 days. When a susceptible individual ingests the oocysts, excystation begins and two sporocysts are excreted (Figure 2). The sporozoites exit the sporocyst as a result of enzymatic digestion when passing through the gastrointestinal tract of the infected individual. Sporozoites then infect the intestinal epithelial cells of the small intestine. A parasitic vacuole is formed, and asexual multiplication occurs. This is followed by sexual multiplication, resulting in the formation of a zygote that undergoes differentiation to

Center for Food Safety, University of Georgia, 1109 Experiment St, Griffin, GA 30223, USA.
* Corresponding author: ortega@uga.edu

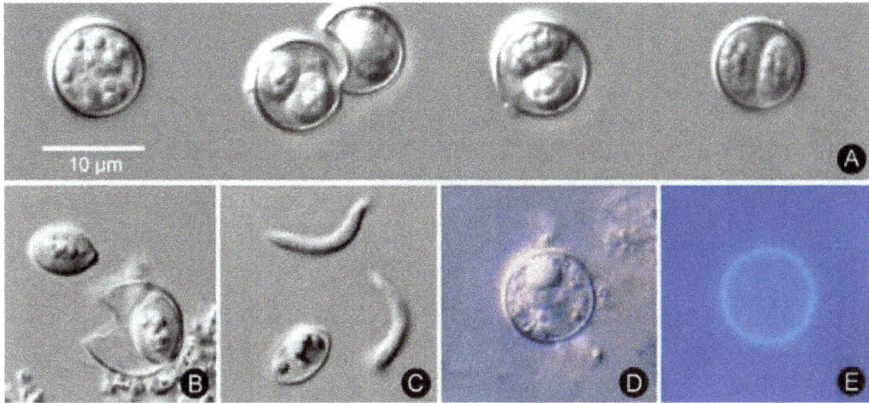

Figure 1. (A) *Cyclospora* oocysts at different stages of sporulation. (B and C) Oocyst excystation (D) Unsporulated oocyst by IDC and (E) epifluorescence microscopy. Image courtesy of DPDx, Centers for Disease Control and Prevention (https://www.cdc.gov/dpdx/).

form the unsporulated oocyst. Oocysts are then excreted in the feces of the infected individual. *Cyclospora* requires at least 7–15 days to fully sporulate and become infectious.

Twenty-one species of *Cyclospora* have been found in animals. Non-human primates, rodents, and snakes can harbor *Cyclospora* spp., but *C. cayetanensis* is the only species identified in humans (Ortega and Sanchez 2010). *C. cayetanensis* is considered anthroponotic, but there have been reports describing oocysts in poultry, ducks, and dogs that are morphologically similar. However, there is no evidence that the parasite infects these animal species. Efforts to infect various animals with *C. cayetanensis* oocysts have been unsuccessful (Eberhard et al. 2000), suggesting that perhaps the identification of oocysts in animal feces may be the result of coprophagia.

Symptoms and Histopathology

Most of the studies on epidemiology and clinical presentation in individuals with cyclosporiasis have been done in endemic locations and confirmed in the US and elsewhere. Symptoms develop between three and seven days after a person is infected. Patients develop acute diarrhea, bloating, cramps, abdominal pain, nausea, weight loss, and in some cases, fever, vomiting, and constipation (Ortega et al. 1997).

Although there are no gross abnormalities such as hemorrhaging or ulcers, the architecture of the intestinal mucosa shows a shortening and widening of intestinal villi. Edema and infiltration of plasma cells and lymphocytes in the villus mucosa have been observed in the intestinal tissues of patients with cyclosporiasis. No parasites have been observed in the lamina propria or submucosa (Ortega et al. 2007; Sun 1996). Parasitic vacuoles are not abundant in intestinal epithelia as is the case of other parasitic infections such as cryptosporidiosis.

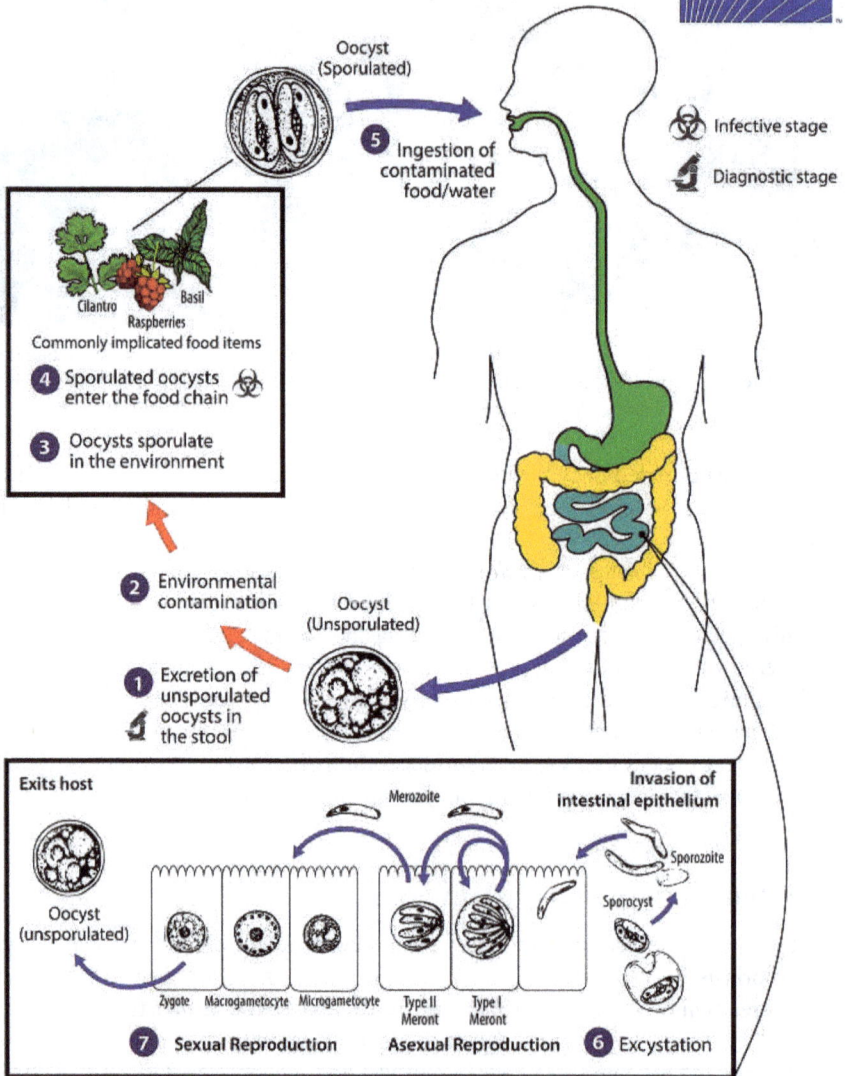

Figure 2. Life cycle of *Cyclospora cayetanensis*. Image courtesy of DPDx, Centers for Disease Control and Prevention (https://www.cdc.gov/dpdx/).

Epidemiology

Cases of cyclosporiasis have been reported worldwide. However, the epidemiology of cyclosporiasis has been studied in only a few countries, most of them in tropical and subtropical areas. Early studies in Nepal were done by expatriates who developed diarrhea while visiting these locations (Hogue et al. 1993). In Peru, cohort studies in children from shanty towns show that a significant proportion of infections are asymptomatic. In many cases, the duration of infection and oocyst shedding is shorter as a result of repeated infections (Bern et al. 2009). Studies in Guatemala and Haiti have shown similar results.

Diagnosis of cyclosporiasis is performed by examining fecal samples for the presence of oocysts (Ortega and Sterling 2018). The disease is most common in tropical and subtropical regions (Hall et al. 2012). A marked seasonal pattern, based on outbreaks and cases in past years, has been observed in several countries. A seasonal pattern of outbreaks in May–August has also occurred in the US (Ortega and Robertson 2017). Hall et al. (2012) observed an increase in stool samples positive for the parasite in June and July.

Sporadic cases and outbreaks of cyclosporiasis have been reported since its discovery in 1994 in the US. Outbreaks prior to 2013 were mostly associated with berries and leafy greens. The most notable outbreaks occurred in 1996 when 1,465 cases were reported in the US and Canada, and in 1997 with 1,012 cases in the US. Both outbreaks were associated with the consumption of raspberries grown in Guatemala. In 1997, an outbreak was associated with basil pesto affecting 308 individuals. Smaller outbreaks associated mostly with salad greens, herbs, and berries were also reported. It was not until 2013 that larger multistate outbreaks were reported in the US. In 2013, 631 cases implicating lettuce and cilantro were reported in various states. In 2015 and 2016, cilantro imported from Mexico was implicated in causing 564 and 384 cases, respectively, of cyclosporiasis in the US. In 2017, 1,065 cases were reported but the implicated food was not determined. In 2018, trays consisting of broccoli, cauliflower, and carrots were implicated in an outbreak affecting 241 individuals. In the same year, another outbreak was associated with salad mixes produced in the US and commercialized by a fast-food chain. A total of 2,999 cases in 33 states were reported. These cases had no travel history, thus suggesting that infection occurred in the U.S. In 2019, the total number of cases of cyclosporiasis in non-travelers was 2,408. Multiple outbreaks occurred that year. One of these outbreaks affecting 241 individuals implicated imported basil from Mexico. In 2021, 1,020 cases of cyclosporiasis were reported. Two outbreaks with 40 and 130 cases, respectively, implicated leafy greens (www.cdc.gov).

Since several outbreaks in the US have been associated with the consumption of imported cilantro from Puebla, Mexico, an Import Alert #24–23 was issued by the US FDA in 2015 and has been in effect in subsequent years from April to August, which coincide with the high season of cyclosporiasis in Mexico. This alert allows for detention without physical examination of fresh cilantro grown in the state of Puebla. The alert covers fresh cilantro, either intact or chopped. Multi-ingredient processed foods that contain cilantro are not covered by this alert. Refusal of admission to the US can result from a product that appears to have been manufactured, processed, or packaged under unsanitary conditions (FDA 2021).

Clinical Detection

Microscopy and conventional ova and parasite examination are used in the diagnosis of cyclosporiasis. Oocysts autofluoresce neon green under ultraviolet light (Long et al. 1991) but, depending on the excitation filter, they can look blue or green (Varea et al. 1998). Differential interference contrast (DIC), phase contrast, and bright field microscopy can be used to view *Cyclospora* (Ortega and Sterling 2018) (Figure 1). Staining methods can also be used. Safranin stains with heating (Visvesvara et al. 1997) and without heating (Maratim et al. 2002) are used, as they provide a more homogeneous red coloration of the oocysts. Acid-fast stains are also used, although oocysts show variability in stain adherence (Ortega and Sterling 2018). Intermittent shedding of oocysts as well as variations in time for sporulation can be challenges for *Cyclospora* detection (Chacin-Bonilla 2017).

Faster methods such as commercial antibody kits for detecting *Cyclospora* are not available but attempts to develop them are in progress (Chacin-Bonilla 2017). There is a commercial BioFire Film Array gastrointestinal panel available that allows for the detection of *Cyclospora* directly from fecal samples with sensitivity/positive percent agreement and specific/negative percent agreement at 100% (Buss et al. 2015). The assay has a run time of 1 hour per specimen (Buss et al. 2015). ELISA has been used to detect *Cyclospora*-specific antibodies in fecal samples in China (Wang et al. 2002).

Detection of *Cyclospora* in Environmental Samples and Foods

Molecular methods are preferred over microscopy for examining environmental samples, as their sensitivity and specificity aid in detecting low counts of *Cyclospora* oocysts in samples (Ortega and Cama 2018), which are often lower than in human fecal samples (Jinneman et al. 1998). Molecular methods also allow for the differentiation of *Cyclospora* from close relatives, e.g., *Eimeria* species (Relman et al. 1996). Microscopic analysis of *Cyclospora* oocysts can be less efficient for examining environmental samples because of challenges concerning variability in autofluorescence and staining (Varma et al. 2003). Molecular methods are performed after purifying and concentrating *Cyclospora* oocysts isolated from environmental samples (Chacin-Bonilla 2017). A major limitation of these methods is that they cannot differentiate between sporulated or unsporulated oocysts, which is needed for infection diagnosis.

Polymerase Chain Reaction Assays

Polymerase chain reaction (PCR) assays for detecting *Cyclospora* were first used for the analysis of clinical samples. Relman et al. (1996) were the first to demonstrate the use of nested PCR targeting the 18S rRNA gene in clinical samples collected from patients infected with *Cyclospora* in Nepal (Relman et al. 1996). When applied to environmental samples, it was observed that this method also amplified DNA from the Eimeridae group (Ortega 2017; Yoder et al. 1996). Jinneman et al. (1998) developed an RFLP method using the same target sequence gene. This assay was able to differentiate some Eimeridae species from *C. cayetanensis*. Jinneman et al. (1999) developed an oligonucleotide ligation assay (OLA) with the intent to simplify discrimination between positive and negative samples. However, this method can be prone to contamination and the process is long and laborious (Varma et al. 2003).

The first qPCR method targeting the 18S rRNA gene sequence to detect *Cyclospora* oocysts was done by Varma et al. (2003). Other methods for detecting *Cyclospora* have also been developed. Sulaiman described a nested PCR method for the amplification of a heat shock protein and the 18SrRNA gene (Sulaiman et al. 2013; 2014). The presence of ITS2 in *Cyclospora* as a diagnostic target has been explored (Lalonde et al. 2008). Further modifications of these methods have resulted in multiple studies in which foods and environmental samples have been tested.

PCR for Detection in Produce

Food and environmental samples can contain PCR inhibitors and high levels of background DNA (Chacin-Bonilla 2017; Lalonde and Gajadhar 2008). This problem has been addressed using exogenous non-competitive internal amplification controls in qPCR assays and tested for efficacy in fresh and frozen berries (Assurian et al. 2020). The US FDA has validated PCR assays for detecting *Cyclospora* in raspberries and cilantro (Murphy et al. 2018) and in shredded carrots, parsley, basil, and romaine lettuce (Murphy et al. 2017).

The first US FDA method for detecting *Cyclospora* in fresh produce was published in 2004 in Chapter 19a of the FDA Bacteriological Analytical Manual (BAM) along with a method for detecting *Cryptosporidium* (Orlandi et al. 2004). The method for *Cyclospora* was later modified and can be found in Chapter 19b of the FDA Bacteriological Analysis Methods Book (2017), which replaced previous methods for detecting *Cyclospora* (Murphy et al. 2017). The 2004 method in Chapter 19a describes isolation and identification steps using microscopy and PCR analysis. Chapter 19b uses a qPCR method. The reagents and washing steps have also changed. Chapter 19a originally used several buffers (e.g., Envirocheck™ elution buffer, NET buffer, and NET-BSA buffer) among other reagents for the produce-washing step. Chapter 19b replaced all reagents with 0.1% Alconox produce wash solution and sterile nuclease-free deionized water.

Assurian et al. (2020) found the U.S. FDA BAM Chapter 19b method able to detect as low as five *Cyclospora* oocysts in fresh and frozen berries. The detection rate increased with an increase in the number of oocysts. C_T values and recovery percentage showed that the method could detect very low numbers in inoculated samples of mixed berries.

Improving Cyclospora Detection in Produce by Washing

Wash solutions for removing *Cyclospora* oocysts from produce can vary in efficacy, so it is imperative that a solution shown to be superior be used for molecular detection. Carefully washing environmental samples may remove inhibitors that interfere with qPCR methods (Assurian et al. 2020). Lalonde and Gajadhar (2008) used 1 M glycine buffer at pH 5.5 to wash basil when they evaluated PCR assays for detecting *Cyclospora*. Shields et al. (2012) demonstrated the efficacy of 0.1% Alconox in removing *Cyclospora* oocysts from inoculated lettuce, basil, and raspberries incubated for 24 hours. The average percent recovery using 0.1% Alconox ranged from 34%

in raspberries to 80.2% in Mesclun/Spring lettuce mix. Chandra et al. (2014) later demonstrated variability in recovery using 0.1% Alconox. They tested the effect of different incubation times on the recovery of *Cyclospora* oocysts inoculated on basil. Six wash solutions (E-pure water, 3% levulinic acid/3% SDS, 1M glycine at pH 5.5, 0.1 PBS at pH 7, 0.1% Alconox, and 1% HCl/pepsin) were evaluated using nested PCR assays. For samples spiked with 1,000 oocysts and incubated for 1 hour prior to analysis, 0.1% Alconox had the same percentage of PCR-positive samples as did washing with 0.1 M PBS at pH 7 (94%), and performed slightly better than 0.1 M PBS at pH 7, resulting in detection of PCR-positive samples of 77.8 and 66.7%, respectively. However, for samples inoculated with only 100 oocysts, washing with 0.1 M PBS at pH 7 performed better than 0.1% Alconox for produce incubated for 1 hour (22.2% vs. 18.5%) or 24 hours (50% vs 37%). Additionally, 0.1% Alconox had the least PCR-positives for recovery of 100 oocysts inoculated in basil and incubated 1 hour compared to all other wash solutions and the second least positive PCR positives, but the same percentage as E-pure water, for samples incubated for 24 hours. These data suggest that 0.1% Alconox performs better for removing *Cyclospora* from basil, inoculated with a larger number of oocysts and with shorter incubation time. This is a concern because environmental samples most likely contain low counts of *Cyclospora* oocysts (Ortega and Cama 2018). Additionally, foam formation occurs when 0.1% Alconox was used in the washing process and this can potentially lead to a reduction in the recovery of oocysts (Chandra et al. 2014). Produce type and incubation times should certainly be considered when choosing a wash solution in methods for detecting *Cyclospora* oocysts.

In another study, *Cyclospora* was detected in 70% of blackberries and strawberries that were inoculated with five oocysts and increased to 90% in mixed berries. For berries inoculated with 10 and 200 oocysts and incubated 2 hours prior to recovery, 100% tested positive (Assurian et al. 2020).

Genotyping

The whole genome of *Cyclospora* was first sequenced by Qvarnstrom et al. (2015). A base length of 44,563,857 was reported. The apicoplast and mitochondrial genomes have also been sequenced (Tang et al. 2015). Multilocus sequence typing (MLST) was developed as a genotyping tool for identifying *Cyclospora* (Guo et al. 2016). The tool was only able to provide useful data for 34 of the 64 specimens due to interference by PCR products (Guo et al. 2016). Hofstetter et al. (2019) evaluated a modified MLST method and found it to have poor discriminatory power. The mitochondrial junction region of *Cyclospora* has been shown to be a potential genotyping marker (Nascimento et al. 2019). Continued work on genotyping tools is valuable, as they can be used to characterize isolates in foodborne outbreak investigations (Assurian et al. 2020). Recent advances have been made in genotyping methodology for *Cyclospora*. One objective has been to determine if *Cyclospora* outbreaks can be discriminated against based on the clusters of cases and second if there are single or multiple isolates involved. This will result in better discrimination among various outbreaks that could be occurring at the same time. These efforts, led by the Division of Parasitic Diseases at the US Centers for Disease Control and Prevention (CDC) and other research groups worldwide are being performed using human fecal samples containing large numbers of oocysts.

The group at CDC has described an assemblage of two similarly-based classification algorithms which infer the relatedness of *Cyclospora* infections. Three SNP-rich loci were selected that captured 4 to 20 SNPs. The clustering of genetically related *Cyclospora* in certain individuals coincided with the epidemiologic clustering of outbreak cases, suggesting that this methodology may be useful for epidemiologic studies (Barratt et al. 2019).

The use of complete mitochondrial genomes and next-generation sequencing is also being investigated to determine if it can be used in outbreak investigations and traceback studies. Whether this methodology can be applied to foods needs to be determined as the number of oocysts in foods is very low.

Treatment

Treatment of patients with cyclosporiasis was determined soon after the description of the parasite. Trimethoprim-sulfamethoxazole (co-trimoxazole 160/800 mg, twice a day) is the treatment of choice. Symptoms, including diarrhea, cease as soon as 1-day post-treatment (Madico 1993). In children, treatment of 5/25 mg/kg/d results in cessation of symptoms by day 2 and oocyst excretion by day 5 (Madico et al. 1997). Ciprofloxacin and nitazoxanide have also been suggested as alternatives but they are not as effective for sulfa-allergic individuals with cyclosporiasis (Zimmer et al. 2007).

Conclusion

Historically, cyclosporiasis was considered an emerging disease that affected individuals from developing countries or travelers returning from *Cyclospora*-endemic locations. More recent reports are showing cases of *Cyclospora* infections in developed countries. Most are associated with foodborne outbreaks. As we study the movement of foods in international commerce, it is clear that this parasite, along with bacterial foodborne pathogens, can be transported around the world. In almost every recent year, the US has experienced outbreaks of cyclosporiasis in travelers and non-travelers. *Cyclospora* continues to be a difficult organism to study and control. This is largely due to the availability of oocysts for research. The lack of an animal model to propagate the parasite has proven to limit studies. However, the use of a surrogate organism may be an alternative to better understand the behavior of this parasite. Traceback studies and analysis of food and environmental samples also present challenges as the number of oocysts present may be too low to be detected using current methods but sufficient to cause infection.

References

Assurian, A., H. Murphy, L. Ewing, H.N. Cinar, A. da Silva and S. Almeria. 2020. Evaluation of the U.S. Food and Drug Administration validated molecular method for detection of *Cyclospora cayetanensis* oocysts on fresh and frozen berries. Food Microbiol. 87: 103397.

Bern, C., Y. Ortega, W. Checkley, J.M. Roberts, A.G. Lescano, L. Cabrera, M. Verastegui, R.E. Black, C. Sterling and R.H. Gilman. 2002. Epidemiologic differences between cyclosporiasis and cryptosporidiosis in Peruvian children. Emerg. Infect. Dis. 8(6): 581–5.

Buss, S.N., A. Leber, K. Chapin, P.D. Fey, M.J. Bankowski, M.K. Jones, M. Rogatcheva, K.J. Kanack and K.M. Bourzac. 2015. Multicenter evaluation of the BioFire FilmArray gastrointestinal panel for etiologic diagnosis of infectious gastroenteritis. J. Clin. Microbiol. 53: 915–925.

CDC. 2018a. Cyclosporiasis Outbreak Investigations—United States, 2017 [Accessed 2018 June 18th] Available from: https://www.cdc.gov/parasites/cyclosporiasis/outbreaks/2017/index.html.

CDC. 2018b. Parasites—Cyclosporiasis (*Cyclospora* Infection) [Accessed 2020 February 6th] Available from: https://www.cdc.gov/parasites/cyclosporiasis/epi.html.

CDC. 2018c. U.S. Foodborne outbreaks of cyclosporiasis—2000–2016 [Accessed 2018 March 23] Available from: https://www.cdc.gov/parasites/cyclosporiasis/outbreaks/foodborneoutbreaks.html.

CDC. 2022. Domestically Acquired Cases of Cyclosporiasis — United States, May–August 2021. https://www.cdc.gov/parasites/cyclosporiasis/outbreaks/2021/seasonal/index.html (Accessed 2022 June 18).

Chacin-Bonilla, L. 2017. *Cyclospora cayetanensis. In*: Rose, J.B. and B. Jimenez-Cisneros (eds.). Global Water Pathogen Project. Michigan State University, E. Lansing, MI, UNESCO.http://www.waterpathogens.org/book/cyclospora-cayetanensis.

Chandra, V., M. Torres and Y.R. Ortega. 2014. Efficacy of wash solutions in recovering *Cyclospora cayetanensis, Cryptosporidium parvum*, and *Toxoplasma gondii* from basil. J. Food Prot. 77: 1348–1354.

Eberhard, M.L., Y.R. Ortega, D.E. Hanes, E.K. Nace, R.Q. Do, M.G. Robl, K.Y. Won, C. Gavidia, N.L. Sass, K. Mansfield, A. Gozalo, J. Griffiths, R.H. Gilman, C.R. Sterling and M.J. Arrowood. 2000. Attempts to establish experimental Cyclospora cayetanensis infection in laboratory animals. Journal of Parasitology 86(3): 577–582.

Erickson, M.C. and Y.R. Ortega. 2006. Inactivation of protozoan parasites in food, water, and environmental systems. J. Food Prot. 69: 2786–2808.

FDA. 2021. Import Alert 24-23. Detention without physical examination of fresh cilantro from the state of Puebla, Mexico. Seasonal (April 1–August 30) https://www.accessdata.fda.gov/cms_ia/importalert_1148.html Accessed: 10/25/2021.

Guo, Y., D.M. Roellig, N. Li, K. Tang, M. Frace, Y. Ortega, M.J. Arrowood, Y. Feng, Y. Qvarnstrom, L. Wang, D.M. Moss, L. Zhang and L. Xiao. 2016. Multilocus sequence typing tool for *Cyclospora cayetanensis*. Emerg. Infect. Dis. 22: 1464–1467.

Hall, R.L., J.L. Jones, S. Hurd, G. Smith, B.E. Mahon and B.L. Herwaldt. 2012. Population-based active surveillance for *Cyclospora* infection—United States, foodborne diseases active surveillance network (FoodNet), 1997–2009. Clin. Infect. Dis. 54: S411–S417.

Hofstetter, J.N., F.S. Nascimento, S. Park, S. Casillas, B.L. Herwaldt, M.J. Arrowood and Y. Qvarnstrom. 2019. Evaluation of multilocus sequence typing of *Cyclospora cayetanensis* based on microsatellite markers. Parasite. 26: 3–3.

Hoge, C.W., D.R. Shlim, R. Rajah, J. Triplett, M. Shear, J.G. Rabold and P. Echeverria. 1993. Epidemiology of diarrhoeal illness associated with coccidian-like organism among travellers and foreign residents in Nepal. Lancet. 341(8854): 1175–9.

Jinneman, K.C., J.H. Wetherington, W.E. Hill, A.M. Adams, J.M. Johnson, B.J. Tenge, N.L. Dang, R.L. Manger and M.M. Wekell. 1998. Template preparation for PCR and RFLP of amplification products for the detection and identification of *Cyclospora* sp. and *Eimeria* spp. oocysts directly from raspberries. J. Food Prot. 61: 1497–1503.

Jinneman, K.C., J.H. Wetherington, W.E. Hill, C.J. Omiescinski, A.M. Adams, J.M. Johnson, B.J. Tenge, N.L. Dang and M.M. Wekell. 1999. An oligonucleotide-ligation assay for the differentiation between

Cyclospora and *Eimeria* spp. Polymerase chain reaction amplification products. J. Food Prot. 62: 682–685.

Lalonde, L.F. and A.A. Gajadhar. 2008. Highly sensitive and specific PCR assay for reliable detection of *Cyclospora cayetanensis* oocysts. Appl. Environ. Microbiol. 74: 4354–4358.

Long, E.G., E.H. White, W.W. Carmichael, P.M. Quinlisk, R. Raja, B.L. Swisher, H. Daugharty and M.T. Cohen. 1991. Morphologic and staining characteristics of a cyanobacterium-like organism associated with diarrhea. J. Infect. Dis. 164: 199–202.

Madico, G., J. McDonald, R.H. Gilman, L. Cabrera and C.R. Sterling. 1997. Epidemiology and treatment of *Cyclospora cayetanensis* infection in Peruvian children. Clin. Infect. Dis. 24(5): 977–81.

Madico, G., R.H. Gilman, E. Miranda, L. Cabrera and C.R. Sterling. 1993. Treatment of Cyclospora infections with co-trimoxazole. Lancet 342(8863): 122–3.

Maratim, A., K. Kamar, A. Ngindu and C. Akoru. 2002. Safranin staining *Cyclospora cayetanensis* oocysts not requiring microwave heating. British J. Biomed. Sci. 59: 114.

Murphy, H.R., S. Almeria and A.J. da Silva. 2017. BAM 19b: Molecular detection of *Cyclospora cayetanensis* in fresh produce using real-time PCR. FDA Bacteriological Analytical Manual Available from: https://www.fda.gov/Food/FoodScienceResearch/LaboratoryMethods/ucm553445.htm.

Murphy, H.R., H.N. Cinar, G. Gopinath, K.E. Noe, L.D. Chatman, N.E. Miranda, J.H. Wetherington, J. Neal-McKinney, G.S. Pires and E. Sachs. 2018. Interlaboratory validation of an improved method for detection of *Cyclospora cayetanensis* in produce using a real-time PCR assay. Food Microbiol. 69: 170–178.

Nascimento, F.S., J.R. Barta, J. Whale, J.N. Hofstetter, S. Casillas, J. Barratt, E. Talundzic, M.J. Arrowood and Y. Qvarnstrom. 2019. Mitochondrial junction region as genotyping marker for *Cyclospora cayetanensis*. Emerg. Infect. Dis. 25: 1314–1319.

Nickerson, D.A., R. Kaiser, S. Lappin, J. Stewart, L. Hood and U. Landegren. 1990. Automated DNA diagnostics using an ELISA-based oligonucleotide ligation assay. Proc. Natl. Acad. Sci. USA 87: 8923–8927.

Orlandi, P.A., C. Frazar, L. Carter and D.T. Chu. 2004. BAM Chapter 19a: Detection of *Cyclospora* and *Cryptosporidium* from fresh produce: Isolation and Identification by Polymerase Chain Reaction (PCR) and Microscopic analysis. FDA Bacteriological Analytical Manual Available from: https://www.fda.gov/food/laboratory-methods-food/bam-19a-detection-cyclospora-and-cryptosporidium.

Ortega, Y. and R. Sanchez. 2010. *Cyclospora cayetanensis*: A food and waterborne parasite. Clinical Microbiology Reviews 23(1): 218–234.

Ortega, Y.R. and C.R. Sterling. 2018. Foodborne Parasites [electronic resource]. Springer International Publishing.

Ortega, Y.R. and L.J. Robertson. 2017. Introduction to *Cyclospora cayetanensis*: The parasite and the disease. pp. 1–7. *In*: *Cyclospora cayetanensis* as a Foodborne Pathogen. Springer, Cham, SpringerBriefs in Food, Health, and Nutrition.

Ortega, Y.R. and V.A. Cama. 2018. *Cyclospora cayetanensis*. pp. 41–56. *In*: Ortega, Y.R. and C.R. Sterling (eds.). Foodborne Parasites. Springer International Publishing.

Ortega, Y.R., C.R. Sterling, R.H. Gilman, V.A. Cama and F. Diaz. 1993. *Cyclospora* species—a new protozoan pathogen of humans. N Engl. J. Med. 328: 1308–1312.

Ortega, Y.R., R. Nagle, R.H. Gilman, J. Watanabe, J. Miyagui, H. Quispe, P. Kanagusuku, C. Roxas and C.R. Sterling. 1997. Pathological and clinical findings in patients with cyclosporiasis and a description of intracellular parasite life cycle stages. The Journal of Infectious Diseases 176: 1584–1589.

Qvarnstrom, Y., Y. Wei-Pridgeon, W. Li, F.S. Nascimento, H.S. Bishop, B.L. Herwaldt, D.M. Moss, V. Nayak, G. Srinivasamoorthy and M. Sheth. 2015. Draft genome sequences from *Cyclospora cayetanensis* oocysts purified from a human stool sample. Genome Announc. 3: e01324–01315.

Relman, D.A., T.M. Schmidt, A.A. Gajadhar, M. Sogin, J. Cross, K. Yoder, O. Sethabutr and P. Echeverria. 1996. Molecular phylogenetic analysis of *Cyclospora*, the human intestinal pathogen, suggests that it is closely related to *Eimeria* species. J. Infect. Dis. 173: 440.

Shields, J.M., M.M. Lee and H.R. Murphy. 2012. Use of a common laboratory glassware detergent improves recovery of *Cryptosporidium parvum* and *Cyclospora cayetanensis* from lettuce, herbs and raspberries. Int. J. Food Microbiol. 153: 123–128.

Sulaiman, I., Y. Ortega, S. Simpson and K. Kerdahi. 2014. Genetic characterization of human-pathogenic *Cyclospora cayetanensis* parasites from three endemic regions at the 18S ribosomal RNA locus. Infect. Genet. Evol. 22: 229–234.

Sulaiman, I.M., M.P. Torres, S. Simpson, K. Kerdahi and Y. Ortega. 2013. Sequence characterization of heat shock protein gene of *Cyclospora cayetanensis* parasites from Nepal, Mexico, and Peru. Journal of Parasitology 99(2): 379–382.

Sun, T., C.F. Ilardi, D.S. Asnis, A.R. Bresciani, S. Goldenberg, B. Roberts and S. Teichberg. 1996. Light and electron microscopic identification of Cyclospora species in the small intestine. Evidence of the presence of asexual life cycle in human host. Amer. J. Clin. Pathol. 105(2): 216–20.

Tang, K., Y. Guo, L. Zhang, L.A. Rowe, D.M. Roellig, M.A. Frace, N. Li, S. Liu, Y. Feng and L. Xiao. 2015. Genetic similarities between *Cyclospora cayetanensis* and cecum-infecting avian *Eimeria* spp. in apicoplast and mitochondrial genomes. Parasit Vectors 8: 358–358.

Varea, M., A. Clavel, O. Doiz, F.J. Castillo, M.C. Rubio and R. Gómez-Lus. 1998. Fuchsin fluorescence and autofluorescence in *Cryptosporidium, Isospora* and *Cyclospora* oocysts. Int. J. Parasitol. 28: 1881–1883.

Varma, M., J.D. Hester, F.W. Schaefer III, M.W. Ware and H.A. Lindquist. 2003. Detection of *Cyclospora cayetanensis* using a quantitative real-time PCR assay. J. Microbial Meth. 53: 27–36.

Visvesvara, G.S., H. Moura, E. Kovacs-Nace, S. Wallace and M.L. Eberhard. 1997. Uniform staining of *Cyclospora* oocysts in fecal smears by a modified safranin technique with microwave heating. J. Clin. Microbiol. 35: 730–733.

Wang, K.-X., C.-P. Li, J. Wang and Y. Tian. 2002. *Cyclospora cayetanensis* in Anhui, China. World J. Gastroenterol. 8: 1144.

Yoder, K., O. Sethabutr and D. Relman. 1996. PCR-based detection of the intestinal pathogen Cyclospora. PCR Protocols for Emerging Infectious Diseases, a Supplement to Diagnostic Molecular Microbiology: Principles and Applications: 169–176.

Zimmer, S.M., A.N. Schuetz and C. Franco-Paredes. 2007. Efficacy of nitazoxanide for cyclosporiasis in patients with sulfa allergy. Clin. Infect. Dis. 44(3): 466–7.

Chapter 9

Advances in the Detection and Diagnosis of Coccidian Parasites Causing Gastrointestinal Disease and Foodborne Outbreaks

Ynes Ortega

Introduction

Foodborne illnesses have occurred and been documented since antiquity. Bacterial outbreaks associated with foods have occurred throughout history. For example, Emperor Leo VI of Byzantine (886–911) announced and ordered that manufacturing blood sausages was forbidden because their consumption was associated with cases of dilated pupils and fatal muscle paralysis. Many centuries later (1895) in a small Belgian village, an outbreak occurred at a funeral dinner where smoked ham was served. This led to a description of the causative agent, *Clostridium botulinum,* by Emile Pierre van Emergem (2004). Now, it is well known that inadequate preservation of foods can result in cases of botulism.

Other bacteria such as *Salmonella, Escherichia coli, Listeria, Chronobacter, Campylobacter*, and *Vibrio* are important foodborne pathogens known to have caused outbreaks of diseases associated with the consumption of contaminated fresh produce, meat, and dehydrated foods. Their detection and identification are based on biochemical characteristics as evidenced when grown in or on selective microbiological media. Molecular tools have been developed and are often used to

Center for Food Safety, University of Georgia, 1109 Experiment St, Griffin, GA 30223, USA.
Email: ortega@uga.edu

identify and genotype bacteria. Validated methods for detecting bacteria associated with foodborne illnesses can be found in the FDA BAM Bacteriological Analytical Manual and the AOAC Official Methods.

Detection of viruses is challenging because *in vitro* cultivation is available for only a few. Diagnostic tools most frequently used are qPCR or RT-qPCR. Two viruses are most often associated with foodborne illnesses. Norovirus, known to cause most cases, is diagnosed using molecular tests (Glass 2009). Various genotypes can infect several animal species; therefore, diagnostic tools should be able to detect genogroups GI, GII, GIV, and GVIII, which are pathogenic to humans and of public health relevance. Hepatitis A, another important viral pathogen most frequently associated with the consumption of imported frozen berries, can also be detected using qPCR and sequencing.

Because most foodborne pathogens can be ingested by the consumption of raw and minimally processed foods, practices are necessary to prevent or reduce the number of illnesses in the U.S. caused by these foods. In 2011, the FDA Food Safety Modernization Act (FSMA) was established, and it is being implemented in various phases to shift disease mitigation emphasis from responding to outbreaks to preventing contamination. Rules and guidance for the food industry address ways that can prevent contamination such as the focus on agricultural water, good agricultural practices, registration of food facilities, transportation, and food supplier verification, among others. In one of these rules, laboratories are subject to an accreditation program for testing and certain analyses. The purpose of this program is to improve the accuracy and reliability of certain food testing. This can be accomplished by insuring uniformity of standards and enhanced FDA oversight of participating laboratories. The main benefit would be to reduce the risk of food-related illnesses by improving the accuracy of analyses of selected samples (FDA 2022).

Parasites

Unlike bacteria, most parasites cannot be enriched or propagated *in vitro*. This makes the methodology for sample processing the most important step in their detection. Parasites are classified into two major groups: protozoa, which are one-celled organisms, and helminths which are multicellular. The adult stage of the helminths can be detected macroscopically and can be found in meats or organs of the infected hosts. However, the eggs and intermediary stages of helminths are small, requiring the expertise of parasitologists to perform meaningful microscopic examinations. Of the many helminths that can infect animal parts consumed by humans and of relevance in domestic and international trade are those causing trichinellosis, anisakiosis, cysticercosis, hydatidosis, and fasciolasis.

Trichinella, an important nematode found in many countries, can be detected in the form of encapsulated larvae in the muscles of infected animals. *Trichinella nativa* and T-6, *T. spiralis, T. britovi,* T-8 and T9, and *T. murrelli* and *T. nelson* can infect mammals and are characterized by having an encapsulated larva. *T. pseudospiralis, T. papuae,* and *T. zimbawensis* can also infect mammalians but are not encapsulated.

Morphological characteristics are traditionally used to determine the *Trichinella* species infecting animals, including humans. Swine can harbor the parasite and it can also be found in wild fauna such as bears, wild boars, and other animals (Figure 1A).

Encapsulated and non-encapsulated larvae can be observed by pressing muscle tissue between two microscope slides. Detection of a larva is dependent on the chance of coinciding slides with the larva. In animals, the sensitivity is low (5 larvae/g of tissue) and is no longer recommended for examining animal tissue. Digestion of animal muscle (diaphragm, tongue, and cheek muscle) with acidified pepsin facilitates observation of a large amount of sample, increasing the sensitivity of the test. In the case of symptomatic patients, muscle biopsies can be examined using conventional histological processing and staining techniques. Serological testing is available and provides a 100% sensitivity in humans infected with *T. spiralis*.

Anisakis, a nematode, can infect humans and result in a severe allergic reaction and some instances death. The most frequent nematodes of public health relevance from this group are *Anisakis* and *Pseudoterranova*. They can be acquired by ingestion of infected finfish and shellfish. Detection of larvae is done by examination of the fish flesh. The larvae can be removed physically or killed by cooking or prolonged freezing. However, in some instances, careful examination of fresh fish is overlooked and can result if ingested, in anisakiosis.

The cestoda, e.g., *Taenia*, has a scolex, and the strobila that consists of proglottids at different stages of maturation. Mature proglottids (containing eggs) are released in feces. *Taenia* species can be detected by examination of the proglottids, the number of primary uterine branches, and whether the scolex has a rostellum, rostellar hooks, and suckers (Figure 1B). The scolex is rarely found in the feces of infected food

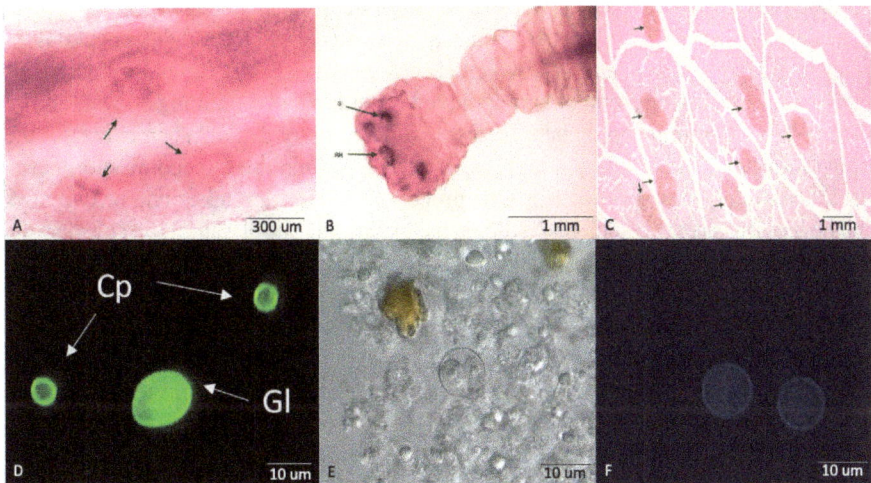

Figure 1. Microscopic observation of foodborne parasites. (A) Histological examination of *Trichinella spiralis*. Larvae (*arrows*); (B) scolex of *Tenia solium,* rostellar hooks (RS) and suckers (S); (C) *Sarcocystis* sp. tissue sarcocysts (arrows); (D) *Giardia lamblia* (Gl) cyst and *Cryptosporidium parvum* (Cp) oocysts; (E) sporulated oocyst of *Cyclospora cayetanensis* by DIC, and (F) *Cyclospora cayetanensis* autofluorescence.

animals and humans, unless they have been treated with antiparasitic drugs. It is also important to examine bovine or porcine tissues for the presence of tissue cyst forms (cysticerci).

In 1950, the life cycle of *Taenia solium* was described, and since the 1970s new methods for the identification of clinical cases have improved and contributed to the implementation of programs to eradicate cysticercosis (the cystic form of *T. solium*). If cysticerci are in the brain or central nervous system, the infection is called neurocysticercosis. In humans, diagnosis of cysticercosis is achieved using serological tests, imaging (CAT scan, MRI, and traditional X-rays), and the patient's clinical history, often associated with epileptic seizures. The location of *T. solium* cysticerci in tissues and organs will determine the symptoms the host will present. In swine carcasses, the diaphragm and organs, like the heart, are examined for the presence of cysts. Consumption of meat containing live cysts will result in teniasis infection in which tapeworms mature and grow in the intestine. Individuals with teniasis shed proglottids and eggs in their feces. If food or water contaminated with these eggs is ingested by swine or humans, it may result in cysticercosis.

The protozoa are another group of parasites that require the use of microscopy to detect their vegetative and environmentally resistant stages. Amoeba, ciliate, flagellate, and coccidia can be acquired by ingestion of contaminated water or food. The following text will address advances in the diagnosis of coccidia in clinical samples as well as in food and water samples.

Coccidia

Coccidia is single-celled, obligate, intracellular parasites that belong to the phylum Apicomplexa. This group is characterized by possessing an apicoplast and an apical complex that is involved in the invasion of host cells. One of the coccidias of relevance is *Cystoisospora belli*, which causes gastrointestinal illness in humans. It can be acquired by ingestion of contaminated water or food containing oocysts and be detected by examining fecal samples from infected individuals. When excreted, the oocyst contains one or two immature sporoblasts that mature to become sporocysts, each containing four sporozoites. This process of maturation takes one or two days in the environment. Diagnosis is done by examining stool samples. The oocysts are oval and autofluorescent. Oocysts contain two sporocysts, each containing four sporozoites. *C. belli* is prevalent in tropical and subtropical areas. Cystoisosporiasis is one of the most common parasitic infections in immunocompromised subjects, resulting in acute diarrhea. Diagnosis can also be done by microscopic examination of wet mounts; however, concentration methods are often recommended. Stains such as safranin, modified acid-fast stain, Giemsa, and auramine-rhodamine can be used to detect *C. belli* oocysts in fecal smears. *Cystoisospora* oocysts autofluoresce; therefore, using epifluorescence microscopy is a method of preference as it is easy to implement. Molecular tools targeting the 18S and ITS genes have also been developed and are used mostly for clinical applications (Dubey and Almeira 2019; Samarasinghe et al. 2008; Esvan et al. 2018).

Sarcocystosis

There are more than 200 species of *Sarcocystis* that can infect mammals, birds, and reptiles. Humans can play a role as definitive hosts for *Sarcocystis hominis, S. herdorni,* and *S. suishominis*. In most cases sarcocystosis is asymptomatic or develops an enteric or muscular infection, depending on the host's immune response and the amount of infected meat consumed. When humans ingest meat containing tissue cysts, these cysts rupture followed by infection of the intestinal epithelium by bradizoites. Bradizoites differentiate into gametocytes and sexual multiplication occurs, resulting in the formation of sporulated oocysts that are excreted in the feces of the infected individual. When sporocysts are ingested by an intermediate host (food animals or humans) the sporozoites will enter endothelial cells of blood vessels and undergo schizogony. Asexual multiplication will occur and eventually infect muscle cells and form sarcocysts that in time fill with bradyzoites (Figure 1C). If humans ingest sarcocysts produced by a different *Sarcocystis* species, the bradyzoites will migrate to muscle tissues and form cysts. Symptoms such as myalgia, muscle weakness, and edema are signs of infection. Muscle biopsies revealing the presence of sarcocysts in tissues will confirm the diagnosis. Detection can be facilitated by staining with hematoxylin and eosin, periodic acid-Schiff, Giemsa, etc. Sarcocysts present in muscle can measure up to 100–325 um in length. Digestion of tissues by enzymes can be useful; however, small sarcocysts may be destroyed by digestion (Spickler 2020). In the case of enteric infections, diagnosis is done by identifying sporocysts or sporulated oocysts in fecal samples. Acid-fast staining can be used to detect sporocysts, but they may be confused with *Cystoisospora*. Oocysts measure 15×20 um and sporocysts measure 6×12 um (Mehrothra et al. 1996). Oocysts can be concentrated using flotation concentration methods that include sucrose, sodium chloride, and zinc sulfate. Microscopic observation does not differentiate between species, but molecular testing has been used to discriminate between species. The use of 18S rDNA, ITS-1, and mitochondrial cytochrome oxidase has been used to detect and understand the parasite diversity (Rosenthal 2021) (Table 1).

Cryptosporidium spp.

Cryptosporidium is a coccidia that can be acquired by ingestion of contaminated food or water. *Cryptosporidium* oocysts are spherical with a thick wall. They range in diameter from 4 to 6 μm (Fayer et al. 2000) and when excreted they have already sporulated and contain four elongated sporozoites (Fayer and Xiao 2007). Of the 30 *Cryptosporidium* species, *C. hominis* and *C. parvum* are the two most common species infecting humans with the latter zoonotic species originating in ruminants (Xiao and Cama 2018). *C. hominis* is anthroponotic, causing infections in humans (Vanathy et al. 2017). The size of *Cryptosporidium* oocysts differs in species affecting the gastric and intestinal tract compared to those that infect the lower gastrointestinal tract, which is larger (Fayer and Xiao 2007).

    Cryptosporidium species that cause illness in humans colonize the gastrointestinal and respiratory epithelium after ingestion of infectious oocysts, causing persistent diarrhea, especially in immunocompromised individuals (Khurana and Chaudhary

Table 1. Some molecular methods for the detection of *Sarcocystis* ssp.

Target	Test	Species	Reference
18S RNA gene	PCR and sequencing	Cattle	Moré et al. 2013
18S RNA gene	Multiplex RT qPCR	Cattle	Moré et al. 2013
18S RNA gene	PCR and sequencing	Foxes, raccoon, dogs	Moré et al. 2016
Cytochrome oxidase subunit cox 1	Multiplex PCR and sequencing	Deer	Abe et al. 2019
Cytochrome oxidase subunit cox 1	PCR and sequencing	Cervids, cattle, and sheep	Gjerde B 2013
Cytochrome oxidase subunit cox 1, 18S RNA gene	Multiplex PCR and sequencing	Cattle	Rubiola et al. 2018
ITS1, cytochrome b, sag2, sag3, sag4	Nested, heminested, double PCR, and sequencing	Opossum	Valadas et al. 2016
18S RNA gene	PCR and sequencing	Cattle, cebu, bison	Fischer and Odening 1998
Cytochrome oxidase subunit cox 1, 18S RNA gene	PCR and sequencing	Moose	Prakas et al. 2019

2018; Xiao and Cama 2018). The infection, cryptosporidiosis, may result in death in developing as well as developed countries, making *Cryptosporidium* responsible for most parasitic diarrheal illnesses, but the prevalence has been reduced as disease management has improved (Cacciò et al. 2005). The immune status of infected individuals is important as the illness can be self-limiting in the immunocompetent, but chronic in individuals who are immunocompromised (Chalmers and Davies 2010). Oocyst shedding is intermittent, lasting an average of seven days and up to 50 days after the symptoms have stopped (Fayer and Xiao 2007). The oocyst is infectious upon shedding. Transmission can be person-to-person, foodborne, waterborne, and by zoonotic routes (Dillingham et al. 2002).

The most notable waterborne outbreak of *Cryptosporidium* infections in the U.S. occurred in 1993 in Milwaukee, Wisconsin. Public water contaminated with *C. hominis* caused illness in approximately 403,000 people (Mac Kenzie et al. 1994). During 2009–2017, there were 444 outbreaks of cryptosporidiosis in 40 U.S. states and Puerto Rico. Sixty-five of the outbreaks were attributed to cattle and 156 were attributed to treated water (Gharpure et al. 2019). Of the 444 outbreaks, 22 were foodborne with vehicles, such as unpasteurized milk, unpasteurized apple cider, and fresh produce (Gharpure et al. 2019). It makes sense that these foods are vehicles as *Cryptosporidium* as oocyst inactivation was not achieved by exposure to heat or desiccation. Foods that are not thermally processed or inadequately heated and with high water content are likely to become vehicles for this parasite (Robertson 2014).

Extraintestinal infections, although uncommon, have been reported. Diagnosis is done by detecting the oocysts in secretions or biopsies. Diagnosis can be done by observation of biopsies of the small or large intestine where the intracellular tissue stages of the parasite can be observed. *Cryptosporidium* infects the gastrointestinal tract and detection is achieved by examination of fecal samples. Concentration either by flotation or sedimentation improves the detection of the parasite. Fecal smears and concentrates can be stained using a modified acid-fast stain (e.g., modified Ziehl-Neelsen stain or safranin) (Garcia 2016). These methods stain all *Cryptosporidium* species but have limited use for traceability during outbreak investigations.

Serological tests are used for increased sensitivity during *Cryptosporidium* screening (Bialek et al. 2002). Immunoassays such as those using direct fluorescent antibodies (DFA), enzyme immunoassays (EIAs), and indirect fluorescent antibodies (IFA) (Figure 1D) for examining stool samples are used in the diagnosis of cryptosporidiosis (Aghamolaie et al. 2016; Garcia 2016). EIA and DFA show similar sensitivity (94% and 91%, respectively) when compared to a modified immunofluorescence assay (64% sensitivity) and nested PCR (97% sensitivity) for detecting *Cryptosporidium* in stool samples from at-risk patients and calves (Bialek et al. 2002).

Differentiation between *C. hominis* (anthroponotic) and *C. parvum* (zoonotic) in clinical samples cannot be achieved by staining oocysts. Other methods for species-specific identification are required (Garcia 2016). Molecular methods have been developed to test for the presence of *Cryptosporidium* in stool samples. These methods have increased sensitivity and accuracy and can be less time-consuming than conventional methods (Fayer et al. 2000). PCR assays, e.g., nested PCR, qPCR, multiplex qPCR, and other molecular methods such as microsatellite analysis, fluorescent *in situ* hybridization, and loop-mediated isothermal amplification can be used to genotype and subtype *Cryptosporidium* (Vanathy et al. 2017) (Table 2). If present in water, *Cryptosporidium* oocysts are likely found in small numbers, thereby requiring a concentration step. Various water filters have been used (Smith and Nichols 2010). The U.S. Environmental Protection Agency (EPA) Method 1623.1 is an updated version of Method 1623 for the detection of *Cryptosporidium* in water (USEPA 2012). Changes made to the method include the requirement for flow cytometer-enumerated *Cryptosporidium* in spiked suspensions, the addition of sodium hexametaphosphate for capsule filter elution, and the requirement for a bead pellet wash step during the immunomagnetic separation (IMS) procedure (USEPA 2012). The basic steps in the method are collection through filtration, elution, separation, and enumeration, which is done by staining oocysts with fluorescently labeled antibodies and examination using an epifluorescent and DIC microscope (USEPA 2012). Other methods have been developed for testing large volumes of drinking and surface waters using filters with high filtration capacity. Real-time PCR is also used to detect and enumerate *Cryptosporidium* oocysts in water. Kimble et al. (2015) modified the DNA extraction process by using the UNEX buffer and a second spin column to evaluate the use of qPCR to detect *Cryptosporidium* in water. They compared the modified procedure to the USEPA method. Oocyst recovery was higher (49%) compared to detection by microscopic analysis (41%). Although recovery of *Cryptosporidium* oocysts from the water was demonstrated, it was noted that additional evaluation using different water sources as well as fewer oocysts is needed (Kimble et al. 2015). Other studies (Adamska et al. 2012; Mahmoudi

Table 2. Some molecular methods for the detection of *Cryptosporidium* spp.

Target	Test	Samples	References
18S	Nested PCR	Human, animal	Xiao et al. 1999
18S	Nested	Animal	Mirhashemi et al. 2015
CPGP40/15	PCR-RFLP and RFLP-SSCP	Human, animal	Wu et al. 2003
COWP		Human, animal	Homan et al. 1999
Gp60	PCR-RFLP	Cattle	Cai et al. 2017
18S, actin loci	qPCR	Fish	Yang et al. 2015
18S	RT-qPCR	Calves	Wang et al. 2021
Multiplex real-time PCR	Gastroenteritis/Parasite Panel I (Diagenode), RIDAGENE Parasitic Stool Panel (R-Biopharm), Allplex Gastrointestinal Parasite Panel 4 (Seegene), and FTD Stool Parasites (Fast Track) real-time PCR	Human	Paulos et al. 2019
Cell imprinted polymer PDMS	Stamping	Human	Sarkhosh T et al. 2021
SSUr RNA, COWP, DNA-J-like protein	qPCR	Human	Weinreich F. et al. 2021
COWP	qPCR	Human	Shin J.H. et al. 2018
GP60	qPCR		Dashti A et al. 2022

et al. 2017) have used qPCR to enumerate *Cryptosporidium* oocysts in water. More research is needed to address the limitations of qPCR and oocyst enumeration before these methods can be adopted. The whole genome of *Cryptosporidium* has been determined, and genotyping of oocysts isolated from water and foodborne outbreaks has been successful.

Methods to detect *Cryptosporidium* oocysts in food are less developed. Foods are washed using an elution buffer and the wash is concentrated by centrifugation. Oocysts are observed either using immunoassays or molecular assays (Fayer and Xiao 2007). An immunomagnetic separation technique (IMS) is recommended for concentrating oocysts in turbid samples (Smith and Nichols 2007). Two methods have focused on specific food matrixes. One is in the U.S. FDA BAM Chapter 19a and the other was published by Cook et al. (2006b). The FDA method describes procedures to isolate *Cryptosporidium* species in fresh produce washes, juices, cider, and milk followed by detection by microscopic analysis or nested PCR (Orlandi et al. 2004). The restriction fragment length polymorphism (RFLP) method is needed for more species-specific analysis (Orlandi et al. 2004).

Cook et al. (2006b) validated and proposed standard methods for the detection of *Cryptosporidium parvum* on/in raspberries and lettuce. This method consists of four steps: oocyst extraction, concentration, staining, and identification by microscopy.

The method has a sensitivity of 89.6% and a specificity of 85.4%. The development of more robust methods is needed to aid in the detection of *Cryptosporidium* species in food.

Cyclospora

Cyclospora cayetanensis, another coccidian parasite, has been associated with foodborne outbreaks since the early 1990s. In the US, sporadic cases, and clusters of cyclosporiasis have been associated with the consumption of raw produce, including herbs and berries. In the past 10 years, lettuce and cilantro imported from Mexico have been associated with cyclosporiasis. Since 2018, produce grown and processed in the US has also been linked with foodborne outbreaks.

Cyclospora oocysts can be detected in clinical samples by microscopic observation. Oocysts measure 8–10 um in diameter (Figure 1E). Oocysts can also be observed using epifluorescence microscopy (Figure 1F). They autofluoresce when exposed to wide UV light (450–490 DM filter) (Ortega et al. 1994). Detection can be improved by oocyst concentration using conventional ova and parasite detection methods, including discontinuous sucrose gradients, ethyl acetate, and cesium chloride. Molecular-based methods are also used to detect oocysts in clinical samples; however, in recent years, culture-independent methods have been successfully used in clinical settings to detect *Cyclospora* (Buss et al. 2015).

The official FDA method 19b and 19c (2022) includes a qPCR method to detect *Cyclospora* in fresh produce, complex food matrixes, and agricultural water. The food testing procedure uses Alconox, a detergent commonly used in laboratories, to remove oocysts from whole vegetable samples. Water samples are being tested using Rexeed filters that were developed for dialysis purposes. These filters can be used to test large volumes of water, not only for concentrating environmental samples but also for water used for irrigation purposes. Other laboratories have used various elution buffers including the one described in the EPA 1963 protocol or glycine buffer.

Molecular tools are used mostly for detecting oocysts in food and environmental samples. Several platforms have described targeting conserved regions of the 18S RNA, mitochondrial DNA, and ITS DNA. The ITS1 gene was developed initially using conventional PCR (Adam et al. 2000) and later as qPCR for environmental surveillance (Temesgen et al. 2019; Lalonde and Gajadhar 2016). Conventional PCR has been used in multilocus sequence typing (MLST) for detecting *Cyclospora* in human fecal samples (Guo et al. 2016) and in the surveillance of berries (Temesgen et al. 2022; Murphy et al. 2018) (Table 3).

Toxoplasma gondii

Toxoplasma is a coccidian parasite whose definite host is felines. Sexual multiplication occurs in the intestine of cats and the unsporulated oocysts are excreted in the environment in their feces. It takes 1 to 5 days for the oocysts to mature and differentiate forming two sporocysts, each containing four sporozoites. The feces of felines can cross-contaminate water and foods. When ingested by a human or other

Table 3. Some Methods for Detection of *Cyclospora cayetanensis*.

Target	Test	Matrix	References
ITS1, MLST	qPCR, PCR, sequencing	Raspberry, blueberry, and strawberry	Temesgen et al. 2022
Mitochondrial DNA	PCR	Cilantro, raspberries, and canal water	Durigan et al. 2022
SSU RNA gene	PCR	Sewer water	Fan et al. 2021
Mitochondria Genome	qPCR	SS	Guo et al. 2019
ITS-1	qPCR	Raspberry, blueberry, strawberry	Temesgen et al. 2019
Mitochondrial junction	PCR	SS	Nascimento et al. 2019
18SrRNA	qPCR	Cilantro, raspberries	Murphy et al. 2018
Microsatellite and minisatellite	nPCR	SS	Guo et al. 2016
Apicoplast	WGS	SS	Cinar et al. 2016
ITS-2	qPCR, SSCP	Soil, water, cucumber, lettuce, fennel, celery, tomato, melon, and herbs	Giangaspero et al. 2015
18SrRNA	PCR	SS	Sulaiman et al. 2014
HSP70	nPCR	SS	Sulaiman et al. 2013
ITS-2	PCR	Basil, parsley, carrot, romaine, blackberry, and shredded cabbage	Lalonde and Gajadhar 2008
18SrRNA	PCR, RT multiplex	Fresh produce wash, juices, cider, Milk	Orlandi et al. 2004
ITS-1	PCR	SS	Adam et al. 2000
WGA	Lectin binding	Mushrooms, lettuce, raspberries	Robertson et al. 2000

SS: stool samples; no food matrixes were used.

animals, the oocysts will excyst and the sporozoites will infect the intestinal cells, migrate to various tissues, and form cysts containing bradyzoites. Cats may become infected if they eat tissue containing cysts or ingest sporulated oocysts. Cysts are most often found in the striated muscle, myocardium, brain, and eyes of infected animals. The cysts can stay dormant for years, but they can reactivate if the host immune system competency decreases. Severe cases are observed when the parasite invades the brain, but symptoms vary depending on the localization of the cyst. Multiple studies show neurological behavioral changes in humans. The association of toxoplasmosis with schizophrenia has been reported in various studies; however, others have not found such an association. More studies are needed to better understand these implications.

Toxoplasma can cross the placenta barrier in animals, including humans, and infect the fetus which can result in miscarriage. Serological diagnosis is done by determining the presence of IgG or IgM antibodies against the parasite. In the case of congenital toxoplasmosis, diagnosis is done using molecular methods to detect the *Toxoplasma* DNA in the amniotic fluid. Birds, rodents, and swine can also serve as intermediary hosts. Humans can acquire toxoplasmosis by consuming raw or improperly cooked pork meat. Detection of tissue cysts can be done by tissue digestion followed by molecular assays.

Animals, including humans, can also acquire toxoplasmosis by ingesting contaminated water or foods containing sporulated oocysts. The detection of oocysts follows the same procedures used for the detection of *Cyclospora*. Oocysts of *Toxoplasma*, like *Cyclospora*, autofluoresce when exposed to UV light. Epifluorescence can be used to detect the oocysts, but this method is not specific when used to examine environmental samples. The number of oocysts present can be low, requiring the use of concentration methods.

Detection in Environmental and Food Samples

Two methods focus on the detection of specific food matrixes. One is in the US FDA BAM (Chapter 19A, 2022) and the other was described by Cook et al. (2006b). The US FDA BAM method isolates *Cryptosporidium* species from fresh produce washes, juices, cider, and milk followed by detection by microscopic analysis or nested PCR (Orlandi et al. 2004). This method dictates that RFLP be used for more species-specific analysis (Orlandi et al. 2004). Cook et al. (2006b) validated the proposed standard method for the detection of *C. parvum* on raspberries and lettuce. This method has four steps: extraction, concentration, staining of the oocysts, and observation by microscopy. The method has a sensitivity of 89.6% and a specificity of 85.4% (Cook et al. 2006b). The capture of *Cryptosporidium* oocysts or *Giardia* cysts from water has been accomplished using immunomagnetic beads followed by observation using fluorescently labeled antibodies. The development of more robust methods is needed to aid in the detection of *Cryptosporidium* species in food.

Multiplex testing of food and environmental samples for the presence of parasites has been done by various researchers. Berries have been tested for the presence of *Echinococcus, Toxoplasma,* and *Cyclospora* (Temesgen et al. 2022). Detection of *Cyclospora, Cryptosporidium*, and *Toxoplasma* in spinach and arugula was done using universal coccidia primers (Lalonde et al. 2015). A new diagnostic platform, the Biofire® FilmArray® System, has been developed as a multiplex diagnostic tool for the analysis of clinical samples. Some laboratories are using this tool because of its simplicity and ease to use. Recently, a new FilmArray® Food & Water Panel was developed for the simultaneous detection of 15 targets in environmental, food, and beverage samples. This panel detects bacteria (*Campylobacter, Salmonella, Vibrio,* and *Yersinia*) diarrheagenic *E. coli/Shigella, E. coli* (EAEC, EPEC, ETEC, STEC, O157, and EIEC) parasites (*Cryptosporidium, Cyclospora cayetanensis,* and *Giardia lamblia*), and norovirus GI/GII. The panel was evaluated using water, deli meats, spinach, and orange and apple juices, among other food matrixes (Buss et al. 2015).

References

Abe, N., K. Matsuo, J. Moribe, Y. Takashima, T. Baba and B. Gjerde. 2019. Molecular differentiation of five *Sarcocystis* species in sika deer (*Cervus nippon centralis*) in Japan based on mitochondrial cytochrome c oxidase subunit I gene (cox1) sequences. Parasitol. Res. 118: 1975–1979.

Adam, R.D., Y.R. Ortega, R.H. Gilman and C.R. Sterling. 2000. ITS1 variability in *Cyclospora cayetanensis*. J. Clin. Microbiol. 38: 2339–2343.

Adamska, M., A. Leonska-Duniec, M. Sawczuk, A. Maciejewska and B. Skotarczak. 2012. Recovery of *Cryptosporidium* from spiked water and stool samples measured by PCR and real time PCR. Vet. Medicina. 57.

Aghamolaie, S., A. Rostami, S. Fallahi, F.T. Biderouni, A. Haghighi and N. Salehi. 2016. Evaluation of modified Ziehl–Neelsen, direct fluorescent-antibody and PCR assay for detection of *Cryptosporidium* spp. in children's faecal specimens. J. Parasitic Dis. 40: 958–963.

Bialek, R., N. Binder, K. Dietz, A. Joachim, J. Knobloch and U.E. Zelck. 2002. Comparison of fluorescence, antigen and PCR assays to detect *Cryptosporidium parvum* in fecal specimens. Diagnostic Microbiol. Infect. Dis. 43: 283–288.

Buss, S.N., A. Leber, K. Chapin, P.D. Fey, M.J. Bankowski, M.K. Jones, M. Rogatcheva, K.J. Kanack and K.M. Bourzac. 2015. Multicenter evaluation of the BioFire FilmArray gastrointestinal panel for etiologic diagnosis of infectious gastroenteritis. J. Clin. Microbiol. 53: 915–25.

Cacciò, S.M., R.A. Thompson, J. McLauchlin and H.V. Smith. 2005. Unravelling *Cryptosporidium* and *Giardia* epidemiology. Trends Parasitol. 21: 430–437.

Cai, M., Y. Guo, B. Pan, N. Li, X. Wang, C. Tang, Y. Feng and L. Xiao. 2017. Longitudinal monitoring of *Cryptosporidium* species in pre-weaned dairy calves on five farms in Shanghai, China. Vet. Parasitol. 241: 14–19.

Chalmers, R.M. and A.P. Davies. 2010. Minireview: clinical cryptosporidiosis. Exp. Parasitol. 124: 138–146.

Cinar, H.N., Y. Qvarnstrom, Y. Wei-Pridgeon, L. Wen, F.S. Nascimento, M.J. Arrowood, H.R. Murphy, A. Jang, E. Kim, R. Kim, A. daSilva and G.R. Gopinath. 2016. Comparative sequence analysis of *Cyclospora cayetanensis* apicoplast genomes originating from diverse geographical regions. Parasit. Vect. 9: 611.

Cook, N., C. Paton, N. Wilkinson, R. Nichols, K. Barker and H. Smith. 2006a. Towards standard methods for the detection of *Cryptosporidium parvum* on lettuce and raspberries. Part 1: development and optimization of methods. Int. J. Food Microbiol. 109: 215–221.

Cook, N., C. Paton, N. Wilkinson, R. Nichols, K. Barker and H. Smith. 2006b. Towards standard methods for the detection of *Cryptosporidium parvum* on lettuce and raspberries. Part 2: validation. Int. J. Food Microbiol. 109: 222–228.

Dashti, A., H. Alonso, C. Escolar-Miñana, P.C. Köster, B. Bailo, D. Carmena and D. González-Barrio. 2022. Evaluation of a novel commercial real-time PCR assay for the simultaneous detection of *Cryptosporidium* spp., *Giardia duodenalis*, and *Entamoeba histolytica*. Microbiol. Spectrum. pp. 00531-22.

Dillingham, R.A., A.A. Lima and R.L. Guerrant. 2002. Cryptosporidiosis: epidemiology and impact. Microbes Infect. 4: 1059–1066.

Dubey, J.P. and S. Almeria. 2019. *Cystoisospora belli* infections in humans: the past 100 years. Parasitology 146: 1490–1527.

Durigan, M., E. Patregnani, G.R. Gopinath, L. Ewing-Peeples, C. Lee, H.R. Murphy, S. Almeria and H.N. Cinar. 2022. Development of a molecular marker based on the mitochondrial genome for detection of *Cyclospora cayetanensis* in food and water samples. Microorganisms 10: 9.

El-Din, A.E. and A. El-Bhairy. 2015. Evaluation of direct fluorescent antibody and enzyme linked immunosorbent assay versus copromicroscopy in diagnosis of cryptosporidiosis. Al-Azhar Med. J. 44: 351–362.

Esvan, R., L.F. Suleková, S. Gabrielli, E. Biliotti, D. Palazzo, M. Spaziante and G. Taliani. 2018. Severe diarrhoea due to *Cystoisospora belli* infection in a good syndrome patient. Parasitol. Int. 67: 413–414.

Fan, Y., X. Wang, R. Yang, W. Zhao, N. Li, Y. Guo, L. Xiao and Y. Feng. 2021. Molecular characterization of the waterborne pathogens *Cryptosporidium* spp., *Giardia duodenalis*, *Enterocytozoon bieneusi*,

Cyclospora cayetanensis and *Eimeria* spp. in wastewater and sewage in Guangzhou, China. Parasit. Vect. 14: 66.

Fayer, R., U. Morgan and S.J. Upton. 2000. Epidemiology of *Cryptosporidium*: transmission, detection and identification. Int. J. Parasitol. 30: 1305–1322.

Fayer, R. and L. Xiao. 2007. *Cryptosporidium* and Cryptosporidiosis. CRC Press. 576p.

FDA. 2022. Laboratory Accreditation for Analyses of Foods, Final Regulatory Impact Analysis. https://www.fda.gov/about-fda/economic-impact-analyses-fda-regulations/laboratory-accreditation-analyses-foods-final-regulatory-impact-analysis. Accessed June 13, 2022.

Fischer, S. and K. Odening. 1998. Characterization of bovine *Sarcocystis* species by analysis of their 18S ribosomal DNA sequences. J. Parasitol. 84: 50–54.

Garcia, L.S. 2016. Diagnostic Medical Parasitology. John Wiley & Sons.

Gharpure, R., A. Perez, A.D. Miller, M.E. Wikswo, R. Silver and M.C. Hlavsa. 2019. Cryptosporidiosis Outbreaks—United States, 2009–2017. Morbid. Mortal. Weekly Rep. 68: 568.

Giangaspero, A., M. Marangi, A.V. Koehler, R. Papini, G. Normanno, V. Lacasella, A. Lonigro and R.B. Gasser. 2015. Molecular detection of *Cyclospora* in water, soil, vegetables and humans in southern Italy signals a need for improved monitoring by health authorities. Int. J. Food. Microbiol. 211: 95–100.

Gjerde, B. 2013. Phylogenetic relationships among *Sarcocystis* species in cervids, cattle and sheep inferred from the mitochondrial cytochrome c oxidase subunit I gene. Int. J. Parasitol. 43: 579–591.

Glass, R.I., U.D. Parashar and M.K. Estes. 2009. Norovirus gastroenteritis. N. Engl. J. Med. 361(18): 1776–85.

Guo, Y., Y. Feng, N. Li, K. Tang, M. Frace, Y.R. Ortega, M.J. Arrowood, D. Roellig, Y. Qvarnstrom, L. Wang, L. Zhang and L. Xiao. 2016. Multilocus sequence typing tool for *Cyclospora cayetanensis*. Emerg. Infect. Dis. 22(8): 1464–1467.

Guo, Y., Y. Wang, X. Wang, L. Zhang, Y. Ortega and Y. Feng. 2019. Mitochondrial genome sequence variation as a useful marker for assessing genetic heterogeneity among *Cyclospora cayetanensis* isolates and source-tracking. Parasit. Vect. 12(1): 47.

Homan, W., T. van Gorkom, Y.Y. Kan and J. Hepener. 1999. Characterization of *Cryptosporidium parvum* in human and animal feces by single-tube nested polymerase chain reaction and restriction analysis. Parasitol. Res. 85: 707–712.

ISO. 2016. ISO 18744:2016. Microbiology of the Food Chain Detection and Enumeration of *Cryptosporidium* and *Giardia* in Fresh Leafy Green Vegetables and Berry Fruits.

Karanis, P., C. Kourenti and H. Smith. 2007. Waterborne transmission of protozoan parasites: a worldwide review of outbreaks and lessons learnt. J. Water Health. 5: 1–38.

Khurana, S. and P. Chaudhary. 2018. Laboratory diagnosis of cryptosporidiosis. Trop. Parasitol. 8: 2–7.

Kimble, G.H., V.R. Hill and J.E. Amburgey. 2015. Evaluation of alternative DNA extraction processes and real-time PCR for detecting *Cryptosporidium parvum* in drinking water. Water Sci. Technol.: Water Supply. 15: 1295–1303.

Lalonde, L.F. and A.A. Gajadhar. 2008. Highly sensitive and specific PCR assay for reliable detection of *Cyclospora cayetanensis* oocysts. Appl. Environ. Microbiol. 74: 4354–8.

Lass, A., B. Szostakowska, P. Myjak and K. Korzeniewski. 2015. The first detection of *Echinococcus multilocularis* DNA in environmental fruit, vegetable, and mushroom samples using nested PCR. Parasitol. Res. 114: 4023–4029.

MacKenzie, W.R., N.J. Hoxie, M.E. Proctor, M.S. Gradus, K.A. Blair, D.E. Peterson, J.J. Kazmierczak, D.G. Addiss, K.R. Fox and J.B. Rose. 1994. A massive outbreak in Milwaukee of *Cryptosporidium* infection transmitted through the public water supply. New Engl. J. Med. 331: 161–167.

Mahmoudi, M.-R., M. Bandepour, B. Kazemi and A. Mirzaei. 2017. Detection and enumeration of *Cryptosporidium* oocysts in environmental water samples by real-time PCR assay. J. Basic Res. Med. Sci. 4: 42–47.

Mehrotra, R., D. Bisht, P.A. Singh, S.C. Gupta and R.K. Gupta. 1996. Diagnosis of human *Sarcocystis* infection from biopsies of the skeletal muscle. Pathology 28: 281–282.

Mirhashemi, M., A. Zintl, T. Grant, F.E. Lucy, G. Mulcahy and T. De Waal. 2015. Comparison of diagnostic techniques for the detection of *Cryptosporidium* oocysts in animal samples. Exp. Parasitol. 151–152: 14–20.

Moréa, G., S. Scharesa, A. Maksimova, F.J. Conrathsa, M.C. Venturinib and G. Schares. 2013. Development of a multiplex real time PCR to differentiate *Sarcocystis* spp. affecting cattle. Veterinary Parasitology 197: 85–94.

Moré, G., S. Schares, A. Maksimov, F.J. Conraths, M.C. Venturini and G. Schares. 2013. Development of a multiplex real time PCR to differentiate *Sarcocystis* spp. affecting cattle. Vet. Parasitol. 197(1-2): 85–94.

Moré, G., A. Maksimov, F.J. Conraths and G. Schares. 2016. Molecular identification of *Sarcocystis* spp. in foxes (*Vulpes vulpes*) and raccoon dogs (*Nyctereutes procyonoides*) from Germany. Veterinary Parasitology 220: 9–14.

Murphy, H.R., H.N. Cinar, G. Gopinath, K.E. Noe, L.D. Chatman and N.E. Miranda. 2018. Interlaboratory validation of an improved method for detection of *Cyclospora cayetanensis* in produce using a real-time PCR assay. Food Microbiol. 69: 170–178.

Nascimento, F.S., J.R. Barta, J. Whale, J.N. Hofstetter, S. Casillas, J. Barratt, E. Talundzic, M.J. Arrowood and Y. Qvarnstrom. 2019. Mitochondrial junction region as genotyping marker for *Cyclospora cayetanensis*. Emerg. Infect. Dis. 25: 1314–1319.

Orlandi, P.A., C. Frazar, L. Carter and D.T. Chu. 2004. BAM Chapter 19a: Detection of *Cyclospora* and *Cryptosporidium* from fresh produce: isolation and identification by polymerase chain reaction (PCR) and microscopic analysis. FDA Bacteriological Analytical Manual Available from: https://www.fda.gov/food/laboratory-methods-food/bam-19a-detection-cyclospora-and-cryptosporidium.

Ortega, Y. and L.J. Robertson. 2017. *Cyclospora cayetanensis* as a Foodborne Pathogen. Springer Briefs in Food, Health and Nutrition, ISBN 978-3-319-53585.

Paulos, S., J.M. Saugar, A. de Lucio, I. Fuentes, M. Mateo and D. Carmena. 2019. Comparative performance evaluation of four commercial multiplex real-time PCR assays for the detection of the diarrhoea-causing protozoa *Cryptosporidium hominis/parvum, Giardia duodenalis* and *Entamoeba histolytica*. PLoS One 14(4): 0215068.

Pineda, C.O., T.T. Temesgen and L.J. Robertson. 2020. Multiplex quantitative PCR analysis of strawberries from Bogota, Colombia, for contamination with three parasites. J. Food Prot. 83: 1679–1684.

Prakas, P., V. Kirillova, R. Calero-Bernal, M. Kirjušina, E. Rudaitytė-Lukošienė, M.A. Habela, I. Gavarāne and D. Butkauskas. 2019. *Sarcocystis* species identification in the moose (*Alces alces*) from the Baltic States. Parasitol. Res. 118(5): 1601–1608.

Robertson, L.J., B. Gjerde and A.T. Campbell. 2000. Isolation of *Cyclospora* oocysts from fruits and vegetables using lectin-coated paramagnetic beads. J. Food Prot. 63: 1410–4.

Robertson, L.J. 2014. *Cryptosporidium* as a Foodborne Pathogen. Springer.

Rosenthal, B.M. 2021. Zoonotic Sarcocystis. Res. Vet. Sci. 136: 151–157.

Rubiola, S., F. Chiesa, S. Zanet and T. Civera. 2018. Molecular identification of *Sarcocystis* spp. in cattle: partial sequencing of cytochrome C oxidase subunit 1 (COI). Ital. J. Food Saf. 7: 7725.

Sarkhosh, T., E. Mayerberger, K. Jellison and S. Jedlicka. 2021. Development of cell-imprinted polymer surfaces for *Cryptosporidium* capture and detection. Water Research 205: 117675.

Samarasinghe, B., J. Johnson and U. Ryan. 2008. Phylogenetic analysis of *Cystoisospora* species at the rRNA ITS1 locus and development of a PCR-RFLP assay. Exp. Parasitol. 118: 592–595.

Shin, J.H., S.E. Lee, T.S. Kim, D.W. Ma, S.H. Cho, J.Y. Chai and E.H. Shin. 2018. Development of molecular diagnosis using multiplex real-time PCR and T4 phage internal control to simultaneously detect *Cryptosporidium parvum, Giardia lamblia*, and *Cyclospora cayetanensis* from human stool samples. The Korean J. Parasitol. 56(5): p.419.

Smith, H.V. and R.A. Nichols. 2007. *Cryptosporidium*. pp. 233–276. *In*: Foodborne Diseases. Springer.

Smith, H.V. and R.A. Nichols. 2010. *Cryptosporidium*: detection in water and food. Exp. Parasitol. 124: 61–79.

Spickler, A.R. 2020. Sarcocystosis. Retrieved from www.cfsph.iastate.edu/DiseaseInfo/Factsheet.php.

Sulaiman, I.M., M.P. Torres, S. Simpson, K. Kerdahi and Y. Ortega. 2013. Sequence characterization of heat shock protein gene of *Cyclospora cayetanensis* parasites from Nepal, Mexico, and Peru. J. Parasitol. 99(2): 379–382.

Sulaiman, I.M., Y. Ortega, S. Simpson and K. Kerdahi. 2014. Genetic characterization of human-pathogenic *Cyclospora cayetanensis* parasites from three endemic regions at the 18S ribosomal RNA locus. Infect. Genet. Evol. 22: 229–234.

Temesgen, T.T., L.J. Robertson and K.R. Tysnes. 2019a. A novel multiplex real-time PCR for the detection of *Echinococcus multilocularis, Toxoplasma gondii*, and *Cyclospora cayetanensis* on berries. Food Res. Int. 125: 108636.

Temesgen, T.T., K.R. Tysnes and L.J. Robertson. 2019b. A new protocol for molecular detection of *Cyclospora cayetanensis* as contaminants of berry fruits. Front. Microbiol. 10: 1939.

Temesgen, T.T., V.M. Stigum and L.J. Robertson. 2022. Surveillance of berries sold on the Norwegian market for parasite contamination using molecular methods. Food Microbiol. 104: 103980.

USEPA. 2012. Secondary Drinking Water Standards: Guidance for Nuisance Chemicals: Environmental Protection Agency [Accessed 2018 March 17th].

Vanathy, K., S.C. Parija, J. Mandal, A. Hamide and S. Krishnamurthy. 2017. Cryptosporidiosis: A mini review. Trop. Parasitol. 7: 72.

Valadas, S.Y., J.I. da Silva, E.G. Lopes, L.B. Keid, T. Zwarg, A.S. de Oliveira, T.C. Sanches, A.M. Joppert, H.F. Pena, T.M. Oliveira, H.L. Ferreira and R.M. Soares. 2016. Diversity of *Sarcocystis* spp. shed by opossums in Brazil inferred with phylogenetic analysis of DNA coding ITS1, cytochrome B, and surface antigens. Exp. Parasitol. 164: 71–78.

Wang, Y., B. Zhang, J. Li, S. Yu, N. Zhang, S. Liu, Y. Zhang, J. Li, N. Ma, Y. Cai and Q. Zhao. 2021. Development of a quantitative real-time PCR assay for detection of *Cryptosporidium* spp. infection and threatening caused by *Cryptosporidium parvum* subtype IIdA19G1 in diarrhea calves from northeastern China. Vector-Borne and Zoonotic Diseases 21(3): 179–190.

Weinreich, F., A. Hahn, K.A. Eberhardt, T. Feldt, F.S. Sarfo, V. Di Cristanziano, H. Frickmann and U. Loderstädt. 2021. Comparison of three real-time PCR assays targeting the SSU rRNA Gene, the COWP gene and the DnaJ-like protein gene for the diagnosis of *Cryptosporidium* spp. in stool samples. Pathogens 10(9): 1131.

Wu, Z., I. Nagano, T. Boonmars, T. Nakada and Y. Takahashi. 2003. Intraspecies polymorphism of *Cryptosporidium parvum* revealed by PCR-restriction fragment length polymorphism (RFLP) and RFLP-single-strand conformational polymorphism analyses. Appl. Environ. Microbiol. 69: 4720–4726.

Xiao, L., U.M. Morgan, J. Limor, A. Escalante, M. Arrowood, W. Shulaw, R.C. Thompson, R. Fayer and A.A. Lal. 1999. Genetic diversity within *Cryptosporidium parvum* and related *Cryptosporidium* species. Appl. Environ. Microbiol. 65: 3386–3391.

Xiao, L. and V.A. Cama. 2018. *Cryptosporidium* and cryptosporidiosis. pp. 73–117. *In*: Ortega, Y.R. and C.R. Sterling (eds.). Foodborne Parasites. Springer International Publishing.

Yang, R., C. Palermo, L. Chen, A. Edwards, A. Paparini, A. Tong, S. Gibson-Kueh, A. Lymbery and U. Ryan. 2015. Genetic diversity of *Cryptosporidium* in fish at the 18S and actin loci and high levels of mixed infections. Vet. Parasitol. 214: 255–263.

Index

www.ingramcontent.com/pod-product-compliance
Lightning Source LLC
Chambersburg PA
CBHW060331220326
41598CB00023B/2673